Seasonal Variability in Third World Agriculture

**Other Books Published in Cooperation with
The International Food Policy Research Institute:**

Agricultural Change and Rural Poverty: Variations on a Theme by Dharm Narain
Edited by John W. Mellor and Gunvant M. Desai

Crop Insurance for Agricultural Development: Issues and Experience
Edited by Peter B. R. Hazell, Carlos Pomareda, and Alberto Valdés

Accelerating Food Production in Sub-Saharan Africa
Edited by John W. Mellor, Christopher L. Delgado, and Malcolm J. Blackie

Agricultural Price Policy for Developing Countries
Edited by John W. Mellor and Raisuddin Ahmed

Food Subsidies in Developing Countries: Costs, Benefits, and Policy Options
Edited by Per Pinstrup-Andersen

Variability in Grain Yields: Implications for Research and Policy in Developing Countries
Edited by Jock R. Anderson and Peter B. R. Hazell

Seasonal Variability
in Third World Agriculture
The Consequences for Food Security

EDITED BY DAVID E. SAHN

Published for the International Food Policy Research Institute
The Johns Hopkins University Press

Baltimore and London

Tallahassee, Florida

© 1989 The International Food Policy Research Institute
All rights reserved
Printed in the United States of America

The Johns Hopkins University Press, 701 West 40th Street, Baltimore, Maryland 21211
The Johns Hopkins Press Ltd., London

The paper used in this publication meets the minimum requirements of American National Standard for Information Sciences—Permanence of Paper for Printed Library Materials, ANSI Z39.48-1984.

LIBRARY OF CONGRESS CATALOGING-IN-PUBLICATION DATA

Seasonal variability in Third World agriculture : the consequences for food security / edited by David E. Sahn.
    p.   cm
"Published for the International Food Policy Research Institute."
Bibliography: p.
Includes index.
ISBN 0-8018-3829-0 (alk. paper)
    I. Sahn, David E.   II. International Food Policy Research Institute.
HD9018.D44S43   1989
338.1'91724—dc20   89-32232
                                                                 CIP

Contents

List of Tables and Figures   ix

Acknowledgments   xiii

PART I   Introduction and Summary

1   A Conceptual Framework for Examining the Seasonal Aspects of Household Food Security   3
DAVID E. SAHN

PART II   The Extent and Implications of Seasonal Food Insecurity and Malnutrition

2   Public Health and Functional Consequences of Seasonal Hunger and Malnutrition   19
PHILIP PAYNE

3   Seasonal Pattern of Activity and Its Nutritional Consequence in Gambia   47
MARK LAWRENCE, F. LAWRENCE, T. J. COLE, W. A. COWARD, J. SINGH, AND R. G. WHITEHEAD

4   The Effect of Seasonality on Intrahousehold Food Distribution and Nutrition in Bangladesh   57
MOHAMMED ABDULLAH

5   Seasonal Demands for Nutrient Intakes and Health Status in Rural South India   66
JERE R. BEHRMAN AND ANIL B. DEOLALIKAR

PART III   Food Acquisition Behavior, Employment, and Labor Market Considerations

6   Understanding the Seasonality of Employment, Wages, and Income   81
HAROLD ALDERMAN AND DAVID E. SAHN

7 Agricultural Wages in India: The Role of Health, Nutrition, and Seasonality   107
JERE R. BEHRMAN AND ANIL B. DEOLALIKAR

8 Seasonal Food Insecurity and Vulnerability in Drought-Affected Regions of Burkina Faso   118
THOMAS REARDON AND PETER MATLON

9 From Seasonal Income to Daily Diet in a Partially Commercialized Rural Economy (Southern Cameroon)   137
JANE GUYER

10 Seasonality in Food Systems: An Anthropological Perspective on Household Food Security   151
ELLEN MESSER

PART IV Grain Marketing and Price Variability

11 The Nature and Implications for Market Interventions of Seasonal Food Price Variability   179
DAVID E. SAHN AND CHRISTOPHER DELGADO

12 Seasonality in Burkina Faso Grain Marketing: Farmer Strategies and Government Policy   196
LYNN ELLSWORTH AND KENNETH SHAPIRO

PART V The Role of Technology and Policy

13 Indigenous versus Introduced Solutions to Food Stress in Africa   209
JON R. MORIS

14 The Impact of Technology and Policy on Seasonal Household Food Security in Asia   235
ROBERT W. HERDT

15 The Effect of Irrigation on Seasonal Rice Prices, Farm Income, and Labor Demand in Thailand   246
DOW MONGKOLSMAI AND MARK W. ROSEGRANT

16 The Impact of Drought and Technological Change in Rice Production on Intrayear Fluctuations in Food Consumption: The Case of North Arcot, India   264
PER PINSTRUP-ANDERSEN AND MAURICIO JARAMILLO

17 The Role of Agricultural Research and Secondary Food Crops in Reducing Seasonal Food Insecurity   285
RICHARD LONGHURST AND MICHAEL LIPTON

PART VI Policy Implications

18 Policy Recommendations for Improving Food Security 301
DAVID E. SAHN

References 317

Contributors 349

Index 355

Tables and Figures

Tables

2.1 Body size and energy expenditure rates, Gambian and Burmese farmers   35
2.2 Seasonal weight changes of Gambian farmers, simulated using different food allocation strategies and rates of postharvest losses   39
2.3 Effect of different food allocation strategies on annual storage losses   40
2.4 Seasonal weight changes of Burmese farmers, simulated using different food allocation strategies and rates of postharvest losses   41
2.5 Simulation of seasonal changes of body mass index, Burmese farmers   42
3.1 Subject profile   48
5.1 Characteristics of seasonal distribution for individual calorie and protein intakes and for weight-for-height, rural South India, 1976–78   69
5.2 Reduced-form seasonal log-linear demand relations for standardized calories, proteins, and weight-for-height for individuals in rural South India, 1976–78   71
5.3 Reduced-form seasonal log-linear demand relations for standardized calories, proteins, and weight-for-height for men, women, boys, and girls in rural South India, 1976–78   75
7.1 Peak and slack labor periods for adults in six SAT villages of peninsular India, 1975–76   112
7.2 Means and standard deviations by season, SAT India, 1976–78   114
7.3 Agricultural semilogarithmic wage equations, SAT India, 1976–78   115
8.1 Percentage of cultivated area sown to all crops, Sahel and Sudano-Sahel village samples, 1984   121

8.2 Household characteristics, Sahel and Sudano-Sahel village samples 124
8.3 Average daily intake per adult equivalent by season and stratum, Sahel village sample 125
8.4 Percentage of food-insecure adult equivalents per stratum per season, Sahel and Sudano-Sahel village samples 126
8.5 Average daily intake per adult equivalent by source, Sahel and Sudano-Sahel village samples, five-season period 127
8.6 Breakdown by food item and source of caloric intake per season, Sahel village sample 128
8.7 Breakdown by food item and source of caloric intake per season, Sudano-Sahel village sample 129
8.8 Chronically insecure versus secure households, Sahel and Sudano-Sahel village samples, five-season period 132
9.1 Economic origin of the diet, by value, according to season 141
9.2 Indices of seasonal variation in cash income 142
9.3 Women's cash income, by source, July and November 1976 146
9.4 Cash expenditure on food, by item, July and November 1976 148
11.1 Selected examples of seasonal range of agricultural prices in the third world 181
11.2 Market town monthly wholesale prices, average of 15 northern Nigerian markets, 1958–65 to 1969–71 182
11.3 Gaps between highest and lowest monthly wholesale price within years of major foodgrain in Nigeria and Bangladesh 185
11.4 Variation in seasonal sorghum price spreads across locations in Burkina Faso, 1981–85 186
12.1 Grain sales and harvest in nine Burkina Faso villages, 1983–84 198
12.2 Grain sales and purchases by quarter in five Burkina Faso villages, 1984 199
12.3 Number of households with their largest volume of sales and purchases in each quarter in five Burkina Faso villages 200
12.4 Market and nonmarket transfers of grain in four Burkina Faso villages 201
12.5 Nonmarket transfers of grain and money by quarter in five Burkina Faso villages 202
13.1 Root crop production in the tropics 214
14.1 Percentage of rice area harvested by month, Indonesia and the Philippines 241
14.2 Patterns of market arrivals in three-month peak season 242
15.1 Irrigated and rice-planted areas, Thailand, 1982 247
15.2 Regression results 251

15.3 Cropping intensity 253
15.4 Production and yield of rice 254
15.5 Farm households' gross cash income by source 256
15.6 Use of inputs for rice cultivation 258
16.1 Daily energy and protein consumption 267
16.2 Mean household-level variance of detrended two-month moving averages of calorie consumption per person per day 271
16.3 Rice production, consumption, share of production consumed, and number of net paddy-buying households 272
16.4 Total annual expenditures and incomes 274
16.5 Mean household-level variance of two-month moving averages of net income and detrended total expenditure 277
16.6 Results from regression analysis aimed at explaining variation in the standard deviation of calorie consumption 278
16.7 Results from regression analysis aimed at explaining variation in the coefficient of variation of calorie consumption 280
17.1 Sources of calories by season in three Zaria villages in Northern Nigeria, 1970–71 290
17.2 Food energy sources in developing countries (1980) and approximate annual allocation of CGIAR centers expenditures (1983) by commodity 292

Figures

2.1 Relationship between body weight and prospective risk of death in Bangladeshi and Indian children 26
2.2 The "model man," showing the tissue compartments and the balancing compartment 33
2.3 The "model farm," with compartments showing the farmer ("model man"), the farm, and the food store 34
2.4 Energy intake, work output, and body weight of Gambian farmers 36
2.5 Seasonal changes in body weight of Gambian farmers 37
2.6 Four patterns of food allocation tested using the model, and actual work output of Gambian farmers 38
2.7 Seasonal changes in body weight of Burmese rice farmers 41
3.1 Seasonal variation in total energy expenditure by rural Gambian women, adjusted for stage of pregnancy or lactation 51
3.2 Seasonal variation in body weight and body fat content of rural Gambian women, adjusted for stage of pregnancy or lactation 54
4.1 Mean daily energy and protein intakes of adults by sex, socioeconomic group, and season 60

4.2 Mean daily energy and protein intakes of children by sex and season 61
4.3 Daily energy intake of women and children as a percentage of that of the household head 62
13.1 Seasonal fluctuations in biomass under different ecological zones 218
14.1 Pattern of total rice area harvested in Java and Madura, Indonesia, 1955, 1968, and 1977 240
16.1 Two-month moving averages of daily calorie consumption per adult equivalent, small paddy farmers, five villages 268
16.2 Two-month moving averages of daily calorie consumption per adult equivalent, landless laborers, five villages 269
16.3 Two-month moving averages of daily calorie consumption per adult equivalent, large paddy farmers, five villages 270

Acknowledgments

This book was made possible through the support and cooperation of many institutions and individuals. The U.S. Agency for International Development (USAID) supported the preparation of many of the papers, and the overall compilation and editing of this volume. Judith McGuire, in her previous capacity at USAID, deserves special recognition for her insights in conceiving this project. Nancy Pielemeier, also from USAID, displayed continued interest in and vital support for the work on seasonality at the International Food Policy Research Institute (IFPRI).

The Food Policy and Nutrition Division of the Food and Agriculture Organization of the United Nations also played an important role in assuming the costs of the workshop, held in Annapolis, Maryland, where the preliminary drafts of the papers in this book were presented and discussed among an impressive group of experts assembled from around the world. Paul Lunven, director of the division, was instrumental in both organizing and conceptualizing the workshop, and ensuring the support that allowed us to bring together such an excellent and diverse group of scientists and policymakers.

Among the individuals not represented by papers in this book who nevertheless attended the meeting as participants and commentators, thanks is offered to Jean Ensminger, Igor de Garine, Brhane Gebrekidan, Natalie Hahn, Jan Hoorweg, Rebecca Huss-Ashmore, Mulumba Kamuanga, Shubh Kumar, Paul Lunven, Judith McGuire, Nancy Pielemeier, Chrisman Silitonga, Dunstan Spencer, Peter Temu, and Lauren Unnevehr. Their discussions and commentaries were stimulating and provocative, and set the tone for a most constructive workshop.

Special acknowledgment is due to all those at IFPRI who contributed to making this book possible. The support of John Mellor, from the outset of the project to the bringing of this book to press, is greatly appreciated. Per Pinstrup-Andersen, whose recognition of the importance of seasonality, both in the process of rural development and in influencing household food security, provided the impetus for the research presented in this volume. His guidance and encouragement were instrumental in bringing this publication to fruition.

Finally, thanks is offered to Ding Dizon, Annie Go, and Vickie Lee, who labored over typing and revising the manuscript, and to Robert Bordonaro and Laurie Goldberg, who organized the Annapolis workshop.

PART I

Introduction and Summary

1 A Conceptual Framework for Examining the Seasonal Aspects of Household Food Security

DAVID E. SAHN

The purpose of this book is to explore the seasonality of household food security. This involves examining the extent, patterns, causes, and consequences of seasonal variations in wages, agricultural earnings, food availability, prices, consumption, and nutritional status. In addition, we are interested in whether seasonal cycles are stable and predictable from year to year, whether and why seasonal fluctuations change with agricultural and economic development, and the implications of such changes for household food security.

A number of excellent and insightful papers are included in this book. The chapters that follow represent a systematic examination of the most salient issues regarding the seasonal dimensions of household food security. This introductory chapter presents an overview to place in context the subjects dealt with in this volume.

Concept of Household Food Security

Food security, at the household level, is defined as adequate access to enough food to supply the energy needed for all family members to live healthy, active, and productive lives. In particular, this book is concerned with whether, and the reasons why, households do not have adequate food resources during certain seasons of the year; and what policy options might address this form of transitory food insecurity.

Country-level aggregate data obscure the fact that even though a country may achieve adequate and relatively stable levels of food supply and prices, there may be great regional and local inequality and seasonal disparities in the distribution of consumption. For example, within a given town or village, only part of the population may face a seasonal shortage of food or display marked deficiencies in its levels of food intake. Similarly, aggregate data do not account for the fact that some members of a household may receive less food than others; thus the data may conceal the fact that some individuals, most likely women or children, may suffer from transitory seasonal declines in food intake while other family members do not.

While the global and national dimensions of food supplies affect the well-being of the household and individual, it is the determinants of the seasonal fluctuations in the levels of household and individual consumption, and the extent and policy implications of these fluctuations, that are the subject of this book.

The concept of food security is based on target levels of consumption. A number of factors determine whether the normative target levels are consumed. These include the availability of food in the market or on the farm, the command over adequate resources to grow or purchase food, and the desire to acquire sufficient food (Pinstrup-Andersen 1985). This book is primarily concerned with the transitory problem of short-term variability, in which consumption falls below trend levels as a consequence of seasonal patterns in the rural economy. In addition, the book includes an examination of how seasonal patterns and fluctuations—in terms of food availability, access to credit, employment, and so forth—contribute to persistent poverty and chronic (i.e., nontransitory) food insecurity. That is, a number of chapters discuss how the seasonal characteristics of the rural economy cause poverty through processes such as disaccumulation of assets during the lean season, or increased demand for labor during certain periods of the agricultural calendar, which discourages school attendance and related investments in human capital.

The Functional Consequences of Seasonal Food Insecurity and Malnutrition

Food insecurity, or inadequate levels of consumption whereby food energy intake is less than the energy required for growth and work, will adversely affect an individual's nutritional status and functional performance. The impact of transitory food insecurity, as manifested in a seasonal decline in access to food and nutrient consumption, on nutrition and human function is difficult to assess. This difficulty arises from the fact that an individual's energy balance is determined by the extent to which food energy (i.e., calorie) consumption is in equilibrium with energy requirements. This balance is seasonally determined by a complex of factors, including variability in work and basal metabolic rate, genetically determined seasonal patterns of growth, and environmentally caused seasonal patterns of disease. The complex relationship among them will determine seasonal patterns of energy balance and the extent to which seasonal fluctuations in food intake translate into malnutrition and related functional impairment.

In order to gain better insight into the interaction between seasonal patterns of food energy intake and energy expenditures, and their relationship to seasonal malnutrition, Philip Payne (chapter 2) reviews current theories of biological regulation and the ability of the individual to adapt to lower-than-optimal levels of food intake. He focuses on the role of seasonal food adequacy, and of disequilibria in nutrient intake relative to energy expenditures, in precipitating or perpetuating malnutrition and changes in functional performance. In doing so, he explores the issue of how well an individual can biologically adapt to seasonal changes in energy balance.

Payne points to the limited inferences that can be drawn concerning the prevalence and implications of seasonal food insecurity based on the traditional indicators of malnutrition and food consumption used by nutritionists and economists. For example, some researchers have presented seasonal reductions in calorie intake as evidence of seasonal food insecurity. In contrast, others have argued that a constant level of food energy consumption does not represent an optimal pattern of intake for preventing malnutrition, since, as pointed out by Mohammed Abdullah in chapter 4, energy expenditure patterns vary from season to season. At a minimum, therefore, information on the seasonality of energy expenditures, as well as consumption, is required. Ideally, one would also know the extent of physiological adaptation to reduced nutrient intake during periods of seasonal shortages.

In light of the theoretical and conceptual discussion of the relation between seasonal patterns of intake and malnutrition, Mark Lawrence and his colleagues draw upon their extensive work in Gambia to address in chapter 3 the functional consequences of seasonal food imbalances. Besides showing the adverse effects of seasonal food insecurity as indicated by reduced birthweight, their research addresses practical policy options for reducing seasonal malnutrition, including the role of food supplementation programs. The potential for timely food aid and health interventions to cope with seasonal food shortages is an issue that is explored in greater detail in the case study from Burkina Faso prepared by Tom Reardon and Peter Matlon (chapter 8).

In addition to the functional consequences of seasonal food insecurity as manifested by malnutrition and disease, the possibility arises that worker productivity is also affected by seasonally reduced food energy intake. In the paper by Jere R. Behrman and Anil B. Deolalikar (chapter 7), the reduced-form wage equation indicates that food energy intake has a differential impact on productivity by season. Specifically, it is during the peak season, when tasks such as harvesting require sustained expenditure of human energy, that reduced calorie intake has its greatest negative impact on worker productivity. Thus the household that experiences seasonal shortages of food in one year faces the prospect that low levels of food energy intake will impair work performance in the future. A possible consequence is a ratcheting effect whereby reduced performance leads to a lower wage offer and farm profits, and thus to even lower food intake in the following year. Similarly, distress sales, placing children with relatives or friends, or migration are strategies employed by households to cope with a period of seasonal dearth (see Ellen Messer, chapter 10). These strategies of coping with the lean season may in fact result in a household having to mortgage land, human capital, and part of its productive assets.

Determinants of Seasonal Malnutrition and Food Insecurity

While the consequences of, and methods to cope with, seasonal stress and shortages are of considerable importance, most of this book focuses on the determinants of seasonal malnutrition and food insecurity, and on policies and programs to prevent the problem. In considering the causes of seasonal malnutrition and food insecurity, one important issue is the intrahousehold distribution of food. The emphasis of the book, however, is on exploring the determinants of access to food at the household level, which are conditioned by a variety of factors, including the desire to obtain food (i.e., the household's preference ordering) and the ability to obtain food (i.e., the household's ability to produce and purchase food).

Intrahousehold Food Distribution

The consumption of food by the individual is related to the level of household calorie intake and the choices of household decision makers regarding the intrahousehold distribution of food resources. An important question is whether and to what extent food allocation within a household varies by season.

Harold Alderman and David E. Sahn (chapter 6) propose a variety of hypothetical reasons for predictable changes in the pattern of intrahousehold distribution by season. For example, households may adjust their distribution in accordance with the variability in energy expenditure that results from seasonal patterns of labor. Similarly, in periods of seasonal stress as measured by household food availability, certain family members are more likely than others to be systematically discriminated against.

A further consideration in the analysis of factors influencing seasonal patterns of food consumption is that income controlled by different household members may be allocated differently. The same may apply to income that takes different forms (e.g., agricultural versus cash income). To the extent that the form and source of income are seasonal, intrahousehold distribution of resources will be influenced.

Two chapters in this volume explore the extent to which there are systematic seasonal changes in intrahousehold food allocation. First, Behrman and Deolalikar estimate calorie and health production functions for children (chapter 5). Their finding of significant seasonal differences in intrahousehold nutrient supply among children suggests that in the lean season, when resources are constrained, parental decision makers allocate in a manner close to a pure investment strategy, favoring productivity over equity. In contrast, during the surplus season, equity concerns prevail in the allocation of nutrients among children. This clearly raises the prospect of vulnerable groups facing the additional risk of discrimination during the season of scarcity.

In another study, from Bangladesh (chapter 4), Abdullah suggests that children, especially females, are discriminated against in the intrahousehold distribution of food energy. This may be partially an artifact of the normative standards employed in the study. However, his study found that during the preharvest (i.e., hungry) season, the share of household resources allocated to young girls is greater than during the postharvest season. In contrast to the findings of Behrman and Deolalikar, it is argued that households display a strategy of protecting the most vulnerable during seasons of greatest stress, in contrast to the general pattern of discrimination evident during more prosperous seasons.

Despite the differing conclusions of the studies by Behrman and Deolalikar and by Abdullah, both underscore the importance of examining intrahousehold dynamics and the need for further empirical research to determine more precisely why, and under what circumstances, the lean season represents a period of heightened risk for certain family members.

Access to Food at the Household Level

While the nutritional status of the individual is partially a function of allocation of available food resources within the household, it is the household's ability and desire to obtain food that determine the level of household food security. A number of the contributors to this book present data on seasonal variability in household consumption. This includes the evidence from Per Pinstrup-Andersen and Mauricio Jaramillo (chapter 16) of contrasting seasonal patterns of household consumption in a drought versus a normal year in North Arcot, India. Similar evidence of strong seasonal variability in consumption is found in the data presented by Reardon and Matlon for Burkina Faso (chapter 8), and by Lawrence and his colleagues for Gambia (chapter 3).

In order to understand better the factors that are responsible for these observed fluctuations, one must consider (1) the nature of household preferences, which are a reflection of the desire to obtain food, and (2) the ability of the household to obtain food seasonally, as conditioned by the patterns of earnings, prices, and savings behavior (including on-farm storage).

DESIRE TO OBTAIN FOOD. Household members allocate their time to work and leisure. The value of time, and thus earnings, is determined by the work being performed, which varies seasonally. One would expect, therefore, the amount of income earned and time dedicated to leisure to be different from one season to the next, and from one household member to the next. Consequently, the source of income and which family members control it vary between the seasons.

Jane Guyer (chapter 9) explores how gender-specific budget control may differ from one season to the next, as may intrahousehold bargaining and exchange relations. Her study highlights (1) the importance of incorporating a seasonal perspective into analyzing how differential control of income affects consumption, and (2) the bargaining process through which the different priorities of household members are reconciled.

Alderman and Sahn (chapter 6) stress that, besides the consideration of who receives income, a related issue that may affect the desire to obtain food at the household level is the changing composition of income by season. There is limited, although contradictory, evidence that income from the household's own production, food stamps, and subsidies may contribute relatively more to food consumption than wages and salaries (see Alderman 1986; Edirisinghe 1987; Garcia and Pinstrup-Andersen 1987; and West and Price 1976). To the extent that income sources are seasonal, this clearly represents an important consideration in examining seasonal patterns of consumption.

Ellen Messer (chapter 10) and Lynn Ellsworth and Kenneth Shapiro (chapter 12) also indicate that the desire to obtain food revolves around seasonal patterns of celebrations and ceremonies. Whether it be during religious holidays or other festivals, there may indeed be regular periods when either food consumption and sharing increase, or changes in the composition of the diet occur that are attributable to traditional beliefs and culturally determined food practices.

ABILITY TO OBTAIN FOOD. The ability of the household to obtain adequate food in a given season is determined by (1) the level and flow of wage and agricultural earnings and food available for home consumption; (2) the seasonality of food prices; and (3) intertemporal savings behavior. The latter includes savings in the form of money, assets, and food. In combination, the level and flow of income and the seasonal changes in food prices and savings define the level of real expenditures and food purchasing power from one season to the next.

The seasonality of earnings. The seasonality of earnings is conditioned by the pattern of income from a variety of sources, including agricultural sales, wages, business profits, remittances, and transfer payments. It is likely that the pattern of agricultural income from field crops, which is a function of the timing of market arrivals and prevailing prices, will be markedly seasonal. Guyer (chapter 9) discusses how production of cash crops, in addition to or instead of traditional food crops, may balance or exacerbate the seasonality of agriculture-related activities; and Jon R. Moris (chapter 13) points out that integrated livestock-grain production systems, which in Africa are largely centered on highland plateaus, are another means of reducing seasonality in agricultural incomes. The anthropological review by Messer (chapter 10) also indicates that a variety of methods to diversify income sources, such as production of traditional crafts, migration, expansion of foraging activities, food processing, and so forth, are also employed to avoid periods of seasonal shortages. In addition, transfer payments within the community are often timed to alleviate seasonal stress. The importance of traditions of reciprocal sharing to avoid seasonal food insecurity is also reported by Ellsworth and Shapiro (chapter 12) for Burkina Faso.

Given that the wage income from agriculture-related activities is a consequence of the pattern of labor demand and reservation and market wages, Alderman and Sahn (chapter 6) review how the functioning of labor markets and wage formation affect the seasonal pattern of earnings. Behrman and Deolalikar (chapter 7) empirically determine the seasonality of wages in India to illustrate the importance of accounting for seasonal fluctuations when studying rural labor markets.

An examination of the seasonality of rural incomes must also take into account the ability of farmers and laborers to smooth out their stream of earnings through changing farming techniques and management practices (e.g., crop choice, water control, seed varieties) and finding nonagricultural income to compensate for the seasonal cycles in farming. Dow Mongkolsmai and Mark W. Rosegrant (chapter 15) explore some of these issues by analyzing the effect of technological change in Thailand on the seasonal demand for labor, seasonal employment, migration, and wages. They find that irrigation stabilizes the seasonal demand for labor, boosts wages for hired workers, and reduces out-migration. Nevertheless, Robert W. Herdt (chapter 14) points out that new productive technologies often in fact aggravate seasonal labor peaks.

Prices. The formation of prices over seasons represents an important theoretical and empirical question that David E. Sahn and Christopher Delgado examine in chapter 11. Economic theory suggests that prices are formed by the intersection of supply and demand curves. However, predicting the changing seasonal patterns of marketed supply and effective demand proves to be more complex than one would expect from a static supply-demand framework. Seasonal prices are a function of the accuracy of expectations concerning supply and demand conditions in the future, and of the extent to which changes in anticipations bring about adjustments in the quantity of food marketed by producers and traders, as well as in the demands of consumers.

Further insight is needed into how expectations are formed and what role government should assume to improve the accuracy of those expectations. An improved understanding of these issues, along with an exploration of the functioning of commodity and futures markets, is required in order to explain the extent and predictability of seasonal price movements.

The major policy questions that are of interest revolve around the appropriateness of, and mechanisms for, reducing the magnitude of seasonal price increases and the interyear instability of these seasonal price spreads. These questions are addressed in the paper by Sahn and Delgado (chapter 11), which also examines who gains and who loses from stabilization, and the explicit and implicit costs of government intervention. The experiences of government interventions to stabilize prices in Asia are discussed by Herdt (chapter 14) and by Ellsworth and Shapiro (chapter 12) for Burkina Faso.

Stabilizing real expenditures and the role of savings. For households that earn their income in cash, the relationship between the seasonal pattern of commodity prices and incomes affects their ability to achieve an adequate diet throughout the year. If movements in prices and incomes are procyclical, purchasing power remains stable from one season to the next. If prices and incomes are countercyclical, the household must resort to savings to smooth out consumption. Thus it is the covariance of prices and income, and the opportunities and ability to save intertemporally at a reasonable cost, that determines the extent of transitory food insecurity.

Farming households have the additional option of saving in kind, rather than in cash. This involves smoothing their consumption stream through storing their own production, gradually drawing down stocks (which in certain cases may be unharvested roots and tubers) as required. Alternatively, they can sell their products, generating cash income that can be saved like all other money income. The potential for and wisdom of storing and consuming one's own production, or selling it in the marketplace and saving the earnings for retail purchases, are determined primarily by the spread between farmgate and retail prices (i.e., marketing costs), by shifts in the relationship between these prices from one season to the next, by the efficiency of on-farm storage, and by the cost of borrowing any needed capital. For example, if the difference between farmgate prices and retail prices is greater than the economic costs of marketing, or if postharvest prices are kept low relative to preharvest prices through exploitative practices of oligopolistic traders, this should clearly encourage on-farm storage and home consumption. Of course, this option arises only if satisfactory on-farm storage facilities are available and if the farmer's need for capital can be met through accessible financial institutions.

Ellsworth and Shapiro (chapter 12) attempt to improve the understanding of the marketing behavior of farm households in Burkina Faso. They found that wealthy farmers were better able to adjust their marketing behavior to take advantage of seasonal price increases. Even small farm households, however, were not adversely affected by the pattern of seasonal prices; they engaged little in market transactions, and they timed their sales to avoid low prices and their purchases to avoid high prices.

Finally, it is worth noting that agriculture-based savings or dissavings can take special forms, such as livestock. Moris points out in chapter 13 the tendency of semipastoralists to keep goats and sheep because they represent a convenient and liquid mode of savings in parts of Africa where banking systems are ineffective. This point is well illustrated by Reardon and Malton (chapter 8), who amply display the importance of livestock sales in mitigating the stress of the lean season in Burkina Faso. Liquid assets in the form of livestock protect the household from dramatic seasonal declines in calorie intake.

Even when credit markets are efficient or other means of savings (e.g., on-farm storage, livestock) are available, two qualifications arise that greatly complicate the intertemporal transfer of resources as a means of preventing seasonal food insecurity. The first is that the poor face a dilemma when confronted by the choice of saving for the next season, even if anticipated to be a period of stress, in the face of hunger in the present. Second, the entire impetus for saving from season to season is derived from a high degree of certainty as to future conditions compared with the present. Although in theory seasonal fluctuations (unlike intertemporal fluctuations) are predictable events, there is considerable evidence introduced by Sahn and Delgado (chapter 11) that this is not the case for prices. There is also evidence from Pinstrup-Andersen and Jaramillo (chapter 16) that the seasonality of income streams is far from predictable. The uncertainty of the seasonal cycles makes more complex, albeit more important, a household's intertemporal budgeting of scarce resources.

The Seasonal Dimensions of Food Availability

The availability of food during any given season is determined primarily by three factors: (1) the timing of production; (2) the timing of what is put into or removed from stocks; and (3) the timing of trade between communities and regions and across national borders. If one is concerned with assuring stable levels of food availability from season to season, this can be accomplished through a combination of production strategies, storage, and trade.

Farm Production and Technological Change

Farming activities revolve around the climate, a seasonal phenomenon over which human beings exert little control. While weather events (e.g., levels of solar radiation, precipitation) are given exogenously, the interaction of the natural ecology with technology and markets determines the seasonal pattern of food production. Although some periodicity in food production is inevitable, there are a number of policy and research issues that are important. These revolve around how farmers can control or respond to climatic events by adjusting production techniques and management practices.

There is little doubt that certain technologies, especially water control and management, change the seasonal pattern of food production. The paper by Herdt (chapter 14) amply illustrates how cropping intensity increases as a result of technological change. Mongkolsmai and Rosegrant (chapter 15) show that in addition to raising dry-season production, irrigation dramatically increases wet-season yields in rain-fed areas of Thailand. Thus the most important motivation for, and result of, irrigation is increasing yearly production, rather than smoothing out the timing of output.

This issue of increasing aggregate production through technological change raises the specter that there may be tradeoffs between strategies to increase output versus altering the timing of production. Nevertheless, the paper by Pinstrup-Andersen and Jaramillo (chapter 16) amply demonstrates that even though new technology exacerbates the seasonality of production, the entire production curve shifts upward sufficiently to ensure that producers and consumers are better off in all seasons.

On the other hand, there appears to be some evidence, as put forward by Jon R. Moris in chapter 13 and Richard Longhurst and Michael Lipton in chapter 17, that abandoning traditional crops (e.g., roots and tubers) and cultivation practices (e.g., diversification) exacerbates seasonal variability in food supplies. This, they argue, has caused increased seasonal stress in parts of Africa. Therefore, those chapters concentrate on the role of coarse grains, roots, tubers, green leafy vegetables, pulses, and tree crops in alleviating seasonal food insecurity. These are often referred to as secondary food crops. In some countries they may be the main staple—for example, the "false banana" in Ethiopia. And during some periods of the year and for certain drought years, they may assume a primary role in the diet. Furthermore, even if they are a secondary energy source, legumes, fruits, green leafy vegetables, and so forth may be the major suppliers of micronutrients and protein. Perhaps most important is that the so-called secondary crops may be seasonal or otherwise of little importance in the diet of the rich, but may play a vital role for the poor, especially during the preharvest period.

A related issue is that the scarcity of factors at various periods in the production cycle (e.g., labor bottlenecks at harvesting) determines the nature of technological change. The implication is that not only does the choice of production technology affect seasonality of production and earnings, but seasonal constraints in production guide the nature of agricultural progress and farming practices. Consequently, when a new technology is developed that does not take into account seasonal bottlenecks in the production process, its adoption will be impeded.

The interface between seasonality in the agricultural cycle and the adoption of new agricultural technology suggests the importance of understanding the seasonal constraints to adoption, along with how adoption affects the timing of demand for factors of production (e.g., labor, fertilizer) and the subsequent response of factor markets. This interaction also has implications for the role of agricultural research on seasonality, a topic addressed by Longhurst and Lipton (chapter 17).

Storage and Processing

The second important factor that determines the seasonality of food availability is the effectiveness of storage operations. Regardless of when crops are harvested, the ability to store intertemporally can smooth out seasonal food availability (see chapter 13 by Moris). This also suggests that the availability of storage infrastructure and the functioning of markets—that is, efficiency and competitiveness of interperiod arbitrage—are essential components in reducing seasonal fluctuations in food availability.

Related to the storage question is the issue of the processing of foodgrain and other agricultural products. Techniques to preserve foods and reduce their perishability, both through the development of germ plasm that yields hearty grains and through postharvest processing, also are important areas of study and are addressed in a number of chapters in this volume.

A further dimension of the storage issue concerns the role of institutions in promoting temporal arbitrage. The need for and proper role of government—state and parastatal marketing boards—remain a controversial subject. The arguments for such institutions arise primarily out of the assertion that markets are uncompetitive and traders exploitive. These points are examined by Sahn and Delgado in chapter 11 and by Ellsworth and Shapiro in chapter 12.

Trade

Just as stocking operations allow a country or region to stabilize food availability, so too can trade operations be timed so that imports arrive during the lean season. To the extent that a country is geographically diverse, like Indonesia, the potential arises for interregional trade to smooth out seasonal variability. Market infrastructure and integration represent the vital ingredients in promoting trade of foodgrains from one region of the country to another. Engaging in international trade to reduce seasonality is a more precarious endeavor, as discussed by Sahn and Delgado in chapter 11. Not only is it difficult to adjust the timing of imports and exports to correspond to the lean period, but there may be other factors, such as seasonal fluctuations in international commodity prices, that have an equally strong influence over the timing of grain purchases and when they arrive at the port.

Considering Seasonality in the Broader Context

This book concentrates on seasonal dimensions of household food insecurity. However, an underlying current in many of the chapters is that variability in agriculture, food availability, earnings, and consumption is not simply a seasonal phenomenon, but also occurs from one year to the next. The larger domain of concern about instability, therefore, encompasses both interyear and intrayear fluctuations.

To amplify, for a given data element—whether it be prices, food production, or weight gain, for example—there is a combination of trends and cycles as well as a stochastic element. Seasonality can be considered a cyclical component, while interyear variability is the stochastic element. To the extent that this is an accurate portrayal, there is a compelling argument that it is easier for consumers, producers, and traders to deal with the predictable seasonal cycles than with random interyear fluctuations, since the latter pose a more significant threat to consumers and producers because of the introduction of greater uncertainty and risk.

In fact the distinction between stochastic interyear fluctuations and seasonality is not straightforward. A number of chapters demonstrate that in certain circumstances, for some variables the pattern of seasonal variability is indeed difficult to predict. Of course the expense of collecting seasonal data at the household level over a series of years limits what can be said on this matter. Nonetheless, since the observed seasonal patterns in, for example, prices or food stocks are not necessarily regular year after year, we should be careful in the conclusions we draw from one or even two years of seasonal data.

A fundamental question regards the causes of instability in intrayear patterns in variables like prices, income, and expenditures from one year to the next. The answer lies in examining the two other components of a data series—trends and stochastic elements. Concerning the former, it is inevitable that a change in general trend levels of variables, such as food production, amount and sources of income, and income exchange relations, will be accompanied by shifts in the seasonal patterns of these variables.

For example, it is shown by Pinstrup-Andersen and Jaramillo (chapter 16) that a rising trend in production will likely be predicated on technological change, which will alter the seasonal timing of output as it increases the surplus, which in turn will affect marketing efficiency and competition. Similarly, Mongkolsmai and Rosegrant (chapter 15) show that there are changes in the seasonal pattern of labor requirements that accompany production increases. These may alter the nature of wage formation or implicit labor contracts, affecting not only the level of income but its seasonal flow. Likewise, as incomes rise, access to and use of capital markets may increase, further affecting the extent of transitory seasonal declines in consumption. And as Guyer (chapter 9) shows, the changing division of labor by gender that accompanies agricultural progress will influence seasonal patterns of food security, just as will the changes in the nature of social relationships that occur with economic development (see also Messer, chapter 10). This suggests that we must consider whether, why, and how trends that accompany agricultural development increase or decrease seasonality and the stability and predictability of seasonal patterns.

Another important explanation for unstable and unpredictable seasonal undulations is the large interyear fluctuations in agriculture and consumption.[1] The stochastic elements of production and food availability, coupled with policy and behavioral changes from one year to the next, contribute to the difficulty of formulating accurate expectations of seasonal cycles and events. To the extent that the farmer, trader, or consumer faces uncertainties and changing circumstances in levels of production, prices, earnings, and other factors from year to year, this will also be reflected in differing seasonal patterns. Accordingly, subsequent chapters address the following types of questions: Is it only during years of low levels of production that households are at heightened risk during the lean season? If interyear instability in production and imports could be reduced, would the pattern of seasonal prices more closely reflect the economic costs of storage?

Finally, the contention that seasonal variations are predictable, and

1. See Anderson and Hazell (forthcoming) for a detailed discussion of the increasing yield variability that accompanies agricultural development; and Mellor and Desai 1985 and Sahn and von Braun 1987 for an analysis of the extent, determinants, and consequences of interyear variations in consumption.

therefore amenable to adaptation without serious consequence, is based on the efficient functioning of price, capital, and wage markets. There is, however, evidence that poor countries are characterized by market inefficiencies, such as rationing of capital, great divergence between the returns to savings and the cost of borrowing, poorly functioning or nonexistent futures markets, and oligopsonistic traders. If this is the case, perhaps the deleterious consequences of seasonality could be overcome through improving market performance and promoting general economic development.

PART II

The Extent and Implications of Seasonal Food Insecurity and Malnutrition

2 Public Health and Functional Consequences of Seasonal Hunger and Malnutrition

PHILIP PAYNE

During the past few years, the many ways in which the seasons of the year affect the lives of peasants have come to provide a common focal point for scientists and practitioners from a broad range of disciplinary backgrounds. For many people, including no doubt (had they been asked) the peasants themselves, it must have been almost beyond comprehension that such an intrinsic aspect of human existence as the cycle of the seasons actually needed to be drawn to the attention of professionals concerned with the problems of food and poverty. However, with some notable exceptions, this clearly was the case. Chambers (1983b) has described the area as a new "professional frontier" and the phenomenon of neglect as "seasonality unseen." As Chambers suggests, the reasons lie in the barriers created by specialization, and in the biases of perceptions (often the results of the adverse effects of the seasons themselves) aggravated by difficulties of access and contact between researchers and peasants. The position in which we now find ourselves is more serious than just blindness toward a certain kind of problem: the neglect, in nutrition certainly, has been so total and has lasted so long that to some extent we lack even the theoretical concepts necessary for analysis. Thus while we can now say that the phenomenology of seasonal effects on health and nutrition is established—indeed, it is beginning to show an astonishing richness and extent—we cannot do much more than sketch the outlines of the theoretical framework that will be needed to take account of these processes in framing policy: specifically, there is an urgent need to develop key indicators for identifying hunger or the risk of it, and for distinguishing the effects of seasonal food shortages from those of infectious diseases. In addition, we need to know how to take account of the dynamic effects of cyclic changes in physical function in assessing the performance of household food production systems.

It has been traditional to assess food adequacy from the standpoint that equilibrium between energy intake and output is a normal and desirable condition, and that temporary disequilibrium can be dealt with by averaging intake measurements over some suitably extended period. The result is then supposed to be compared with the mean value of a set of normative standards. However, the added interest that a concern for seasonality has brought, for example, to the study of the energy requirements of reproduction (Roberts et al. 1982; Lawrence et al. 1984) has shown that in Gambian women, normal strategies for dealing with the combination of seasonal food availability and the changing requirements of pregnancy and lactation are complex and continuously changing over time. They include adjustments of energy expended on work as well as of basal metabolism. Lawrence et al. (1984) conclude, very reasonably, that "FAO-WHO-recommended intakes during pregnancy and lactation are 'inappropriate' for this community."

The problem, of course, is that the whole armamentarium of recommended intakes, and of techniques for measuring and comparing consumption patterns with respect to these standards, is premised not only on the desirability of the levels chosen but also on the maintenance of a steady state of equilibrium. If the "normal" state is one of a dynamic disequilibrium, with repeated cyclic adjustments over a range of values, then we shall need more than just "appropriate" levels of recommended energy intake in order to establish whether seasonal effects contribute to malnutrition and, if so, when.

Apart from food intake, nutritionists for the most part rely on growth or changes in body weight as an index of nutritional status. Billewicz and McGregor (1982) in Gambia and Black et al. (1982) in Bangladesh have described the influence of seasonality on growth patterns of children and on adult body weight. These and many other studies illustrate the "normality" in rural people of dynamic seasonal changes in the measurements conventionally used to assess nutritional status. In the Bangladesh studies in particular, in addition to cyclic changes over time, the two measurements—height and weight—are shown to be affected differentially with respect to the phase of the seasons, so that even the so-called age-independent ratios such as weight-for-height and weight/height^2 display regular seasonal cycles. Again, there is currently a dispute among nutritionists about the appropriateness of international reference standards for growth. But even if local standards were to be adopted, how would we interpret fluctuations in magnitude in time and in phase? It is perhaps only partly coincidental that just as we have reached the stage of wanting to know the significance of seasonality for malnutrition, we discover that a good many professionals are uncertain about what is meant by the term *malnutrition* anyway.

Thus it is almost universal practice to refer to anthropometry as a technique for measuring nutritional status, and to equate poor growth in children with malnutrition. However, although the rate of growth of small children is certainly sensitive to changes in health and nutritional status, with the passage of time it has become less clear exactly what, if any, normative standards for growth can be applied to measure the incidence of malnutrition; even less clear is the precise etiology of malnutrition as manifested in growth retardation (Grantham-McGregor 1984).

In addition, there is a major controversy about the significance of adaptive changes in food intake and body size. One faction asserts that consumption patterns and growth rates achieved generally in developed countries reflect genetically determined optimum performance, and that lower intakes consumed by smaller people in poor countries must be suboptimal and therefore constitute a nutrition problem. This leads to views like that of Gopalan (1980), who says that despite the fact that child mortality and morbidity are strikingly lower in the state of Kerala than in the Punjab, their generally smaller stature indicates that there is more malnutrition in Kerala than in the Punjab. Gopalan rejects the opposing view, which is that people can adapt to different environments so as to be "small but healthy," and that it is the general levels of quality and risk imposed by the different environments that constitute the problem for policymakers, rather than smallness per se. However, both factions agree that there must be some degree of growth retardation that is so severe as to be beyond the limits of safe adaptation so that individuals do suffer increased risk of early death or permanent disability. These conflicts and contradictions have been building up for some time, and the recognition of the hitherto "unseen seasonality" serves only to highlight the uncertainties.

For the purposes of this chapter, this means that we have to begin with a discussion of current theories about nutritional status, nutritional adequacy, and malnutrition. The most critical issue that divides the theoreticians is that of individual adaptability: to what extent can the adjustments in metabolism, body size, and behavior that individuals make in accommodating to changes in their environment be regarded as cost-free or adaptive, rather than pathological? Obviously, this is a crucial question if we are concerned with distinguishing "malnourished" from "normal" individuals in populations whose situation is relatively static; it is even more so if those adjustments are being made repeatedly in response to regular seasonal changes.

As we shall see, the resolution of this apparently straightforward question is by no means easy. Indeed, it raises some philosophical issues that do not perhaps permit a general solution. Thus the notion of the cost of adjustments or accommodations turns out to have two alternative meanings. In a biological sense, the cost of a change would need to be measured by the extent to which it reduces inclusive fitness—that is, the overall chances of survival of the genotype as distinct from the individual. In the sense of social valuation, however, the costs would need to be assessed either in economic terms, as the value of total output forgone, or in relation to the quality of life, in terms of individual or social values, and these will of necessity vary among different economic and cultural contexts.

Even bearing in mind that malnutrition, like many other aspects of health, is likely to emerge as being as much a social construct as a biological state, it is still necessary to develop a dynamic framework for analysis if the further implications of seasonality are to be included.

Although many other disciplines have a long tradition of the use of simulation models as aids to research and application, this is not the case so far with nutrition, and there is very little on which to build. What is described later in this chapter, therefore, should be seen as a very tentative first step in linking together a few of the concepts of nutrition, work physiology, and seasonal relationships in food production. The results of doing this are so far of heuristic value only—they are not predictive in the sense of providing policy-relevant statements about the future of particular real production systems. They do, however, provide a new perspective from which to look at some case studies of actual situations and, it is hoped, to obtain some clues as to where to go next.

Theories of Malnutrition

The following is a description of two alternative models of the processes of biological regulation that maintain the state of nutrition of an individual. These are presented in starkly contrasting fashion, in order to convey more effectively the nature of the theoretical assumptions that underlie them, and to bring out clearly the different implications for policy. This should not be interpreted as excluding the possibility that the truth lies somewhere between the two. The two theories have been described by Payne and Cutler (1984) as the "genetic potential" and "adaptability" models.

The genetic potential model probably represents the conventional and certainly the most widely held view about the causes of malnutrition. This is the logical development of an assumption that the level of health and functional performance of an individual is a reflection of the degree to which environmental factors such as food supply, the demands for physical work, or the effects of disease have caused the body to depart from its "normal" or preferred state.

The body is seen as a system that is not only self-regulating but self-optimizing. For each individual there exists a preferred state characterized by a unique set of values for a number of variables such as food intake or body size and composition. This state is an optimal one in the sense that all of the functions and capacities of the person taken together maximize fitness. People differ, one from another, in the values of the component variables that characterize their particular preferred state, and in the levels of performance of which they are capable when at their own particular optimum. These interindividual differences are inherent and result from differing genetic constitutions. Individuals who are prevented from returning to their preferred state because of dietary or other constraints are regarded as malnourished in the sense of being in a nonoptimal condition. Malnourished states can be detected and graded for severity by the extent of deviation of one or more of the state variables (intake, body weight, etc.) from those that characterize the preferred state.

There are two important implications of this model: First, nutritional status is made consonant with the now widely held view of health as a positive condition, and not as simply the absence of recognizable disease. It is important to note that the performance of a malnourished individual either could be affected currently with respect to some existing challenge, or could imply a reduced potential for dealing with some future situation, such as an increased demand for work, for resisting an infection, or for a psychosocial response. That is to say, malnutrition results in suboptimal responses both to current stresses and to increased risk of failure at some future time.

Second, the procedure for quantifying malnutrition in a population is (in theory) straightforward. We may count the number of people whose status indicator values depart so much from the normal reference values as to imply that those concerned are more likely to have had to adjust to an environmental stress than simply to have unusually low preferred values (e.g., take as being malnourished all people who have body weights less than two standard deviations below some reference mean). Alternatively, we could measure food consumption and count as malnourished all those people whose intakes are so low as to be more likely the result of food restriction than unusually low food requirements for the maintenance of their preferred states. In terms of energy, this involves the additional complication that the reference level or requirement has to include all allowance for whatever is considered to be a "desirable" level of physical work output.

In contrast, the adaptability model is based upon the view that an individual's genes do not in fact uniquely determine his or her characteristics in any simple way; rather, they define the range and nature of adaptive adjustments that can be made in response to changes in the environment, food, work, and so on. A young child will respond in a relatively reversible fashion to dietary changes and may, for example, follow a period of reduced growth during the lean season by a fast catch-up. Adjustments to changes can be both short-term and reversible, but also long-term and relatively irreversible. There are, however, limits to the extent to which physiological and behavioral adjustments can be made. In the adaptability model, malnutrition is the consequence of adjustments that exceed those limits.

The model for *adequate nutrition* that results from these concepts contrasts with the genetic potential theory of *malnutrition* in the following way. First, the body is still seen as a homeostatic self-regulating system whose state can be measured in terms of such variables as food intake, body size, and growth. However, there are no preferred values for any of these variables, nor is there any internal reference that defines a normal or an optimum state to which the system will always tend to return.

Second, it is accepted that individuals are likely to vary one from another in some important respects. The first of these is their capacity to adapt—that is, the magnitudes of the adjustments they are able to make in response to seasonal stresses without risk of physiological breakdown and subsequent death or disability. The second is the efficiency with which they convert food to work. Thus some people may need more or less food than others in order to take up the same production possibilities.

The adaptability model suggests that the differences in body size among children (and for that matter adults) from different socioeconomic environments are to be interpreted in the first place as indicators of the overall effects of those environments, since they show how much adaptive adjustment has been made by individuals living in them. The model does not say anything predictive about malnutrition. This is to be regarded as the result of failure to maintain homeostatic control, and happens when an individual is forced beyond the limits of his or her adaptive range. Malnutrition is thus the end point of a more or less catastrophic transition or system failure, rather than simply the condition of being in a range of suboptimal states.

The following statements probably represent most dramatically the different policy implications of the two models. The first is by Gopalan (1980, p. 129):

> We have therefore to conclude that during the last 30 years, in spite of all our advances in the agricultural, industrial, and technological fields, and in spite of several applied nutrition programmes and supplementary feeding programmes, we have not really made any significant dent on the problem of malnutrition in our children. We have reduced infant and child mortality to some extent and we have thus saved many children whom, otherwise, the merciful hand of death would have removed. There is thus an increasing pool of survivors who have escaped death but who exist in a substandard state of health and nutrition with permanent impairment of functional competence and productivity. The result is a progressive erosion and deterioration of the quality of our human resources.

Gopalan goes on to say that on the basis of physical size, only some 15 percent of children born in India are destined to become "truly healthy, physically fit, productive, and intellectually capable citizens of the country." The contrasting statement is from Seckler (1982, p. 135): "Except where people are in clear and present danger of functional impairment due to malnutrition, interventions should be targeted to poor environments and not to poor individuals. Once the environment of small people is improved, their size will take care of itself. Until their environment is improved, their size is best left to take care of itself." Thus Seckler's assessment of malnutrition as a public health problem for India's young child population is opposite to Gopalan's. One says that 10-15 percent are severely growth-affected and therefore actually malnourished, the other that only 15 percent are nutritionally healthy and free of the risk of functional impairment.

Seasonal effects on growth obviously assume quite different significance depending upon which of these views is accepted. On the basis of their longitudinal studies of child growth, Black et al. (1982) suggest that because the repeated seasonal checks to growth imposed on a child become progressively less reversible with age, this ratcheting effect is an important mechanism that results in an accumulation of growth deficits throughout the whole of childhood. For Gopalan, this might well mean that a reduction in seasonal growth checks ought to be a major objective of public health policy as a contribution toward dealing with the problem of the 85 percent who exist in a substandard state of health. On the other hand, Seckler might argue that even if seasonal episodes of food shortage contribute to short children, this represents a healthy adaptive process that has no serious functional consequences.

It is important to say that there is little evidence so far to refute either theory. Almost certainly there are elements of truth in both. However, it is not possible to make any effective commentary on the general public health implications of seasonal retardation of child growth, and whether seasonal growth retardation is an indicator of seasonal malnutrition or simply an inconsequential temporary response to seasonal stress, while still appearing to remain indifferent as to the likely significance of adaptive responses. In favor of the adaptability perspective are (1) its general consistency with the rest of biological theory and (2) the evidence tying growth in children and body dimensions in adults to functional capacity, which generally seems to point to small, or negligible, functional losses accompanying a certain range of adjustments in growth or size. In contrast, very significant penalties result from growth deficits beyond a certain degree. In other words, the relationships are found to be markedly nonlinear and may therefore be the result of threshold effects.

The clearest evidence in this regard is the threshold effects on survival. Figure 2.1 shows the relationship between prospective risk of death and body weight as a percentage of standard weight-for-age in young chil-

FIGURE 2.1 Relationship between body weight and prospective risk of death in Bangladeshi and Indian children

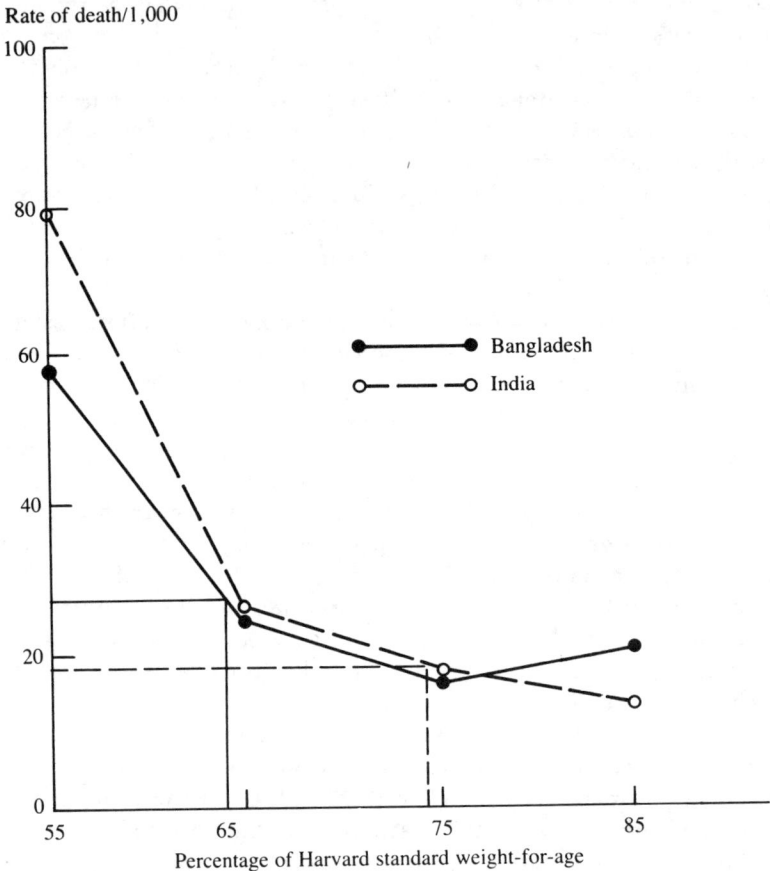

SOURCE: Data on Bangladesh from Chen, Chowdhury, and Huffman (1980). Data on India from Kielmann and McCord (1978).

dren in Bangladesh (Chen, Chowdhury, and Huffman 1980) and India (Kielmann and McCord 1978). Furthermore, as Lunven (1985) notes, it has been observed recently that determining the low limit of adaptability is not as simple as it seems because individuals start losing muscular mass well before their fat has been exhausted.

The highly complex nature of the interactions that define the relationship among disease, nutrition, and growth emerges most clearly from the research in this area. Thus although growth may adapt to match food availability, at least up to a certain point, this seems not to exacerbate the effects of disease. However, when infections do occur (as they inevitably will), they also affect growth. This effect may be direct, through reduced absorption of food (or changes in the metabolic efficiency with which food is used), or it may be indirect, through reduced appetite.

In other words, the effects of the seasonal pattern of disease on growth can be indistinguishable from those of a seasonal reduction in the amount of food supplied. Even more complex than this is the fact that the extent to which food consumption and growth are affected by infections will depend on the level of knowledge and seasonal variations in how much time a mother can devote to child care, which are in turn components of what Sen (1981) describes as the endowment and exchange entitlements of the household. Thus the notion that seasonal fluctuations in body size are a measure of adaptation to the combined effects of a number of different components of the environment in itself reveals little, if anything, about the relative importance of any one of those components.

Anthropometric Indicators and Seasonality

Poor general environmental conditions, with high levels of exposure to diseases, coupled with limited command over economic resources, may provide as an additional effect of seasonality the increment of stress needed to precipitate severe growth depression or loss of body weight. A further complication arises from the fact that many of the infectious diseases are themselves linked to seasonal factors in the environment. Black, Brown, and Becker (1984a) have described how in Bangladesh the seasonal incidence of diarrhea and of corresponding checks to growth is related to reduced food intake coincident with the peak of incidence of infection. It is, however, impossible to say which is the initiating event.

One consequence of the neglect of the subject of seasonality is that of the large numbers of studies on the growth of children in relatively affluent societies, very few have taken account of possible seasonal effects. However, this has not always been the case. During the 1920s and 1930s, the effect of seasonal factors on child growth was a subject of great interest. Reviews by Nylin (1929), Orr and Clark (1930), and Palmer (1933) all agree on the following:

1. Seasonal changes in growth rates occur in children at all ages and in many different countries. Palmer says of her own study of 2,500 children living in Hagerstown, Maryland, United States, that the magnitude of these changes is "such as quite completely to dominate any change in growth rates which may be expected to occur in successive calendar years."
2. Growth in weight proceeds at maximal rates during the late summer

and early autumn, and declines during winter to a minimum during spring and early summer, with velocities of weight gain frequently falling to zero. The same pattern has been found in Australian children, not at the same time, but at the corresponding seasons for the Southern Hemisphere (Fitt 1924).
3. Palmer's study of schoolchildren showed that the effect was not due to cyclic patterns of infectious diseases.
4. Growth in height is also affected, but with a difference in phase such that maximum height increase comes at the time when weight gains are at their minimum, just as Black et al. (1982) found in Bangladesh.

Dugdale and Eaton-Evans (1985) studied a sample of 1,000 infants born in Brisbane and compared the summer and winter weight gains during the first to third months of age. They found significantly slower growth during the winter months. Griffiths (1985) has recorded daily weights for two-year periods for a group of preschool and early school-aged children of professional parents living in London. All shows signs of a marked staircase effect of the season of the year on the course of weight gain. In some individuals, growth virtually ceased for three-month periods, resulting in crossing of a number of percentiles of standard National Center for Health Statistics growth charts—an effect just as great as observed by Billewicz and McGregor (1982) for Gambian children.

This is not the place for an extensive review of the direct biological effects of temperature and day length on growth and metabolism. However, it is well established that a number of animal species—for example, deer and sheep, as well as some smaller animals—show regular seasonal cycles of food intake, resting rates of energy metabolism, and growth. Thus growth in sheep and deer over the first few years *normally* has a staircase pattern—rapid in summer and slow in winter. These effects are not mediated through food supply—they persist even under controlled conditions when food composition is constant and access is unlimited. They can, however, be shown to be set in train by changes in day length—longer summer days produce more rapid growth, higher rates of energy metabolism, and greater voluntary food consumption than short winter days, the organ responsible being the pineal gland.

The picture is thus incomplete and inconsistent with respect to timing and, therefore, likely causes. Nevertheless, there remains the strong likelihood that a significant component of the seasonal growth disturbances we see in developing countries may be found to be due to factors other than lack of food or the incidence of disease.

Despite their shortcomings and uncertainties of interpretation, anthropometric indicators remain the most practical and certainly the most widely used means of assessing the health status of children and adults.

Over the range of adaptive adjustment—so-called mild to moderate malnutrition—growth can provide a proxy for the extent of fluctuations of environmentally caused seasonal stresses, plus, in more seriously affected people, evidence of the timing and magnitude of stresses that cross a critical threshold of risk and result in frank malnutrition. The major problem in interpreting anthropometric indicators remains that of distinguishing the effects of seasonality in disease from those of primary food shortage.

This is particularly the case with child anthropometry, because of the complexity of the interaction between disease and growth. Until about the mid-1970s it was commonly accepted—more because of vigorous advocacy than on the basis of any objective evidence—that small size in children was sufficient prima facie evidence of current deprivation and distress. Moreover, the need for improved food availability was stressed as being the major policy implication. To the extent that disease was acknowledged to be an element in small size, it was asserted that this would also be countered by improved nutrition. Thus there were many who argued that combating "marginal" and "moderate" malnutrition through increasing food consumption would be a sufficient strategy for improving child health—albeit one to be assisted and accelerated wherever possible by improving immunization and basic health provisions.

The evidence now is that infectious disease, or rather the lack of means to combat it, needs to be treated as a primary deprivation in its own right, and not one that can be dealt with by presuming that the benefits of improved food consumption can substitute for those of disease control. Indeed, if anything experience points in the reverse direction: in many cases, disease not only overlays but extensively distorts the effects that seasonal variations in food availability would otherwise have. The implications of this for understanding seasonality are that interpretations of fluctuations and changes in growth and body size must include an assessment of the pattern and character of the infectious diseases to which the particular population is exposed. Comparison of the weights or heights of children in different locations or at different times without reference to the level of exposure to malaria, measles, or diarrhea can tell us little, if anything, about the extent to which their size is conditioned by the food available to them as individuals, and even less about the food entitlement of the households to which they belong.

With respect to the health status of adults, the situation is probably very similar though much less well studied. Seasonal infections are probably an important cause of loss of work time, or of reduced work capacity during the peak season of labor demand. A study by Maxwell, Stutley, and Bojanic (1982) of six Bolivian farmers showed that only one was not ill during the study year, four lost between 3 and 20 days, and one lost more time sick (128 days) than he spent on farm work. Most of this was during the main agricultural season, and there were additional days lost because of illness of wives and children. One of the main effects of seasonality on hunger may well be mediated by the influence of disease on production, and we might even speculate that in some circumstances, reduced growth of children in a household could as well be seen as a measure of the general level of disease risk to which all of the household's members are exposed, and hence an indication of the cause rather than the outcome of impaired food entitlement.

Food Consumption Indicators

There are methodological problems that will probably always severely limit the utility of quantitative estimation of nutrient or energy consumption, except for highly controlled research situations. Intakes *can* be measured in individuals, but precision is bought at the expense of invasiveness as well as money and time: it is possible to know accurately how much food a person has eaten over a period, but only with so much observer interference as to cast doubt on the result as an estimate of normal habitual intake. This does not, of course, mean that measures of household food consumption, or of changes in patterns of food availability, or of critical levels of food stocks, and so on are useless; simply that the widespread belief that nutritionists can "tell what a man is by what he eats" falls sadly short of the truth. Even if we could measure intakes of individuals, the problems of constructing indices of nutritional adequacy are essentially the same as those of interpreting differences in growth and body size. In fact it was Sukhatme's (1961) critical attack on the conventional wisdom about interpreting household intake data, and his advocacy of an approach that takes account of adaptive responses, that largely initiated the current controversy. With respect at least to energy, the problem of interpretation is made even greater by the very wide range of adaptive responses that are made possible through altered physical activity levels; it is of course precisely those changes that are a main feature of the effects of seasonality.

It seems unlikely that these difficulties will be resolved simply by greater precision in the prescription of requirements. The latest FAO-WHO-UNU (1985, p. 12) report on protein and energy defines energy requirements as follows: "The energy requirement of an individual is that level of energy intake which will balance energy expenditure when the individual has a body size and composition and level of physical activity consistent with long-term good health, and which will allow for the maintenance of economically necessary and socially desirable physical activity." This is based on the understanding that people can adjust to changes in energy intake in a number of ways, some physiological and some behavioral. Thus a reduction in energy intake can be compensated for by reducing expenditure either at work or during leisure time. Besides this, however, changes in body weight or body composition and metabolism can result in an altered requirement for maintaining the body, again offsetting a reduced intake. These different kinds of adjustments can take place either singly or in various combinations, so that in practice high or low levels of physical activity can be found associated with either large or small body size, and with marked differences in composition.

This definition of requirements does move a little toward accepting the concept of adaptation. Thus there are said to be upper and lower limits to the range of body weights that are consistent with good health; the same is true for changes in body composition or metabolic regulation. But at the same time it is implied that the requirement should be such as to allow for some minimum necessary and desirable levels of physical activity. The concept of energy requirement is no longer one of a single, fixed level appropriate for a given person, but rather one of a set of safe limits, defined in terms not just of physical health but of an acceptable level of economic and social performance. A person who adjusts to seasonal changes in intake in such a way as to remain within those safe limits could be said to have successfully adapted to the seasonal fluctuations.

Changing the unit of measurement from the individual to the household clearly adds to the problem, since we cannot aggregate the individual requirements of family members without making assumptions about distribution in addition to those that have to be made in order to fix the requirements of each of the members.

The questions still remain: How can food-deficit households be identified? How can chronic or seasonal hunger be measured? For all practical purposes, the only consumption data we have for most of the rural populations of developing countries are household-level assessments, for the most part made at unknown times of the year and usually relating to periods of a very few days. Are these of any value for assessing household food security? Before attempting an answer, we will need to return to the point made in the introduction about the current lack of an adequate theoretical basis for analyzing energy balance in dynamic situations.

A Simulation Model of Seasonal Changes in Energy Balance

In populations subject to seasonality of work and food supply, energy equilibrium is the exception rather than the rule. This is particularly true of regions where peasant farmers harvest one staple crop a year, which has to be stored either in granaries or in the ground until needed. Food in storage inevitably suffers spoilage: the longer it is kept, the greater the loss. In addition, the work load of such farmers is variable. Immediately after harvest there is commonly a rest period of several months. Then the heavy physical work must be done to prepare the ground for the next crop, followed by planting, weeding, and finally harvesting. The period of greatest energy expenditure is therefore several months removed from the last harvest. The farmer and his family could in theory use their stored food in a variety of different ways. If they were to eat large amounts soon after harvest, they might reduce the overall storage losses but would incur greater metabolic losses because of their increased body weights. They might also run the risk of having insufficient food to balance the high energy expenditures needed to cultivate and harvest the next crop, and would thus experience large weight losses. At the other extreme, they might conserve and ration food throughout the year so as always to balance current demands for energy and so maintain a constant body weight. In this case, the mean storage time of food would be longer and year-round losses would be greater. In other production situations, factors such as the use of animal or other power assistance, multiple cropping, or seasonal off-farm employment opportunities would of course complicate the range of strategy choices.

The Model Man

Payne and Dugdale (1977) have described a model that simulates the regulation of energy balance in adults. In this model, the tissues of the body are represented as four compartments linked together as shown in Figure 2.2

The model is operated through a computer program, which calculates day-by-day changes in body weight and composition. After specifying initial values for body size and consumption, the inputs to the computer consist of daily schedules of food energy intake and physical work output. At the end of each day, the energy needed for maintenance is calculated from the sizes of the body compartments, and hence the energy balance for that day can be computed. If the balance is positive, the surplus energy is stored in the different compartments as lean tissue and fat tissue; the ratio of lean to fat stored is fixed for any particular individual. If the energy balance for the day is negative, the deficit is made up by depleting the compartments of lean and fat tissues, again in the same ratio. The model does not have any preferred value of body size, but if the average level of food intake or of work output is changed, compartment sizes and hence maintenance energy requirements will adjust upward or downward until equilibrium is reestablished. If intake and expenditure vary over time—for example, because of seasonal effects—then the output of the model will be a description of the pattern and magnitude of seasonal weight fluctuations.

FIGURE 2.2 The "model man," showing the tissue compartments and the balancing compartment

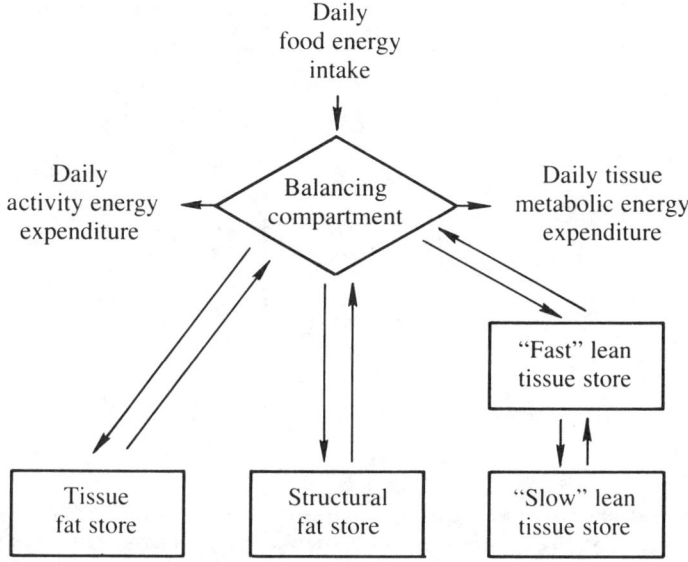

SOURCE: Reprinted with permission from Payne and Dugdale 1977.

The Model Farm

The model can be extended by linking the "man" to a "farm" on which he can grow food in exchange for making work inputs. The harvest from the farm is placed in a food store from which the farmer can take the food every day over the subsequent year until the next harvest comes in.

Figure 2.3 shows how these components of the model farm system are linked. The box containing the farmer represents the model man already described. In what follows, the farm is restricted to a single crop production unit that yields the same amount of food energy every year. The seasonal patterns of work input to the farm and food withdrawn from store can be specified. At harvest, the total annual crop yield is transferred to the food store, after which it becomes subject to a storage loss that can be specified in the program as the daily fractional rate at which the food energy contents of the store are degraded by microbiological spoilage, insect or rodent attacks, and so on. The farmer can take food from the store each day in various ways so as, for example, to meet periods of heavy work requirements or postharvest feasting. A number of runs of the model have been made using different food allocation patterns and different rates of storage loss, all of which have been designed so that the total amount of

FIGURE 2.3 The "model farm," with compartments showing the farmer ("model man"), the farm, and the food store

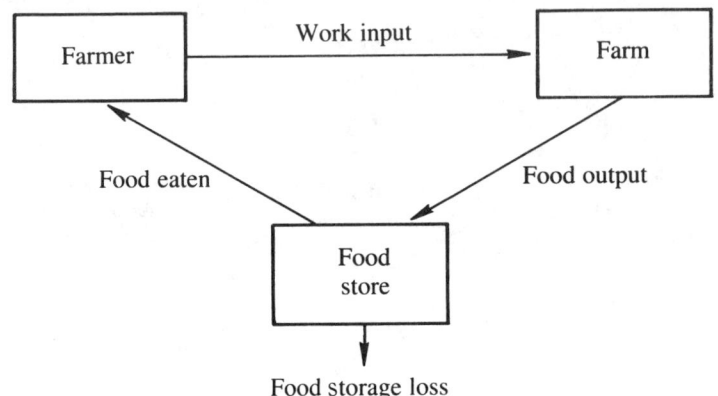

food available each year—that is, farm production minus storage loss—is completely used up. There is never a surplus carried over from one year to the next. In this form the model is self-stabilizing and after a number of simulated years (usually 10–15) settles down to an exactly repeating cycle of weight changes.

In what follows, the model is used to show the effects of different patterns of food use using data reported by Fox (1953) for Gambian farmers and Tin-May-Than and Ba-Aye (1985) for Burmese farmers. Table 2.1 shows a summary of the measurements of body weight and height, body mass index (weight/height^2), and the seasonal range of energy expenditures.

At the time when Fox described them, the Gambian men were growing a single annual crop of millet, for which they prepared the ground by manual labor entailing peak work expenditure rates of a little over 1,000 calories per day. During the postharvest period, they were described as extremely sedentary, with maintenance energy expenditures barely above basal rates. The Burmese farmers were also producing a single crop (rice) and had very high daily rates of energy expenditure during the ploughing season, despite the use of draft animals. Expenditure rates were also high during the harvest period, and some off-farm employment was taken when there was no farm work to be done. These differences in work pattern are reflected in very much higher annual rates of energy expenditure by the Burmese, 210 percent of basal metabolic rate (BMR) compared with 140 percent of BMR by the Gambian men. Despite this, the Burmese are shorter and lighter and have lower body mass indices than the Gambians.

TABLE 2.1 Body size and energy expenditure rates, Gambian and Burmese farmers

	Gambia			Burma		
	Mean	Max	Min	Mean	Max	Min
Weight (kg)	58.47	59.80	56.80	53.50	54.80	52.10
Height (cm)	168.30	—	—	164.70	—	—
Body mass index (BMI)	20.60	21.10	20.10	19.70	20.20	19.20
Seasonal energy expenditure (kcal/day)	2,100	2,600	1,600	3,003	3,840	2,230
Energy expenditure (kcal/kg)	35.90	—	—	56.10	—	—
Energy expenditure as percentage of basal metabolic rate (BMR)[a]	140	—	—	210	—	—

SOURCES: Data on Gambia from Fox 1953; data on Burma from Tin-May-Than and Ba-Aye 1985.
[a]BMR has been calculated using the DuBois and DuBois monogram.

FIGURE 2.4 Energy intake, work output, and body weight of Gambian farmers

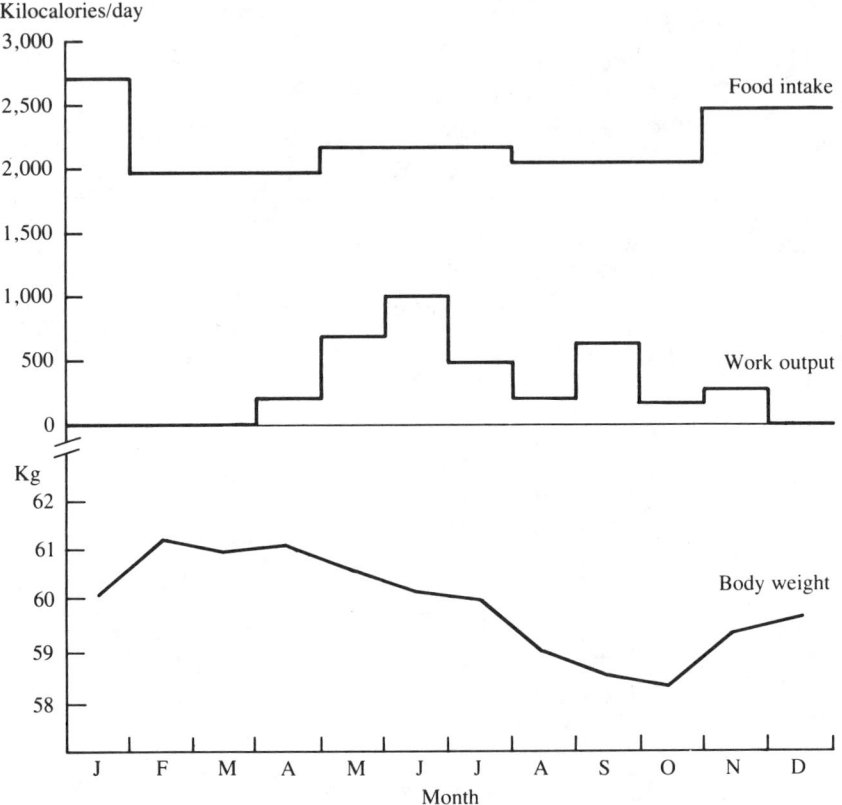

SOURCE: Adapted from Fox 1953.

The Gambian Farmer

The Gambian farmers do not eat the largest amounts of food at the time when work on land preparation and harvesting is heaviest. Instead, they eat most during the immediate postharvest period (see figure 2.4), when their work energy expenditure has fallen to a very low level. This traditional pattern of postharvest feasting, which is not an uncommon feature of traditional peasant communities, is in part responsible for the seasonal weight fluctuations, and it is interesting to see how this chosen strategy compares with other possible ones in terms of weight changes and storage losses.

Four different strategies of food allocation have been compared:

1. The time pattern of energy intakes actually observed by Fox for the whole Gambian farming community.

2. Exaggerated postharvest feasting, falling off rapidly to a much reduced level—this is intended to show what could be the maximum effect of early food use on postharvest storage losses.
3. A constant daily intake over the whole year.
4. A feeding pattern adjusted to maintain constant body weight—that is, intakes always balancing expenditures.

The results of simulations using these four strategies are shown in figures 2.4, 2.5, and 2.6.

Figure 2.4 shows the observed patterns of expenditure and intake and also the actual changes in mean weight of a group of Gambian men. When these observed weights are compared with those predicted by the model, the actual and simulated weights are similar (see figure 2.5). The model shows larger seasonal swings, but these are well within the range of individual variation for men from the same region, as reported by Billewicz and McGregor (1982).

Figure 2.6 shows the three other allocation patterns. Table 2.2 shows the average year-round body weights and the maximum seasonal weight changes predicted by the model and how these are affected by the rate of storage losses.

When storage losses are set at zero, the best strategy for food allocation seems to be the one that sustains body weight at a constant level. However, as the rate of storage loss is increased toward more realistic values, the advantages of high postharvest consumption rates become greater, with average weights highest of all for the exaggerated postharvest feeding schedule and lowest for the constant weight strategy.

FIGURE 2.5 Seasonal changes in body weight of Gambian farmers

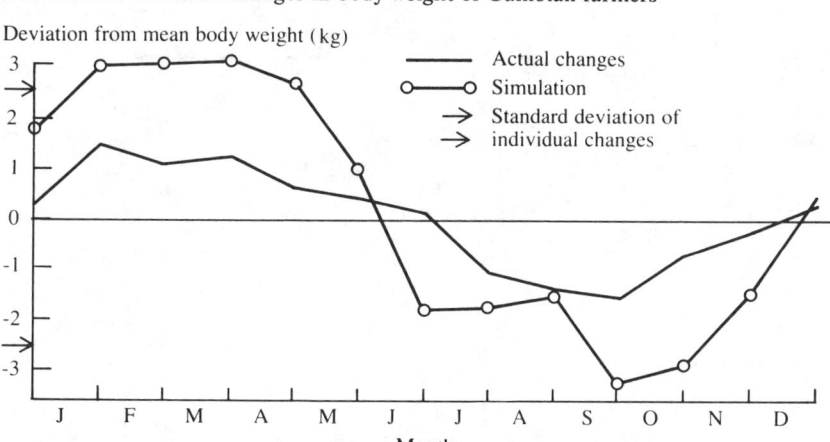

SOURCE: Actual data from Fox 1953.

38 *Philip Payne*

FIGURE 2.6 Four patterns of food allocation tested using the model, and actual work output of Gambian farmers

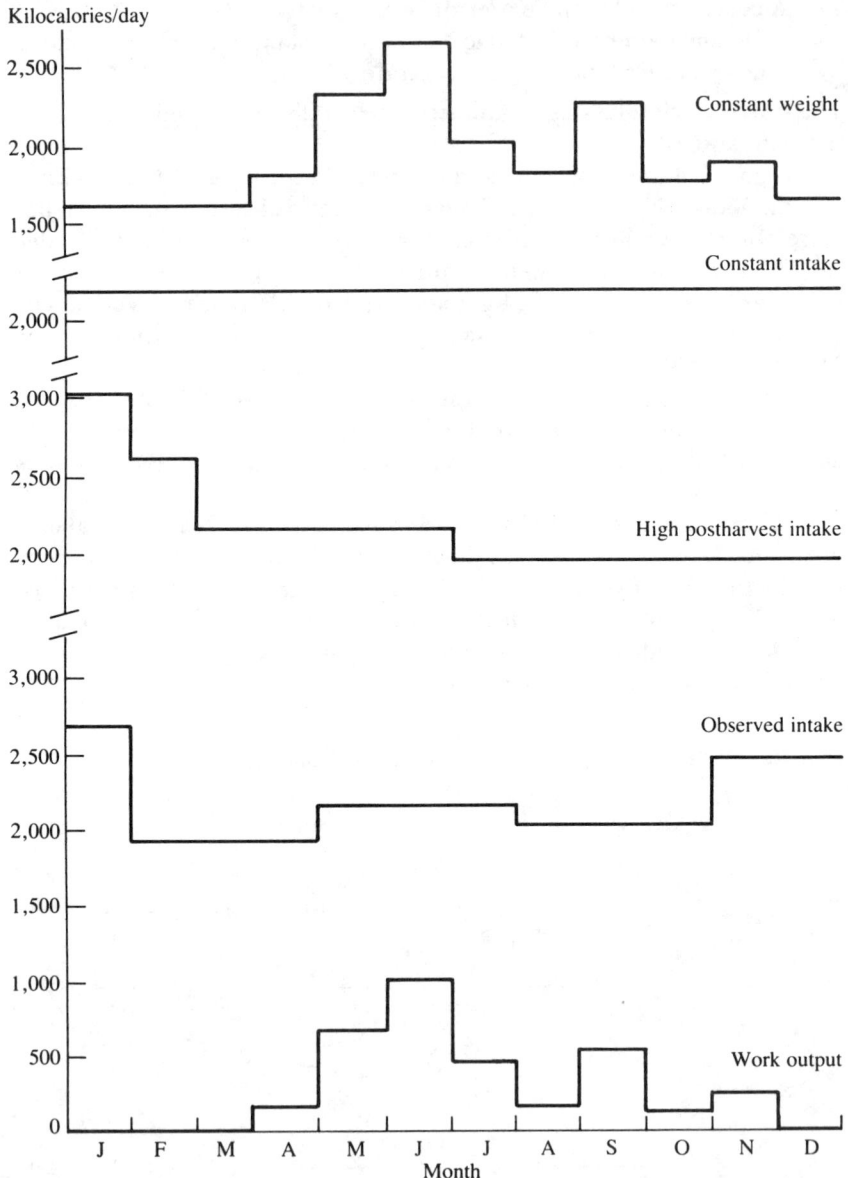

SOURCE: Actual data from Fox 1953.

TABLE 2.2 Seasonal weight changes of Gambian farmers, simulated using different food allocation strategies and rates of postharvest losses

Body Weights at Different Rates of Storage Loss		Food Allocation Strategies			
% Loss/ Day	Weight (kg)	Observed Intake Pattern	High Post-harvest Intake Pattern	Constant Intake Pattern	Constant Weight Intake Pattern
0	Max	64.2	65.1	65.1	65.0
	Min	57.7	57.4	59.9	65.0
	Mean	61.5	60.8	62.2	65.0
	Range	6.5	7.7	5.2	0
0.03	Max	59.4	62.0	58.9	55.4
	Min	52.9	54.4	53.7	55.4
	Mean	56.8	57.7	56.0	55.4
	Range	6.5	7.6	5.2	0
0.10	Max	51.8	54.1	51.3	47.2
	Min	45.4	46.7	46.0	47.2
	Mean	49.1	49.9	48.3	47.2
	Range	6.4	7.4	5.3	0

Average year-round body weight is probably less important than the maximum body weight, which is attained at the start of the heavy work season. As table 2.2 shows, the effect of different allocation patterns on maximum weight is even greater than the effect on average weight. What is particularly striking is the relative disadvantage of attempting to maintain a constant body weight, which is, of course, the result of eating according to the conventional advice of matching intake with "requirements"!

The observed intake pattern shows clear advantages over both constant intake and constant weight strategies, and is indeed quite close in efficiency to the exaggerated postharvest feasting pattern. These advantages derive partly from the fact that high body weight is sustained at just the time when it is likely to be most crucial for work output, but is then allowed to decline and thus reduce maintenance costs during the rest of the year. But as table 2.3 shows, storage losses are significantly lower when using the high postharvest intake strategies, and this of course results in higher overall consumption.

As table 2.3 also shows, the model simulates very well the effect on annual storage losses of withdrawals of food for consumption throughout the year. Food left uneaten in store and exposed to daily losses of 0.03 percent or 0.10 percent would be degraded as much as 10 percent or 30 percent per year, respectively.

TABLE 2.3 Effect of different food allocation strategies on annual storage losses

Percentage of Stored Food Lost per Day, and Theoretical Annual Losses	Percentage of Food Lost during One Year with Different Food Allocation Strategies			
	Observed Intake Pattern	High Post-harvest Intake Pattern	Constant Intake Pattern	Constant Weight Intake Pattern
Gambian farmers				
0.03 (10% per annum)	5.1	4.9	5.3	5.5
0.10 (30% per annum)	16.3	15.9	16.9	17.5
Burmese farmers				
0.03	5.4	5.1	5.3	5.4
0.10	17.2	16.4	16.9	17.4

The Burmese Farmer

The data given by Tin-May-Than and Ba-Aye (1985) suggest that the annual energy intake for Burmese farmers was 17 percent higher than the annual expenditure. An attempt at simulation using these values in the model gave initial weight gains of 11.0 kilograms (kgs) per year, with average body weight plateauing at 100 kgs over the course of five years.

This implausible value suggests that there is a systematic error in the reported intake data, and the total annual intake has been scaled down accordingly while maintaining the relative pattern of seasonal variation. Subject to making this adjustment, the pattern and extent of seasonal weight changes generated by the model are in very good agreement with those observed (figure 2.7).

The same range of strategies of food allocation has been studied as with Gambian farmers. As before, each of these strategies has been tested with various rates of storage loss. As table 2.3 shows, the proportion of food lost during a year depends again primarily on the extent to which food is consumed early in the storage cycle. The difference between the Gambian and Burmese systems, however, is that the observed pattern of food use is far from being the optimum from the point of view of food conservation. Table 2.4 shows the outcomes in terms of body weight and the magnitudes of weight changes.

The contrast with Gambian farmers is that the observed pattern of food use is not the one that would sustain the highest possible body weights, either on average or at the start of the heaviest work period. In effect, the choice seems to be directed much more toward constancy of body weight, even at the cost of somewhat greater food losses.

Table 2.5 shows the effect of the different strategies on body mass index at different times of the year. Compared with either the exaggerated

Seasonal Hunger and Malnutrition 41

FIGURE 2.7 Seasonal changes in body weight of Burmese rice farmers

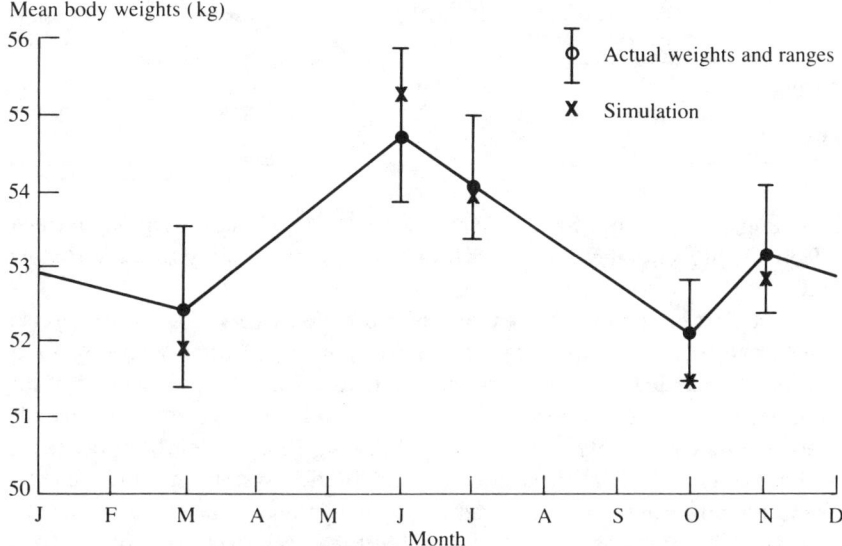

SOURCE: Actual data from Tin-May-Than and Ba-Aye 1985.

TABLE 2.4 Seasonal weight changes of Burmese farmers, simulated using different food allocation strategies and rates of postharvest losses

Body Weights at Different Rates of Storage Loss		Food Allocation Strategies			
% Loss/ Day	Weight (kg)	Observed Intake Pattern	High Post-harvest Intake Pattern	Constant Intake Pattern	Constant Weight Intake Pattern
0	Max	59.5	66.2	62.4	65.5
	Min	57.1	54.0	53.7	65.5
	Mean	58.5	61.6	59.6	65.5
	Range	2.4	12.2	8.7	0
0.03	Max	54.0	58.7	56.9	55.3
	Min	50.2	46.7	48.1	55.3
	Mean	52.9	54.1	54.0	55.3
	Range	3.8	12.0	8.8	0
0.10	Max	43.8	47.5	47.0	43.9
	Min	40.2	35.7	38.0	43.9
	Mean	42.4	42.9	43.9	43.9
	Range	3.6	11.8	9.0	0

TABLE 2.5 Simulation of seasonal changes of body mass index, Burmese farmers

	Observed Pattern	High Post-harvest	Constant Intake	Constant Weight
Mean	19.5	19.9	19.9	20.4
Max	20.4	23.2	22.5	20.4
Min	19.0	17.9	17.8	20.4

early consumption or the constant intake strategy, the observed pattern effectively prevents the body mass index (BMI) from falling to values below 19.0.

The first points to emerge from this admittedly very simplistic exercise are that attempts to assess adequacy of food supply on the average, or food stocks at a particular time, can only be made from the perspective of an understanding of the particular farm and livelihood systems concerned; there are no completely general rules. Looking first at Gambian farmers, we have a relatively simple single-crop production system that (at the time when the data were collected) was still largely one of subsistence farming. Immediately after the harvest, when food was relatively plentiful, there was a period of several months with little agricultural work (for men at least), but a fair number of traditional festive occasions. Food eaten in excess of requirements produced very rapid weight gain, so that with the advent of the rains when the heavy work of land preparation had to be done by human labor, body weight was maximal. Food stocks, however, were then at their lowest and intakes much less than requirements, leading to weight loss. At first sight, the picture is one of people constantly trying, and apparently always failing, to produce enough food to last them over the year so as to sustain nutritional status, or at least body weight, at a constant level. The model shows, however, that the strategy followed for rationing out food and effort was in fact a conservative one, minimizing postharvest storage losses while ensuring maximum body weight during the short period when peak work output was needed. The result was a livelihood system that supported people at a level of year-round energy expenditure of only 1.4 times their basal rate of metabolism—a figure very little above that considered to be the absolute minimum for survival (FAO 1974).

The Burmese farmers rely on rice as a main source of food energy and grow this for a restricted period of the year, but are more engaged in the market. They generally produce surplus rice for sale and in addition take off-farm employment; there is no time of the year spent in total inactivity. Although their daily rates of energy expenditure are very high during the whole of the four months of the ploughing season, the character of the work is likely to be different, since they use draft animal power. They are able to accomplish this work with smaller body weights than the Gambian men, even allowing for their smaller height. Fairly high rates of daily work output are sustained for much of the rest of the year, and the overall production system is maintained by year-round energy expenditures of 2.1 times their basal metabolic rate. The simulations show that although there might be advantages of reduced storage loss and of higher maximum body weights as a result of concentrating consumption of food early in the storage cycle, these would be offset by much greater seasonal swings of body weight. Their higher level of overall production and ability to exchange cash for food probably means that they are less sensitive to the relatively small changes in food entitlement that these reduced storage losses represent. However, the very high rate of energy turnover is combined with the low maintenance requirements that go with small body size and hence small body reserves of energy. Consequently, the factor that is likely to be critical for them is that of maintaining weight or BMI above some minimum value below which their capacity for long, sustained (as opposed to peak-intensity) work output might be threatened. The model simulations show that compared with the other strategies, the observed pattern of eating ensures that BMI does not in fact fall below 19.0.

The model has made it possible for the first time to quantify the effects of food withdrawals on annual storage losses. The literature contains two kinds of estimates of loss. Until recently, most have been based on short-term degradation rates with no adjustment for food withdrawal (Krishnamurty 1970). These studies suggest cumulative annual losses of between 15 percent and 25 percent. More recent estimates of 4–5 percent (Boxall et al. 1978) take account of the fact that in practice food is continually withdrawn from storage for consumption and is therefore not all exposed to a full year of spoilage. The model shows quantitatively the extent to which this withdrawal of food reduces the theoretical losses from spoilage, and demonstrates also how different seasonal patterns of consumption can alter the overall annual losses.

Policy Implications of the Model

As it stands now, the model only offers an exceedingly simplified version of the real-life situation of peasant farmers. Nevertheless, it extends understanding of seasonality by demonstrating the effects of interactions between production and nutritional status, and between food storage and consumption. It is to be hoped that this kind of analysis will underline the limitations of current conventions about how to assess the adequacy of food consumption. Up to now, this has been based on static comparison of intakes and requirements, both of which are usually assumed to have been averaged over a sufficient length of time to obscure or remove the effects of fluctuations. The model suggests the need to take a much more dynamic view of the relationship, and one that recognizes the human subject as part of a system that includes the fluctuating demands of his or her annual work pattern. An important outcome is likely to be a replacement of the simple notion of an adequate or inadequate food supply with a capacity to identify critical situations and points in the calendar when the viability of the system may be at risk. The Gambian type of livelihood depends on attaining the sufficient body weight, hence muscle mass, needed to enable intense but relatively short periods of work to be performed while using up body stores of energy in order to fuel that work.

By contrast, the Burmese situation seems to demand endurance rather than a high peak-power output, and to do so over lengths of time so that body energy stores are of relatively smaller significance. For such people, food intakes must be sustained at a level that will prevent already low body weights from falling below the point at which the capacity for sustained work suffers. At the moment we know very little about these critical weights and muscle masses, and much more research will be necessary before it is possible to predict the probability of failure of farm systems due to nutritional inadequacy. However, this points in the direction in which more research is needed.

In principle, the model could readily be elaborated. The constraints such as a fixed level of food production, no carryover of food stocks from one year to the next, could be removed. Food production could be made variable and dependent on the level of work input to the farm. A model household could be assembled taking account of male, female, and child contributions of labor and requirements for energy. The household could then be exposed to year-by-year fluctuations of production to test the sensitivity of the system and the different strategies of food use (including intrafamily food distribution) to the effects of bad years. However, the constraints are not so much in the complexity of the programming or the computing capacity involved as in the limitations of existing data.

Very few studies of traditional farming communities have been reported that give figures for food intake, energy expenditure, and body weights throughout a complete annual production cycle. Teokul, Payne, and Dugdale (1986) have reviewed the literature and have found only 10 descriptions of energy turnover in farmers and pastoralists. Of these, only 5 provide the full set of data needed. Even such a small sample, however, makes it clear that people adopt quite different strategies for coping with the risks generated by seasonal food insecurity.

Thus it seems probable that at one extreme there is a type of strategy adopted by populations dependent on a single main food crop, grown at a particular and restricted time of the year, which is dictated by the rainfall pattern. If such people are not to any appreciable extent engaged in cash cropping and have no opportunities for off-farm employment, then their pattern of food use and weight changes will follow the Gambian model. With increased opportunities for filling in the gaps of the family farm production calendar, either with extended varieties of food or cash crops or by selling their labor, the seasonal changes in energy balance become less marked, as among the Burmese farmers. This trend may continue to the point where seasonality of food intake can no longer be measured. According to Bidinger, Nag, and Babu (1986), this applies in some parts of India; and Flores and Flores (1984) say the same is true of many populations in Central America.

Taking into account the methodological difficulties of the measurements, it does not seem likely that greater availability or precision of intake and expenditure patterns will be of much assistance in identifying high-risk populations or for defining policy options that would specifically relate to seasonal insecurity. Direct observation of regular episodes of indebtedness or low household stocks seems to be a much more realistic approach.

The simulations do, however, underline the importance of adult anthropometry as a way of identifying the times of the year when failure to sustain work performance may be critical for the viability of particular livelihood systems, despite the many uncertainties in the interpretation of adult anthropometry data. It is evident, however, that we need to know much more about the relationship of body size and build and ability to produce high rates of work output for shorter times or to sustain long hours of work throughout the year. Also, as in the case of children, infectious diseases will complicate the picture, and these will always need to be assessed separately in any research studies of community health monitoring systems.

As far as children are concerned, the most important features of seasonal effects will be those that operate through the nutrition-infection interrelationship. The climatic aspects of infectious disease transmission, particularly for those conditions that adversely affect young children, have been reviewed in detail by Chambers, Longhurst, and Pacey (1981).

Among the vector-borne diseases, malaria, guinea worm, and sometimes dengue and other insect-borne viruses have the greatest seasonal impact. Schistosomiasis and sleeping sickness may show seasonal transmission but the consequences are drawn out, and this is sometimes also true of kala-azar and filariasis. In the savanna areas—for example, Gambia—malaria transmission increases when the rains begin following the rise in mosquito population. In some parts of the Indian subcontinent, malaria is also a disease of the rains, but in others mosquito breeding is restricted to the dry season when the streams are not in spate.

Diarrheal diseases show marked seasonal rises but against a continuously high background level in many areas. The peak often comes in the hot season or during the early rains, but detailed analysis shows that differing causes of diarrhea have staggered peaks. Respiratory infections may resemble the pattern for diarrhea in their peak season, but this is variable. Cerebrospinal meningitis is spread by the respiratory route. The immense epidemics of the African savanna tend to occur near the end of the dry season and in a highly predictable manner in any one locality.

The greatest regular seasonal impact on rural society, then, is often due to malaria, diarrheal diseases, and guinea worm. The brunt of the first two falls upon infants, though all age groups are affected. There are also seasonal variations in human behavior that have a large effect: the contact between people indoors during a cold, dry season will facilitate the spread of some respiratory infections.

Some of the implications of this are obvious—for example, the management and timing of programs for the extension of immunization and for the control of vector-borne infections, and the priority in many populations for an improvement in housing and access to fuel.

The less obvious, but probably equally important, links are those that make the connection between the food and health entitlement of the household and the actual condition of the dependent children. These relate not only to the strategies and resources needed for reducing the risks of infection in the first place, but also to the successful management of diseases such as diarrhea and respiratory infections once they have been caught. In both of these aspects, it is likely that if seasonality has an effect, it will be through intensifying the conflict between the claims of food production and the claims of other work on the time and other resources of adult household members—that is, between the maintenance of household food security on the one hand and the competing demands of child care on the other. It goes without saying that, in most instances, it is the time and resources controlled by the woman that are of key importance to the survival of the child.

It is difficult to draw conclusions about these situations in such a way as to provide general guidelines for public health strategies that would specifically address the impacts of seasonality. However, the general line of argument of this chapter has been toward the view that although the impacts of the seasons on many people are adverse and serious, the nature of the process through which they operate is both complex and varied.

3 Seasonal Pattern of Activity and Its Nutritional Consequence in Gambia

MARK LAWRENCE, F. LAWRENCE, T. J. COLE,
W. A. COWARD, J. SINGH, AND R. G. WHITEHEAD

Gambia is a small country in the extreme west of Africa, a 30-mile-wide strip of territory on the north and south banks of the river Gambia. The three villages that formed the subject of this study—Keneba, Kantonkunda, and Manduar—lie in the western half of the country, in an area somewhat isolated by brackish river and salt marsh. The majority of Gambians are subsistence farmers growing millet, sorghum, and rice for food, and groundnuts for consumption at home and for sale as a cash crop. These crops are grown during a relatively short monomodal rainy season from July to October. The problems of a short growing season have been compounded recently by several years of low rainfall. Although total crop failure has not yet occurred in Gambia, partial crop failure is to a greater or lesser extent an annual event, and this results in regular food shortages prior to each harvest at a time known locally as the hungry season.

Study Design and Analysis

We have recently completed in Keneba the Gambian component of a multinational study of energy balance during pregnancy and lactation. This involved longitudinal measurements, over a period of up to three years, of changes in maternal adipose tissue stores, changes in various aspects of total energy expenditure, and changes in total food intake. These data have been analyzed using within-subject multiple linear regression to investigate the influences of (1) stage of pregnancy or lactation and (2) season. Of these two factors, only the effect of season will be discussed in this chapter.

A further aspect of the study was to assess the impact of a maternal dietary supplementation program. The supplement consisted of an energy-dense groundnut biscuit and a vitamin-enriched tea drink prepared from locally available ingredients and aid foods. Between 1978 and 1982, all lactating women resident in Keneba received the supplement as part of a study to assess its impact on lactational performance. Starting in 1980 the supplement was also distributed to pregnant women in order to assess the impact on birth outcome. Consumption of the supplement took place six days a week throughout the year in a supervised purpose-built center. Maximum daily intake was restricted to three 60-gram (g) biscuits and 380 milliliters (ml) of tea in the dry season (1,140 kilocalories [kcal]) and four biscuits and 380 ml of tea during the rains (1,420 kcal). This generally exceeded the appetite of at least the pregnant subjects, and consequently intake was somewhat less than the amount offered. Average daily supplement intake was further reduced because the supplement center was closed on Sundays and public holidays. The program was very popular. Levels of attendance in excess of 95 percent were reported previously (Prentice et al. 1983b); and similarly high levels were maintained throughout the present study. In pregnant women, daily supplement intake varied between about 600 kcals/day postharvest and 800 kcals/day at the peak of the annual hungry season in August (Prentice et al. 1983b); in lactating women, intake varied between 700 kcals/day postharvest and 950 kcals/day in August (Prentice et al. 1983a).

For the purposes of the present study, women recruited in Keneba received the supplement from before conception and throughout pregnancy and lactation. For comparison with this supplemented group, women from the two nearby villages of Kantonkunda and Manduar were recruited; they were offered medical care similar to that for women from Keneba but did not receive the supplement. Almost all women resident in the three villages and conforming to the following selection criteria were studied: those aged 20–35 years who were expected to conceive during the period of the study and were not primigravidae. A brief subject profile is given in table 3.1.

TABLE 3.1 Subject profile

	Mean ± Standard Deviation
Age (years)	25.90 ± 4.40
Height (mm)	1,576 ± 65
Weight (kg)	51.00 ± 6.70
Parity (live plus stillbirths)	3.80 ± 1.70
Length of gestation (weeks)	39.20 ± 1.30
Pregnancy weight gain (kg)	8.30 ± 3.20
Birthweight (kg)	2.98 ± 0.32

NOTE: Details of the 50 women studied longitudinally during pregnancy and lactation are given, unsupplemented and supplemented subjects combined.

Seasonality of Agricultural Work and Its Impact on Energy Expenditure in Rural Gambian Women

Women throughout Gambia have an important role in agriculture, and in Keneba they cultivate both rice and groundnuts. Agricultural work is not restricted to the rainy season (July–October) but extends into several months before and after. Work begins in April with the preparation of vegetable gardens. In early June the fields are cleared, and in late June, as the first rains soften the hard ground, the fields are dug and seeds are planted. This work continues throughout July. In August and September the main task is weeding. In October the rice harvest starts, and this is followed in November and December by the groundnut harvest. All these tasks are carried out by hand, using cutlasses, hoes, and a variety of other hand-held implements.

Seasonal variation in activity pattern related to agricultural work could influence maternal nutritional status through an effect on total energy expenditure. The magnitude of any change in total energy expenditure will vary according to the amount of time devoted to agriculture, the level of energy expenditure during agricultural work, and, equally important, the energy cost of the activities that the agricultural work replaces.

In order to investigate seasonal variation in activity pattern, we used activity diaries kept by observers. The observer accompanied the subject throughout the day, starting at 0630 (just before the subject woke up) and finishing when the subject went to bed (2100–2200). The observer was equipped with a small "beeper" that prompted him or her every 2.5 minutes to record the activity being performed at that moment. This method was equivalent to taking a photograph every 2.5 minutes throughout the day.

In January, when almost no agricultural work was being performed, the 14.4-hour observed day was divided among lying, sitting, and standing, that is, resting or chatting and so on (7.4 hours or 52 percent of the day); food preparation activities (1.8 hours or 13 percent); household activities such as drawing water and washing clothes (2.1 hours or 15 percent); and walking (2.3 hours or 16 percent). The remaining 4 percent of the observed day was divided among miscellaneous activities, child care, and any agricultural work left from December.

Between April and July, agricultural work accounted for an increasing proportion of the observed day, until by July 3.5 hours/day were actually spent at work (largely digging) in the fields. A further 50 minutes a day were spent walking to and from the fields. This massive increase in agricultural work was achieved by a reduction in the time spent performing other activities: 1 hour less in food preparation, 1 hour less in household work, 1.5 hours less in resting, and a slight reduction in the time spent walking in the village. Finally, the length of the observed day increased by about half an hour a day at this time of year. In August and September, slightly less time was spent in the fields than in July, and thereafter the burden of agricultural work declined considerably to about an hour a day in the harvest months of October, November, and December.

A lot of time is devoted to agriculture in some months of the year, but do these tasks demand a high level of energy expenditure? While full details of the energy costs of individual activities are given elsewhere (Lawrence et al. 1985), the most strenuous activities we measured were clearing the fields and digging in June and July, which required 3.5–4.5 kcal/minute (min). This level of energy expenditure however, would be classified not as heavy but only as moderate, according to the widely accepted classification of Durnin and Passmore (1967). To a casual observer, this would come as a surprise, since outwardly these tasks seem to be quite hard work. One possible mechanism for coping with energy shortage is of course to maximize the efficiency with which any particular task is performed, and to some extent this may be happening here. Certainly these women work with considerable economy of effort; that is, there are few superfluous, energy-wasting movements. Only one other task, digging groundnuts in November, required even a moderate level of energy expenditure (about 4 kcal/min), and many other activities, such as weeding and harvesting rice (August–November), required really quite low rates of energy expenditure (about 2 kcal/min).

We divided the day according to the number of hours spent at activities corresponding to five levels of energy expenditure (<1.2, 1.2–1.7, 1.8–2.5, 2.6–3.5, and >3.5 kcal/min). For this analysis, the unobserved part of the day was assumed to have been spent sleeping at expenditure level 1 (<1.2 kcal/min).

By integrating the results of the hourly record of activity patterns that correspond to the five levels of energy expenditure, it was possible to show the substantial effect of season on total energy expenditure (figure 3.1). The lowest level of daily energy expenditure was apparently achieved in March (2,250 kcal/day), with expenditure rising to a peak of 2,650 kcal/day in July before falling once again to about 2,350 kcal/day between August and December. Some caution must be exercised in interpreting these figures, however, as a number of fairly crude assumptions have been made—for example, that metabolic rate during the night averaged 0.95 kcal/min and that the average cost of each level of expenditure remained constant throughout the year. Neither of these assumptions is likely to be entirely correct. Furthermore, it is evident that these figures for expenditure are substantially higher than the food intakes previously measured in this population. Some seasonal discrepancies might be expected, but the overall average expenditure for the year (about 2,350 kcal/day) should ap-

FIGURE 3.1 Seasonal variation in total energy expenditure by rural Gambian women, adjusted for stage of pregnancy or lactation

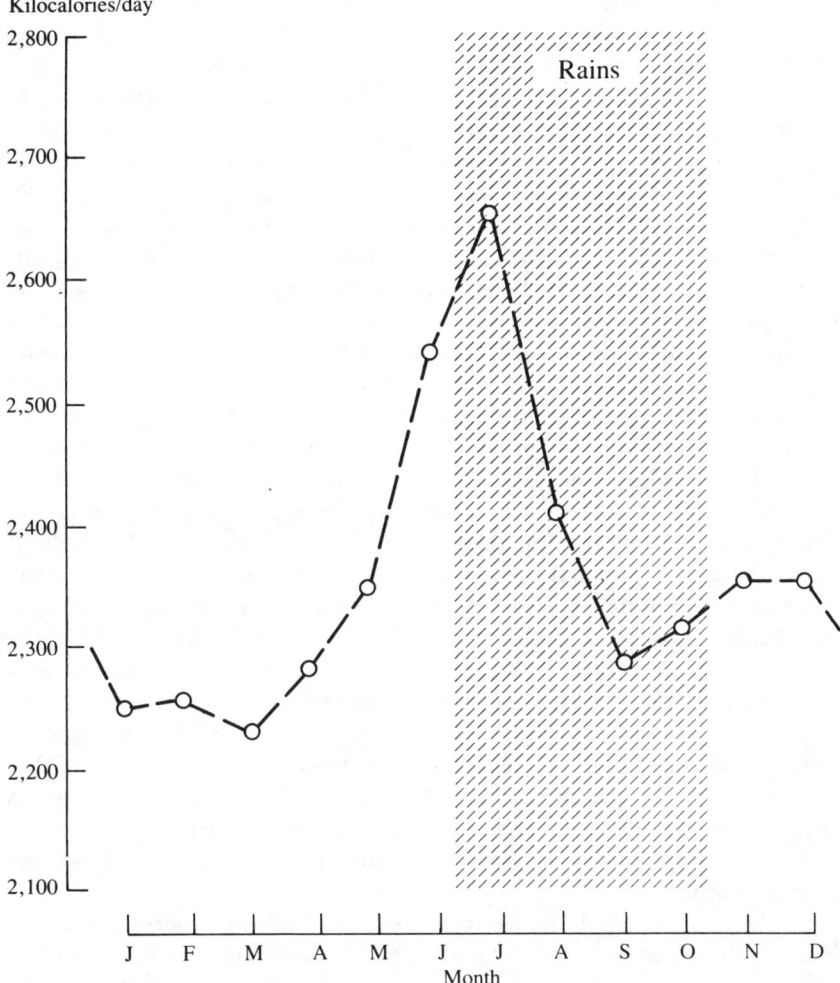

proximate average energy intake. In pregnancy, however, average food intake in Keneba, measured by direct weighing of the quantities consumed, was 1,470 kcal/day rising to 1,900 kcal/day after supplementation (Prentice et al. 1983b), while in lactating women the corresponding figures are 1,570 kcal/day and 2,290 kcal/day (Prentice et al. 1983a). Preliminary analyses of the food intakes measured in the women studied here yield values comparable to these.

Expenditure therefore appears to exceed food intake by as much as 800–900 kcal/day, even before the additional requirements of pregnancy (synthesis of the products of conception) and lactation (synthesis of milk) are taken into account. How can a discrepancy of this magnitude arise? First, the figures for expenditure represent a combined value for unsupplemented and supplemented women in the nonpregnant, nonlactating state, and so are not directly comparable with the figures for food intake. There are reductions in some components of energy expenditure during pregnancy and lactation (e.g., in resting metabolic rate and in total activity during pregnancy), and we have also measured reductions in unsupplemented compared with supplemented women (Lawrence et al. 1984; unpublished data), but these differences are not large enough to explain the discrepancy. Nor is it possible that the women made up the shortfall in energy intake by drawing on adipose tissue reserves, since a deficit of 800–900 kcal/day would be equivalent to a weight loss of 40–50 kilograms (kg) over the whole year. The difference almost certainly arises from methodological errors, the nature of which is not at present clear.

Estimates of total expenditure depend at present on combining information on activity patterns with data on the energy costs of individual activities. It is possible that the presence of an observer altered the habitual pattern of activity, or that the energy cost of individual activities was overestimated. The latter measurements were necessarily of short duration, and the tempo of work could have been temporarily increased during the measurement. We were, however, unable to detect any increase in work rate (e.g., in pounding rate) as a result of measurement. The method of choice for estimating total 24-hour energy expenditure in future will almost certainly be the doubly labeled water method (Coward et al. 1984), a noninvasive technique employing stable isotopes that permits the subjects to go about their business quite freely. This method is currently being used in Keneba and should enable us to evaluate the diary technique used in the present study.

There are also potential errors in the estimation of energy intake. In Keneba, food intake is usually measured on a one-day-a-week basis (with alteration of the day from week to week). The subject is visited on three or more occasions during the day for weighing the main meals, subtracting the weight of leftovers, and recording snack food intake by recall. The energy value of the food consumed is determined using food tables prepared from direct chemical analyses of local dishes and individual foods. Possible sources of error, however, include (1) a change of dietary habits on the day of study and (2) the difficulty of estimating snack food intake accurately. These possibilities are currently being investigated. It should be borne in mind, however, that any errors in the measurement of food intake have so far evaded successive investigators in Keneba, who have consistently produced similar results.

Seasonal Energy Imbalance in Pregnant Gambian Women

A further component of the study was to estimate, in pregnant women, seasonal changes in body weight and body water (TBW) and thus body fat. TBW was measured using a deuterium oxide dilution technique. After adjusting for stage of pregnancy, the impact of season on body weight and body fat in unsupplemented women was highly significant ($p < .001$), with over 4 kg of fat being lost between March and November (figure 3.2), by which time mobilization of body fat was providing over 10 percent of daily energy requirements, or about 250 kcal/day. The rate of fat loss was greatest from the end of June onward, that is, sometime before food shortages began to bite in August and September, and this indicates that the initial stimulus to body fat mobilization was the seasonal increase in activity documented above, rather than any decline in food intake.

Although there was also an effect of season on body fat in supplemented women, supplementation appears to have diminished the magnitude of seasonal energy imbalances. In the supplemented group, no individual monthly increment in fat was significantly different from December–January, while in unsupplemented women, both the February–March and October–November increments were significantly different. In addition, the overall significance of the seasonal effect was greatly reduced, to $p < .02$ in the supplemented group, compared with $p < .001$ in unsupplemented women.

Seasonal Variations in Resting Metabolic Rate

The large seasonal imbalance between energy intake and output has other effects besides that on body weight. Seasonal variation in resting metabolic rate (RMR), after adjustment for stage of pregnancy or lactation, was examined. In unsupplemented women, RMR was significantly lower between August and November than it was in February–March, by about 50 kcal/day. (Note that the calculated expenditures in figure 3.2 take no account of this.) This drop occurred at the time of year when the lowest body weights and lowest food intakes were also recorded.

One possible cause of a drop in RMR could be the loss of lean tissue that normally accompanies the depletion of adipose tissue reserves. However, in this study we found no significant seasonal variation in TBW, an index of lean body mass. Moreover, we used multiple regression analysis to investigate the effect of body weight change on RMR and found that the

FIGURE 3.2 Seasonal variation in body weight and body fat content of rural Gambian women, adjusted for stage of pregnancy or lactation

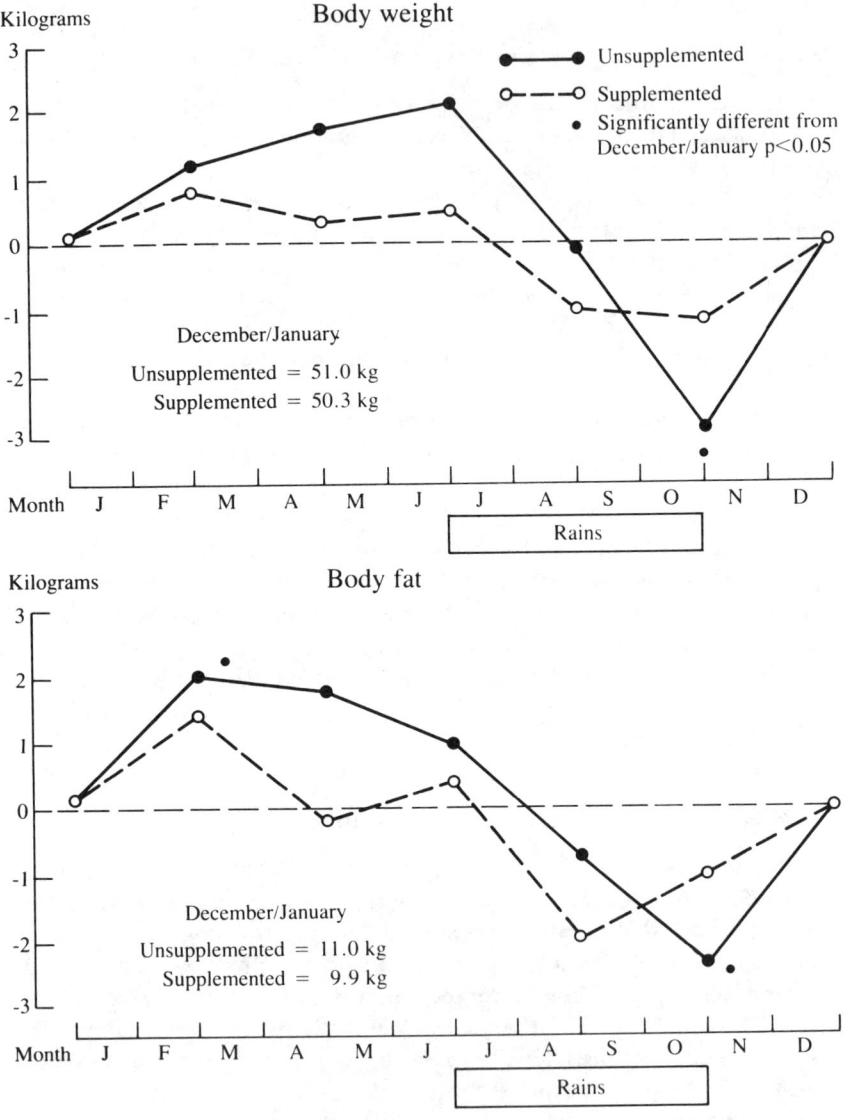

effect of season was still significant even after adjustment for body weight. Alternatively, the RMR per kg lean tissue may have been reduced, as has been observed in studies of semistarvation (James and Shetty 1982).

As with body fat, supplementation appears to have diminished the effect of season on RMR change. The drop observed in supplemented women during August–September did not by itself reach significance, and the overall effect of season was less significant in supplemented than in unsupplemented women ($p < .05$ compared with $p < .01$).

The Effect of Season on Pregnancy Outcome and Lactational Performance

Previous work in Keneba carried out by Dr. A. M. Prentice and others has provided considerable information on the effects of season and supplementation on both pregnancy outcome and lactational performance. The initial study investigated the effect of supplementation on breast-milk output, measured by test-weighing over a 12-hour period. There was evidence of a marked decline in breast-milk output during the rainy season in women in the second to sixth month of lactation (Prentice 1980). However, the results of the supplementation trial were unequivocal, with no improvement in breast-milk output being observed (Prentice et al. 1983a).

In contrast to this was the marked effect on birthweight of supplementation during pregnancy. Prior to the introduction of the supplementation program in Keneba, average birthweight, adjusted for gestational age, sex, and parity, was 2.74 ± 0.05 kg during the rains (defined here as July to January) and 2.97 ± 0.05 kg during the dry season. This seasonal decline in birthweight was eliminated by supplementation, with weights of infants born July to January improved by an average 224 g (Prentice et al. 1983b). There was also a marked reduction in the incidence of low birthweight (<2.5 kg) at this time of year, from 28 percent to 5 percent of births. The supplement was without effect on birthweight during the dry season.

Conclusion

In this chapter we have concentrated on the seasonal demands of agricultural work and their effect on the nutritional status of pregnant and lactating rural Gambian women.

June–July is the period of most intensive agricultural work as fields are cleared, dug, and planted at the beginning of the rains. More than 4 hours/day are spent in tasks directly related to agriculture, and for about half this time, women work at a rate in excess of 3.5 kcal/min. We estimate that total energy expenditure rises by more than 300 kcal/day at this time, with most of this extra requirement having to be met from adipose tissue energy reserves. Total expenditure is somewhat reduced in August–September, the peak of the annual hungry season, but fat mobilization continues at a high rate because of the reduction in food intake. These large seasonal imbalances of energy intake and expenditure cause serious depletion of the relatively small maternal adipose tissue reserves, and also result in reductions in resting metabolic rate, the consequences of which are not at present clear. Perhaps the most important effect of seasonal food shortage is, however, the reduction in average birthweight that occurs at this time of year.

Seasonal food shortages occur because both stored food supplies and the cash derived from the sale of groundnuts in December are exhausted before the next harvest comes in. This situation has been exacerbated by several years of low rainfall and poor harvests, and perhaps also by a change in agricultural and dietary practices. Over the past 10 years, an increasing amount of land has been used to grow groundnuts for sale rather than to grow millet and sorghum, the traditional staple foods. This has increased dependence during the dry season on the purchase of imported rice, supplies of which have often been irregular since 1983. Prior to this date, rice was freely available in village shops, but it has since disappeared from the open market. The local government rice store has also run out of rice on occasions, and this has caused local and sometimes acute food shortages from time to time.

One possible means of alleviating seasonal food shortage is to supplement at-risk groups such as pregnant or lactating women. The Dunn Nutrition Unit has developed a low-cost energy-dense biscuit supplement that can be baked locally and makes constructive use of aid foods, such as dried skim milk, that may otherwise go to waste. The supplement is effective in preventing seasonal reductions in birthweight (Prentice et al. 1983b) and may also reduce seasonal fluctuations in body weight, total body fat, and resting metabolic rate.

It may be undesirable to supplement the diet of lactating women, however, not only because supplementation fails to improve milk output, but also because of its effect in reducing the duration of postpartum lactational amenorrhea (Lunn et al. 1980). Supplementation thus speeds the return to fertility and may potentially increase birthrate, a clearly undesirable change. Any supplementation program should probably be allied with a simultaneous program of birth control.

4 The Effect of Seasonality on Intrahousehold Food Distribution and Nutrition in Bangladesh

MOHAMMED ABDULLAH

In Bangladesh, seasonal differences in the availability and intake of food and the effect of seasonality on the nutritional status of people are well recognized. Conclusive evidence is available from Bangladesh on the seasonal variations in the intake of food and nutritional status (US-DHEW 1966; INFS 1977 and 1983; Chen, Huq, and D'Souza 1981; Chowdhury et al. 1981; Brown, Black, and Becker 1982; Brown et al. 1985; Abdullah and Wheeler 1985). Intrahousehold distribution of food generally favors the productive adult male members of the household (INFS 1977 and 1983), but there is little evidence on how food varies with the seasons and on the availability of food at the household level. It has been suggested elsewhere that intrahousehold distribution of food discriminates against women and young children, and that in most developing countries this discrimination increases at times of shortage (Schofield 1974; Safilios-Rothschild 1980).

The purpose of this chapter is to examine seasonal variations in the intrahousehold distribution of food, following a brief discussion of seasonality in food availability and an overview of the general pattern of intrahousehold food distribution in Bangladesh.

Seasonal Pattern of Food Availability

Food in Bangladesh traditionally means cereals, especially rice, the availability of which is highly seasonal. Relative abundance and shortage of food, particularly rice, occur cyclically in relation to the postharvest and preharvest periods.

The principal rice crop harvested in November to mid-January is *aman,* which constitutes 60 percent of total rice production. *Aus* rice, harvested in June–August, accounts for 26 percent and *boro* rice, harvested in mid-March–May, for 14 percent. The availability of rice peaks after the principal rice *(aman)* harvest and thereafter reaches its lowest level in late

September to nearly November. This period is traditionally known as the hungry season in Bangladesh.

Seasonal variations in the availability of foods other than rice play an important role in adding variety to the diet. However, they contribute very little to the aggregate intake of macronutrients (that is, calories and protein). Pulses are harvested from February through March and are more abundant early in the year. Wheat is harvested in March-April. White and sweet potatoes are harvested from January through May. Fruits are highly seasonal, with peak production occurring in May-June. Vegetables are available throughout the year, although the availability diminishes in September-November. This is the period when the monsoon floods recede and fish supply increases.

Methods

Data for this in-depth analysis of seasonal variations in the intrahousehold distribution of food were generated in a longitudinal study (Abdullah 1983) of individual food intake and nutritional status, conducted in a typical central-western Bangladeshi village during 1981-82.[1] The purpose of the study was to examine the differential allocation of food in relation to estimated requirements and whether the differential allocation put women or young children at extra risk during times of shortage.

Food intake was measured for three consecutive days in four different seasons of the year—March-April, June, September-October, and December. In seasons 1 and 2 (March-April and June), people were doing intensive agricultural work. There was also a food-for-work program in the area in season 1, providing additional employment to the landless laborers. During season 2, the second *aus* rice crop was harvested. Season 3 (September-October) is the preharvest hungry season before the main rice *(aman)* harvest. This is the period of least agricultural work, lowest food stocks, and increased food prices. Season 4 (December) is just after the main rice *(aman)* harvest.

Female enumerators with special training were employed for the dietary survey. All foods that were going into the household cooking pots were weighed using Salter Duet Scales of 1-kilogram capacity. Each individual's portion of cooked food was then weighed at each meal using the same scale. Any food left over or given to others or to household pets was also weighed. Intake of snack foods was estimated at the end of each day by questioning. Individual intake, in terms of raw food ingredients in the cooked food, was determined. Individuals who had eaten any meal away

---

1. Data from this study were published in the *American Journal of Clinical Nutrition* (Abdullah and Wheeler 1985).

from home were excluded from the analysis. This resulted in a reduction of about 5 percent of all person-days studied.

After a preliminary census, 55 households were found to have one or more children under 5 years of age, and all were asked to join the study. Two refused, leaving 53 study households. On the basis of socioeconomic information—such as access to land, expenditure on and income from agriculture, cash income, and changes in assets—obtained by the author, 31 households (mostly landless) were classified as "poor" and 17 as "better off" (small farmers and sharecroppers). The remaining 5 households were relatively wealthy (big landowners). Because of large differences in income, these 5 households were omitted from the analysis.

Results

Seasonal Variations in Food Intake (Energy and Protein)

Energy intakes by seasons and by age groups and sex have been plotted in figures 4.1 and 4.2 to show the pattern of seasonality in food intake. Figure 4.1 clearly demonstrates the seasonal variations in the energy intake of adults (14–55 years). Analysis of variance showed that there is a highly statistically significant difference in the energy intakes (kilocalories per kilogram body weight per day) of both male and female adults between the first two seasons and the second two, the intake being higher in March–April (season 1) and June (season 2) than in September–October (season 3) and December (season 4) ($p < .001$). In all four seasons, energy intakes of adult men were significantly higher than those of adult women ($p < .001$). The difference between men's and women's energy intakes can be accounted for by differences in requirements and levels of activity. In seasons 1, 3, and 4, the intakes of better-off women were also significantly higher than those of poor women. Except in season 2, the better-off women had a 15–18 percent higher energy intake per unit of body weight.

Figure 4.2 shows a similar pattern of seasonality in the energy and protein intake of children—that is, intakes are higher in seasons 1 and 2 than in seasons 3 and 4, except for 1–4-year-old girls, who had their highest intake in season 3. Analysis of variance did not, however, show any seasonal difference in the intake among children in any age group. Between sexes, there is no difference among the 5–14-year-olds, but in the 1–4-year-old group, boys' intake was significantly higher than that of girls in all four seasons ($p < .04$).

Superimposed on figures 4.1 and 4.2 are protein intakes (grams per kilogram body weight). It is evident that the pattern of protein intake is almost the same as that of energy.

Figure 4.3 shows the energy intakes by adult women and children as a proportion of the intake by male household heads. It is an indicator of the

FIGURE 4.1 Mean daily energy and protein intakes of adults (14–55 years) by sex, socioeconomic group, and season

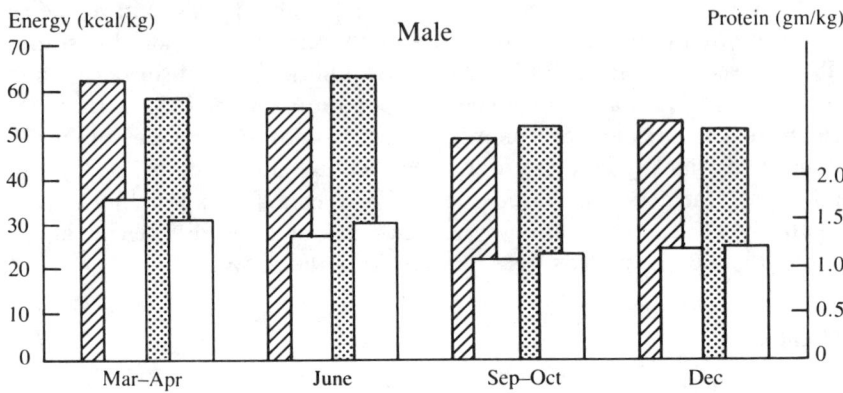

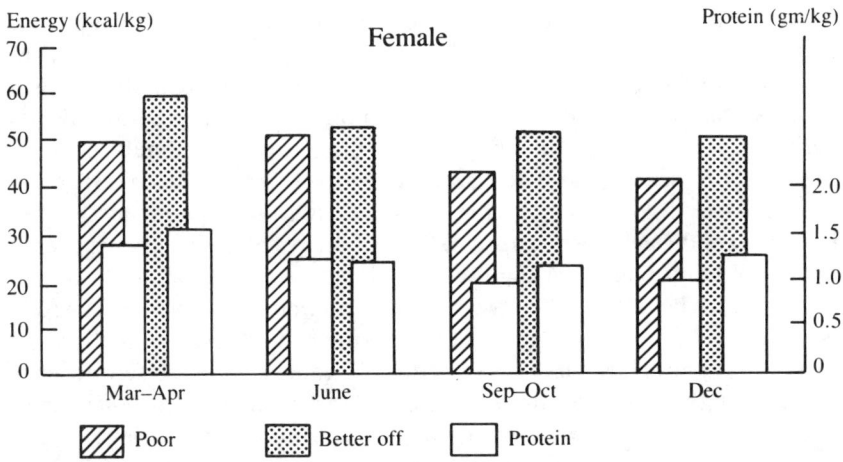

intrahousehold distribution of available food resources: the higher the percentage, the larger the share received by the women or children. By examining these percentages from one season to the next, it is possible to determine whether the intrahousehold distribution of food resources to various members changes.

Also included on the graphs are solid and broken horizontal lines. They represent two normative values of the proportion of the energy intake of the male household head that the women or children should be consuming, based on their needs relative to the head of the household. So, for example, regardless of the total calories available to the household, if the head of the household is engaged in moderate activity, a male child aged

FIGURE 4.2 Mean daily energy and protein intakes of children by sex and season

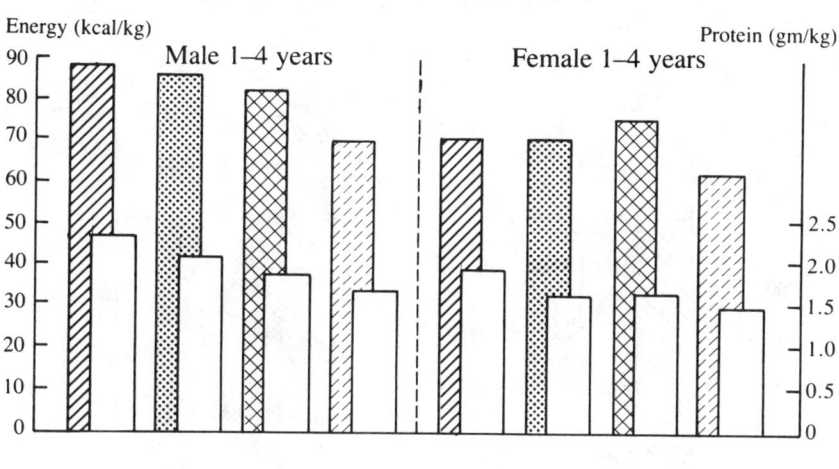

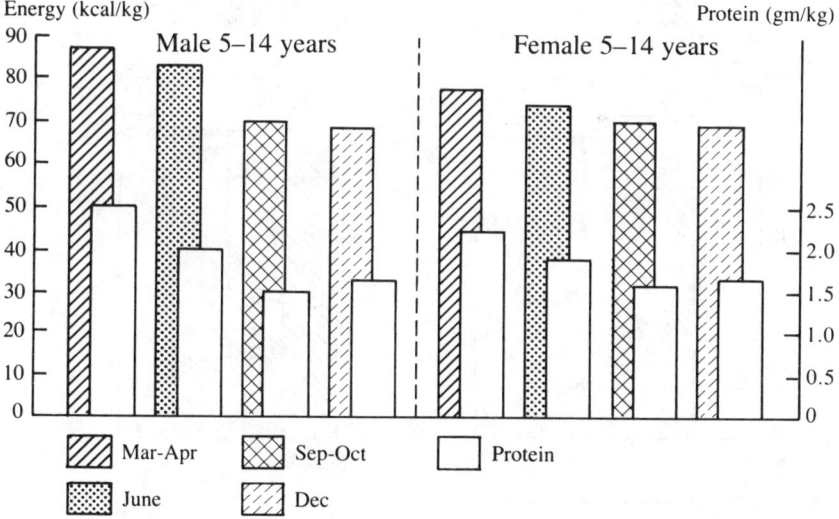

5–14 years in the sample should, on the average, be consuming 54 percent of the calories that are being consumed by the household head. If food is distributed equitably, according to the normative standards, then the proportion of the calories consumed by the population groups should fall somewhere between the solid and dashed lines.

These normative standards were calculated based on the actual average body weights of the population and their age, in accordance with the dietary requirements recommended by the Food and Agriculture Organi-

FIGURE 4.3 Daily energy intake (kilocalories) of women and children as a percentage of that of the household head

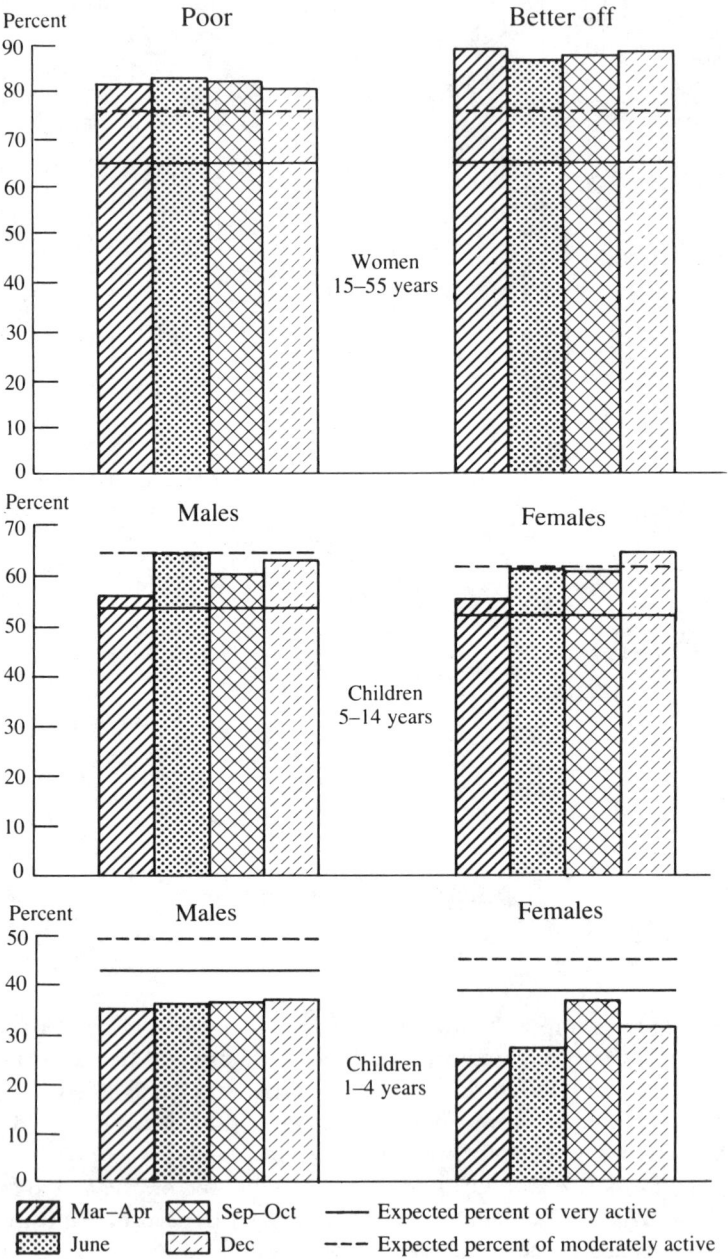

zation (FAO). To the extent that the FAO requirements are incorrect, so too are the normative standards used in this analysis.

Figure 4.3 demonstrates that, relative to the energy intake of household heads, the energy intake of women is remarkably constant throughout the year—80.0-82.8 percent of the male household head's intake for poor women and 86.2-88.3 percent for better-off women. In all seasons, women's proportional intake was within the expected range based on the normative values.

The energy intake of children aged 5-14 years as a percentage of that of the household head is also within the expected range, and no seasonal trend is noticed. Relatively lower proportional intake among children in season 1 (March-April) reflects a redistribution of household food resources to the head of household. This is to be expected in view of the fact that adult males had their highest absolute intake during that period, coincident with increased energy expenditure in agricultural activities and food-for-work programs. The proportional intake by both male and female children in the 1-4-year-old group is below the expected range, given their requirements relative to the adult head of household. Also, the proportional intake by girls is lower than that by boys for their age group. This suggests, if the FAO dietary requirements are correct, that young children are discriminated against in the intrahousehold distribution of food, and this is especially the case for girls. But of greater interest is that while no seasonal difference is seen in energy consumption by boys, the proportional intake by girls increased considerably in season 3 (September-October). In season 4 (December) also the proportional intake by girls (1-4 years) was higher than in seasons 1 and 2. It is important to note that the increase in the proportional intake among young girls that occurs in season 3 is coincident with the preharvest hungry season.

Conclusion

Bangladesh is predominantly an agrarian country, and the availability of food at the household level is primarily determined by the cropping pattern, because most rural households are directly or indirectly dependent on agriculture for their food supply. The household food situation improves after rice harvests. Stocks peak after the principal *aman* rice harvest in November-January and are at their lowest levels in September-October. Overall food intakes have highly significant seasonal correlations, intakes being generally higher during postharvest periods.

Seasonal variations in food energy intake among adult men and women (14-55 years) are clearly demonstrated. Figure 4.1 illustrates seasonal differences in intake per unit of body weight. Intakes during the first two seasons are significantly higher than in the second two ($p < .001$).

Adult intake also correlates with activity pattern. High intake in March-April and June may be partly accounted for by high-energy-demanding activities characteristic of the season and partly by higher availability of food. Decreases in consumption during the hungry season in September-October are obvious. The season is also characterized by the least agricultural activity. Relatively low intake in the postharvest season in December is probably due mainly to slack agricultural activities during this time. Another explanation, not investigated in this study, is that people voluntarily ration their intake to stretch out the amount of available food to cover a longer period.

Male-female differentials in food energy intake among adults remained almost constant in all four seasons studied. Figure 4.3 clearly demonstrates this. Average proportional intake by women was above the expected range calculated on the basis of FAO-recommended intake and actual body weights. Evidence does not support the theory that the pattern of intrahousehold distribution of food is affected by seasonal fluctuations in the availability of food insofar as adult men and women are concerned. The same is true of older children in the age group of 5-14 years. Minor seasonal differences are not statistically significant, and there is no sex differential. It is only among the younger children (1-4 years) that there is a sex differential ($p < .04$). In absolute terms, boys generally received more food than girls, although there is no sex differential in the FAO recommendations.

Chen, Huq, and D'Souza (1981) observed similar discrimination against female children in the 1-4-year-old group in a cross-sectional dietary intake study conducted in four Matlab villages. The overall intake by both boys and girls in the 1-4-year group was well below the recommended level throughout the year. The reason might be that the exceptionally high food requirements of young children per unit of body weight are not fully appreciated. The pattern of intrahousehold distribution of food shows that the proportional intake by boys remained almost constant year-round, but that by girls increased during the hungry season in September-October. During this period, their absolute intake was also the highest of the year. The seasonal difference was not, however, statistically significant. Overall, despite significant seasonal variations in the intake in absolute terms, the pattern of intrahousehold distribution of food remains more or less the same across seasons except in case of young female children in the age group of 1-4 years.

Young girls tend to receive a larger share of family food at times of shortage. Most of the parents interviewed in the study think that young children are unable to withstand hunger and must be fed first when family food is in short supply, even if the parents themselves are required to go hungry. Frequently, parents would say that since the girls would eventually leave their parental homes after marriage, they should be given preferential treatment in the allocation of choice foods in order to increase their attractiveness in the marriage market. It is tempting to speculate that this kind of parental attitude might have resulted in the increased allocation of family food to young girls at times of shortage, despite the prevailing attitude during the peak season, when young boys received a larger share of family food resources than young girls.

Several unresolved questions remain: Why do young children not receive the share of family food that the theoretical estimates of requirements would prescribe? Why do young girls receive a lower share than young boys? Is it because of parental misperceptions about the food needs of young children? Although there is no sex discrimination among adults as far as energy intake is concerned, it also remains to be explored whether there is qualitative discrimination that affects micronutrient intake.

More longitudinal studies under differing ecological and sociocultural conditions are needed for precise quantification of the effect of seasonality on intrahousehold distribution of food and nutritional outcomes in order to identify appropriate counterseasonal measures.

5 Seasonal Demands for Nutrient Intakes and Health Status in Rural South India

JERE R. BEHRMAN AND ANIL B. DEOLALIKAR

Seasonal variations are substantial in many rural areas of developing countries. There are often great seasonal fluctuations in environmental conditions, food availability, food prices, and labor demands. Schofield (1974), Chambers, Longhurst, and Pacey (1981), and others have suggested that such variations may have substantial impact on nutrition and health status.

However, even if seasonal variations in environmental conditions, income, and prices are substantial, seasonal savings and dissavings patterns of income or food stocks may preclude there being much impact on nutrient intakes. Moreover, even if there are seasonal patterns in nutrient intakes, adjustments in energy expenditures may mitigate the effect on indicators of even fairly short-run health status.

Whether seasonal variations in such variables as prices substantially affect nutrition and health status is thus an empirical question. Our contribution in this chapter is to explore such impacts for a rural South Indian sample by investigating questions such as these: Do nutrient intakes and health status vary seasonally? Do demands for nutrient intakes differ significantly across seasons? Do price and asset elasticities differ seasonally? Are they different for small versus large cultivator households? Does health status differ significantly across seasons, and between small and large cultivator households? Do comparisons of the nutrient and health status results indicate similar or different responses to prices and assets? Do parental preferences vary across seasons regarding the equity-productivity tradeoff and the favoring of children by birth order or gender in the intrahousehold allocation of nutrients?

We begin by discussing briefly alternative farm household models of demand for nutrient intakes and health status and why seasonality might matter in these demand relations. We then present and summarize our data, and finally present and discuss our empirical estimates.

Seasonality and Reduced-Form Demand Functions for Health and Nutrient Status

The determination of the allocation of nutrients to individuals and of individual health status occurs substantially within households, given the prices that households face and the assets that they control. Thus reduced-form demand functions for individual nutrients and health status are employed in this study. The important implications of the farm household literature is that the reduced-form demands depend upon both consumption and production prices and assets, and that the income effect of production prices opposes the demand effect of consumption prices for goods that are both produced and consumed by farm household units. We do not need to be able to look into the "black box" mechanisms of intrahousehold decision making to test empirically whether the coefficients in the particular (log-linear) approximation of these relations that we use are or are not different across seasons.

Whatever the origin of these reduced forms, there is the question of why there might be differences in them across seasons. Such differences would seem to be large only if it is costly to transfer resources—for example, purchasing power or food stocks—across seasons. For instance, in the lean season, most cultivators may be net *buyers* of food, so that the income effect of higher food prices on nutrient consumption would be negative (reinforcing the substitution effect). In addition, it is possible that the lower real incomes in the lean season might even result in a more negative substitution effect. (Timmer and Alderman [1979] and Timmer [1981] have provided evidence on the basis of cross-sectional data that low-income consumers have a stronger [i.e., more negative] substitution effect than high-income consumers.) On the other hand, in the surplus season, a larger number of cultivators may be net *suppliers* of food, so that the income effect of a food price increase would be positive (and hence opposite in sign to the substitution effect). The substitution effect, in addition, might be weaker.

In this study, we were not able to explore such processes directly because of data limitations. However, we are able to provide some insight into the possible importance of seasonality for health and nutrition by estimating reduced forms for nutrient and health status, and asking whether the price and asset effects differ significantly and substantially between seasons.

Data from Rural South India

We use the ICRISAT VLS (International Crops Research Institute for the Semi-Arid Tropics Village Level Studies) data set to estimate reduced-form individual nutrient intake and health status relations. The ICRISAT VLS data are panel data that have been collected at regular intervals since 1975 on production, expenditure, time allocation, prices, wages, and socioeconomic characteristics for 240 households in six carefully selected "typical" villages in three different agroclimatic zones in SAT India. Within each village, households are a random sample stratified by size of landholdings. For the 1976-77 and 1977-78 crop years, four nutrition surveys were taken that recorded individual nutrient intakes in the past 24 hours and anthropometric measures of health status. These surveys were timed to obtain observations from each village on both the surplus season, during which food is relatively available, and the lean season, during which food is relatively scarce. We use these seasonal definitions for our analysis below. We use the average value for each season for each individual as an observation, which gives us a total of 1,786 individual-season observations.[1]

The basic dependent variables of interest for this study are measures of individual calorie and protein intakes and of individual health status.[2] For short-run health status, we use weight-for-height—an anthropometric measure that is widely assumed to pertain to short-run health status.[3] Each of the nutrient variables was standardized for age and sex by using the Indian recommended daily allowances in Gopalan, Sastry, and Balasubramanian (1971). Weight-for-height was standardized by modified Harvard standards (see Ryan et al. 1984). As further controls for possible systematic deviations from the standards, we include among the right-side variables the age and gender of the individual.

Table 5.1 gives the means, standard deviations, standard errors of the means, and coefficients of variation by seasons for these nutrient intake and health status variables for all sample individuals. The means are significantly below the standards for calories and weight-for-height, and sig-

---

1. For further details concerning the ICRISAT VLS data, see Binswanger and Jodha 1978 and Ryan et al. 1984. The same individual may be included in the original data set up to four times. In another study, we have found for nutrient estimates across crop years that controlling for fixed effects does not make much difference (Behrman and Deolalikar 1987). Since each household has multiple individuals, we also have been able to explore the possibility of additive unobserved household fixed effects. These estimates do not differ substantially from those presented here.

2. We also have considered the seven other nutrients for which data are available in the ICRISAT VLS surveys: β-carotene, riboflavin, vitamin C, calcium, iron, thiamine, and niacin. However, the general results do not differ substantially from those for calories and proteins presented below, so we focus on these two often-emphasized nutrients to preclude the presentation of an avalanche of estimates.

3. Often such anthropometric measures are referred to as indicators of nutritional status, but since nutrients are but one of the inputs in the production of health status, the labeling of the output as if it reflected only one or only some of the inputs may be misleading (see Behrman and Deolalikar 1988a).

TABLE 5.1 Characteristics of seasonal distribution for individual calorie and protein intakes and for weight-for-height, rural South India, 1976-78

	Mean	Standard Deviation	Standard Error of Mean	Coefficient of Variation
Lean season				
Calories	93.4	35.0	0.8	37.5
Protein	137.2	64.0	1.4	46.6
Weight-for-height	84.5	9.8	0.2	11.7
Surplus season				
Calories	97.0	32.0	0.7	33.0
Protein	138.8	52.2	1.2	38.4
Weight-for-height	83.9	10.0	0.2	11.9

SOURCE: Calculated by authors from ICRISAT VLS data tape.
NOTE: All variables are relative to standards of 100.

nificantly above the standards for proteins.[4] For calories, the overall mean is 93.4 percent of the standard for the lean season and 97.0 percent for the surplus season. Mean calorie consumption is significantly greater in the surplus than in the lean season. For proteins, the lean season mean is 137.2 percent and the surplus season mean is 138.8 percent of the standards. For weight-for-height, the lean season mean is 84.5 percent and the surplus season mean is 83.9 percent of the standards. The means for both proteins and weight-for-height are not significantly different at the 5 percent level between the lean and surplus seasons.[5]

Thus with respect to the dependent variables, sample individuals are significantly below the standard, especially for calorie intakes. For calories—but not for proteins or for weight-for-height—there are significant differences in the means between the seasons. Of course, whether there are significant differences in the means does not indicate whether the reduced-form relations differ seasonally. For the nutrients, finally, there are substantial interpersonal variations around the means that tend to be systematically larger in the lean season than in the surplus season.

The independent variables of primary interest for this study pertain to consumption and to production prices and assets. Because of limited price variation in our sample (due to limited variation in observations both over

4. Unless otherwise qualified, we use *significant* to refer to the standard 5 percent level for a *t*-test throughout this chapter.
5. Ryan et al. (1984) also conclude that their detailed study of these data reveals little in the way of seasonal effects by the comparison of means of distributions between seasons. However, when they disaggregate considerably—that is, by village or region, by three landholding classes, and by seven gender-age groups—they find significant differences in the means between seasons in about a quarter of the cells. They do not discuss the reasons for such patterns.

time and across space), multicollinearity precludes including a large number of prices. We focus on four key prices: for rice, sorghum, pulses, and labor. Though individual households vary, the average household both produces and consumes all of these items. We do not include nonfood prices for goods and services that are directly related to health status because we do not have observations on such prices.[6] We use monthly prices to calculate seasonal prices for each of these four items for each village. We use village-level mean prices in all cases because the prices that individual households pay or receive depend on the quality of their demands or supplies. With regard to assets, we focus on total asset value[7] but also distinguish between small and large cultivator households, since the primary form of physical wealth in these communities is access to land, and one might conjecture that the seasonal nutrient and health status relations vary depending on such access. Since all households in the sample hold some land, we use an arbitrary cutoff of 7.0 hectares—which is not a large holding in the semiarid tropics—to distinguish between small and large cultivator households; about 70 percent of the sample falls below this cutoff. Those that are below the cutoff tend to depend on earnings from the rural labor markets for relatively large shares of their income.

Estimates of Reduced-Form Demand Relations for Nutrient Intakes and Health Status for All Individuals

Table 5.2 presents the basic reduced-form log-linear estimates of demands for the two individual nutrient intakes and the weight-for-height indicator of health status for all 1,786 individual-season observations. In these estimates we include among the right-side variables the four prices, total asset value, a dichotomous variable with a value of one if the household is a large cultivator (and zero otherwise), and age and gender of the individual of relevance for each observation. Because of the log-linear specification, each of the coefficients of the right-side variables is an elasticity giving the percentage change in the dependent variable for that column relative to a given percentage change of the designated right-side

---

6. Pitt and Rosenzweig (1985) report that such prices in the form of the availability of community health care have significant impacts. If the unobserved prices are correlated across the ICRISAT villages with our four included prices or with assets, our estimates of the coefficients of our observed variables may be biased.

7. The data set includes information on schooling, but schooling levels are quite low and do not seem to be significantly associated with nutrient intakes and health status—see, for example, Behrman and Deolalikar 1988b and Ryan et al. 1984. Therefore, in the interests of a parsimonious specification, we have not included schooling in the estimates presented below. We also do not include other assets and endowments, many of which are unobserved. If such unobserved variables are correlated with our included variables, our coefficient estimates may be biased.

TABLE 5.2 Reduced-form seasonal log-linear demand relations for standardized calories, proteins, and weight-for-height for individuals in rural South India, 1976–78

Right-Side Variables	Calories	Proteins	Weight-for-Height
Price of sorghum	−2.16	−2.16	0.22
	(8.4)	(7.3)	(2.5)
Price of sorghum × season	1.63	2.02	−0.15
	(5.8)	(6.3)	(1.5)
Price of rice	−1.77	−1.17	0.37
	(4.6)	(2.7)	(2.9)
Price of rice × season	1.54	1.85	−0.24
	(3.9)	(4.1)	(1.8)
Price of pulses	−1.20	−1.31	0.20
	(4.3)	(4.1)	(2.2)
Price of pulses × season	1.41	1.39	−0.19
	(4.9)	(4.3)	(2.0)
Wage rate	0.58	1.05	−0.004
	(2.8)	(4.4)	(0.1)
Wage rate × season	−0.24	−0.74	0.004
	(1.1)	(2.9)	(0.1)
Large cultivator household	−0.06	−0.03	−0.003
	(1.8)	(0.8)	(0.2)
Large cultivator household × season	0.08	0.07	−0.01
	(1.7)	(1.4)	(0.4)
Value of assets	0.02	0.03	0.01
	(1.5)	(2.9)	(2.1)
Value of assets × season	0.01	−0.01	−0.001
	(1.0)	(0.4)	(0.1)
Age	0.08	−0.02	−0.05
	(6.1)	(1.6)	(11.7)
Age × season	−0.01	−0.004	0.001
	(0.9)	(0.2)	(0.2)
Male	−0.28	−0.41	−0.02
	(3.1)	(3.9)	(0.7)
Male × season	0.07	0.27	0.01
	(0.07)	(2.4)	(0.2)
Intercept	6.97	6.70	3.94
	(12.1)	(10.2)	(20.2)
Season dichotomous variable	−3.22	−2.87	0.46
	(5.4)	(4.2)	(2.3)
R^2	0.17	0.20	0.21
F-ratio	22.47	26.18	28.17

NOTES: The season dichotomous variable has a value of zero in the lean season and a value of one in the surplus season. The male dichotomous variable has a value of one for males and zero for females. In parentheses beneath the point estimates are the absolute values of t-statistics.

variable. Each coefficient for each right-side variable is allowed to vary by seasons: the elasticity in the lean season is just the coefficient of the variable itself, but to obtain the elasticity in the surplus season, to that value must be added the coefficient immediately below the variable times the seasonal variable—for example, the lean-season elasticity of calories with respect to the price of sorghum is -2.2 and that for the surplus season is -0.6. We discuss first the nutrient determinants and then those for health status.

Reduced-Form Individual Nutrient Demand Relations

Significant price effects are pervasive for both calories and proteins, as are significant differences between price elasticities for the lean versus the surplus season. The sign patterns of the price elasticities for the lean season and for the adjustment for the surplus season are the same for both calories and proteins, without significant differences between the relations for these two nutrients. Note that this pattern occurs even though the seasonal differences in the means for proteins in table 5.1 are not significant while those for calories are.

What is the nature of the seasonal patterns in the nutrient price elasticities? For all the prices, the lean-season price elasticities are fairly large in absolute magnitude, implying considerable price responsiveness when food is relatively scarce. For all but labor, the signs are negative. As prices of any of the three farm production/household consumption goods—that is, sorghum, rice, and pulses—increase during the lean season, intakes of both nutrients decline at least as much in proportional terms. For labor, in contrast, the positive elasticity suggests that on the average these households are net suppliers in the lean season, and the positive impact on nutritional intakes of the income gained from labor services outweighs negative price effects on own production when the wage rises.

For the surplus season, all the nutrient price elasticity estimates are significantly closer to zero, except for the caloric response to wages—that is, except for labor, they are algebraically more positive, presumably because the income effects are relatively more important in these cases during the surplus season, when these households on the average are selling their marketed surplus. Thus the seasonal changes in the price elasticities seem generally consistent with the seasonal patterns of production and market involvement. The fact that the seasonal differences in the price elasticities generally are significant and fairly large suggests that interseasonal transfers are costly.[8]

---

8. Per Pinstrup-Andersen has suggested to us that the difference may reflect the fact that households on the average are on the buying side of the market in the lean season and on the selling side during the surplus season, so the relevant price to the average household relative to the village average switches between seasons. Such a pattern might imply that each

What about the impact of household assets and individual characteristics? Whether a household is a large cultivator does not have a significant additive impact except in the case of calorie demand (in which case large cultivators *ceteris paribus* have lower calorie intake in the lean season but greater intake in the surplus season than small cultivator households); we return to this variable in a later section. The asset value is a significant determinant of proteins, though not of calories, and without any significant difference between seasons.[9] Protein elasticity with respect to assets, however, is quite small (0.03). Since part of the total asset effect is represented by the values of the products that the assets help produce and, therefore, the product prices, these results should not be interpreted to mean that assets are unimportant in nutrient determination. Age has a positive significant impact on caloric intake, but with no seasonal difference or any significant effect on proteins. Males are estimated to receive significantly fewer calories in both seasons (with no marked difference between seasons) and significantly fewer proteins in the lean season (with a marked difference between seasons). Thus there is some limited additional support in the asset and individual characteristic elasticity estimates for significant seasonal differences that may reflect high costs of transferring resources between seasons in the case of the assets, and seasonal differences in nutrient requirements in the case of the individual characteristics. Finally, the significant additive terms indicate additional seasonal impacts beyond those captured by our variables for prices, assets, and individual characteristics.

Reduced-Form Individual Health Status Relation

Price effects are somewhat less pervasive and less substantial for the indicator of health status than for nutrient intakes: wages do not have significant effects on weight-for-height; the absolute magnitudes of the product price elasticities are 0.4 or less, rather than unitary or greater; and only one of the seasonal differences is significantly nonzero at the 5 percent level (though two others are at the 10 percent level). Such results are not surprising, given the probable lags in health response to nutrient intakes. Efforts to explore possible lags and cross-seasonal effects with this data set were not helpful in providing enlightenment on such lags because of the high multicollinearity and limited price variance.

What is surprising, however, is that the sign pattern of the estimates for the product prices for weight-for-height are opposite to those in the

---

price is multiplied by a marketing margin factor that is dependent on the season. This would not affect the price elasticity estimates, however, but only the constant estimate and the additive seasonal variable estimate.

9. Though the seasonal difference coefficient estimate is significant at the 10 percent level and equal in magnitude and opposite in sign to the lean-season estimate.

nutrient estimates. One possible explanation is that in the lean season for the weight-for-height relation, the product prices are serving as inverse proxies for the amount of energy expenditure by farmers, a variable not included in the regressions. The story might be that in the lean season, say, sorghum prices are high, so fewer nutrients are demanded. But at the same time there is less agricultural activity or less human heat generation, so energy expenditures are relatively even less, and people gain weight. If this theory has validity, then what may appear from the estimates to be a seasonal price effect reflecting high costs of transferring resources across seasons may be in fact a seasonal energy expenditure effect. We do not find such a possibility completely implausible, but we do wish to emphasize that we consider this explanation to be very speculative.

What about the impact of household assets and individual characteristics? The total asset coefficient in the health determination relation is significantly positive, but quite small. For age, the impact is negative and significant. There is no significant effect for gender. And for none of these asset and individual characteristic variables are the seasonal effects significantly nonzero, though there is a significant additive seasonal effect.

In summary, there appears to be much less evidence of significant seasonal effects in the reduced-form relation for health status, as contrasted with the substantial seasonal effects in the reduced-form relations for nutrients. Moreover, those seasonal estimates that are significant in the weight-for-height relation are difficult to interpret.

Estimates of Reduced-Form Demand Relations for Nutrient Intakes and Health Status for Small versus Large Cultivator Households

In a society in which the major form of production depends on access to land, as noted above, the capacity for adjusting to seasonal market and environmental changes may also depend on access to land. Those with limited access to land, for example, tend to be much more dependent on rural labor markets for their income, which may result in seasonal adaptive capacities different from those of cultivators who depend primarily on their own agricultural production. The estimates in table 5.2 indicate that there is no significant seasonality effect of an additive variable distinguishing between small and large cultivators. However, there still may be differences in the prices and asset elasticities for the two groups.

Table 5.3, therefore, gives separate estimates for small and large cultivators for the same three relations as in table 5.2. The lean-season product price elasticities and the seasonal adjustments in these elasticities do differ significantly between the two groups in a number of cases for nutrient demands, but not for weight-for-height. In all of these cases of significant differences, the absolute values of the lean-season nutrient elasticities

TABLE 5.3 Reduced-form seasonal log-linear demand relations for standardized calories, proteins, and weight-for-height for men, women, boys, and girls in rural South India, 1976-78

Right-Side Variables	Small Cultivator Households			Large Cultivator Households		
	Calories	Proteins	Weight-for-Height	Calories	Proteins	Weight-for-Height
Price of sorghum	−2.52	−2.58	0.19	−1.04	−1.00	0.32
	(8.2)	(7.3)	(1.9)	(2.2)	(1.7)	(1.7)
Price of sorghum × season	2.10	2.49	−0.12	−0.01	0.64	−0.26
	(6.3)	(6.6)	(1.1)	(0.0)	(1.0)	(1.3)
Price of rice	−2.06	−1.51	0.33	−0.61	−0.31	0.47
	(4.7)	(3.0)	(2.3)	(0.8)	(0.3)	(1.5)
Price of rice × season	1.94	2.24	−0.18	−0.20	0.61	−0.44
	(4.3)	(4.3)	(1.2)	(0.2)	(0.6)	(1.4)
Price of pulses	−1.30	−1.50	0.18	−0.75	−0.82	0.28
	(4.0)	(4.1)	(1.6)	(1.3)	(1.2)	(1.3)
Price of pulses × season	1.56	1.58	−0.15	0.73	0.94	−0.30
	(4.7)	(4.2)	(1.4)	(1.3)	(1.4)	(1.4)
Wage rate	0.61	1.16	0.04	0.34	0.67	−0.11
	(2.6)	(4.2)	(0.4)	(0.8)	(1.4)	(0.7)
Wage rate × season	−0.27	−0.85	−0.04	−0.04	−0.35	0.11
	(1.1)	(2.9)	(1.7)	(1.7)	(1.9)	(1.1)
Value of assets	0.02	0.04	0.01	0.03	0.04	0.01
	(2.0)	(2.9)	(1.7)	(1.7)	(1.9)	(1.1)
Value of assets × season	0.01	−0.01	−0.00	−0.02	−0.02	0.00
	(0.5)	(0.1)	(0.4)	(0.8)	(0.6)	(0.4)
Age	0.09	−0.02	−0.05	0.06	−0.02	−0.05
	(5.7)	(1.3)	(10.1)	(2.6)	(0.8)	(6.2)
Age × season	−0.01	0.00	0.00	−0.05	−0.05	−0.01
	(0.7)	(0.1)	(0.3)	(1.5)	(1.2)	(0.5)
Male	−0.28	−0.46	−0.03	−0.22	−0.26	0.01
	(2.7)	(3.8)	(1.0)	(1.2)	(1.2)	(0.1)
Male × season	0.09	0.33	0.03	−0.10	0.02	−0.08
	(0.8)	(2.5)	(0.7)	(0.5)	(0.1)	(1.0)
Intercept	7.17	7.09	3.98	5.36	5.39	3.85
	(10.9)	(9.5)	(18.2)	(4.4)	(3.7)	(8.4)
Season dichotomous variable	−3.61	−3.29	0.40	−0.43	−1.09	0.68
	(5.4)	(4.3)	(1.8)	(0.3)	(0.7)	(1.4)
R^2	0.19	0.21	0.20	0.19	0.18	0.26
F-ratio	22.62	24.61	23.83	6.04	5.58	8.63

NOTE: See note to table 5.2.

are larger and the seasonal adjustments in these elasticities are greater for small cultivators than for large cultivators.[10] Thus there is evidence that the small cultivators have greater price elasticities and more seasonal dependence than large cultivators. The greater price elasticities apparently reflect their greater dependence on product and labor markets. Since the small cultivators also tend to have lower incomes, these results are consistent with those in other studies that report higher nutrient demand price elasticities for lower incomes—for example, Behrman and Deolalikar (1988c), Behrman and Wolfe (1984), and Pinstrup-Andersen (1985). The greater seasonal difference in such elasticities apparently reflects the fact that small cultivators have more difficulty in transferring resources across seasons than large cultivators.

For the asset and individual characteristics, however, there is a little more evidence of differences between these two cultivator groups. For proteins, there is evidence of a significant seasonal adjustment for males in small but not in large cultivator households; that this effect is significant only for males is consistent with the evidence in Behrman and Deolalikar (1989), which shows that only for males is there evidence of seasonal patterns in labor market productivity.

Conclusion

This sample from rural South India indicates that on the average there are seasonal differences in mean standardized intakes of calories, but not for proteins or for the weight-for-height short-run indicator of health status. However, such averages do not in themselves reveal whether there are significant seasonal differences in the *determinants* of nutrient intakes and short-run health status—that is, in the responsiveness to prices and assets. We use a farm household model to estimate—for the first time, to our knowledge—seasonal reduced-form demand relations for individual nutrient intakes and health status.

Our estimates indicate that there are significant and substantial seasonal differences in these relations for nutrients. There is a systematic pattern of price responses being relatively large in absolute magnitude in the lean season when food is relatively scarce. They change substantially for the surplus season, with the adjustment in an algebraically positive direction for the product price responses. This is consistent with the implications of the farm household framework if it is costly to transfer resources across seasons, since there is a large positive income effect in the surplus

---

10. The differences between small and large cultivators are significant in the lean season for the prices of sorghum and rice and the intercept for both calories and proteins and for the wage rate for calories. The differences in seasonal adjustments are significant for the prices of sorghum and rice and the intercept for both calories and proteins.

season when these farm households tend to sell their products. The nutrient elasticities with respect to some major prices and the seasonal adjustments in these elasticities are significantly greater in absolute value for small than for large cultivator households. These patterns reflect the often-observed tendency for price elasticities to be higher for lower-income households and the fact that apparently lower-income households are less easily able to transfer resources across seasons. For small cultivator households alone there also is a seasonal shift in proteins consumed by males.

For the determinants of short-run health status as represented by weight-for-height, in contrast to the determinants of nutrient intakes, we find less evidence of significant seasonal differences in the responses to prices or to total assets for both small and large cultivator households. The pattern of signs in the price responses, moreover, contrasts with those for the nutrient relations, which we speculate may be due to seasonal patterns in unobserved energy expenditures.

These results have several important implications for analysis and policy, all subject to qualifications because of the multicollinearity associated with the limited price variation in our sample and conditional on our choice of functional forms.

First, the analysis of seasonal nutrient and, perhaps, health patterns is enriched by analysis of seasonal differences in reduced-form determinants and in underlying structural relations, apparently because of the high cost of transferring income and food stocks across seasons, in particular for small cultivator households. There may be important seasonal differences in the responses to such determinants, even though there are no significant seasonal differences in means, as for proteins in this sample.

In order to analyze the effects of market changes or of proposed policies, it is necessary to understand the extent of responsiveness in the underlying relations, and to what extent there are seasonal variations in such responses. Of course, policies might be directed toward lessening the extent of seasonality in the nutrient and health elasticities by facilitating transfer of resources across seasons. For the villages of the current study, policies directed toward public storage are not likely to be efficient, because private storage is low cost and efficient in these regions. There may be some gains, however, from policies to improve local goods and financial markets by better integration into regional and national markets.

Second, for nutrient intakes, the price responses dominate, though asset responses also are significant for the small cultivator households. Therefore, analysis of nutrient determinants and of related policies must be very sensitive to market environments and changes in these environments, including seasonal changes.

Third, in contrast, at least over the range of nutrient and health variations experienced by this sample, there is less evidence of significant seasonal differences in price and asset responses for health status. Though substantial adjustments apparently occur in the determinants of nutrient intakes, we do not find evidence of related adjustments in the health relations. If indeed weight-for-height is a good indicator of short-run health status, this result may reflect the pattern of unobserved seasonal adjustments in metabolism and energy expenditures.

Fourth, if one is interested in nutrient intakes, attention to the lean season seems to be important for this sample. The combination of relatively great price responsiveness in the lean season means that the more vulnerable children are likely to be particularly exposed to malnutrition risk when food is scarcest.

PART III

Food Acquisition Behavior, Employment, and Labor Market Considerations

6 Understanding the Seasonality of Employment, Wages, and Income

HAROLD ALDERMAN AND DAVID E. SAHN

Persephone ate six pomegranate seeds in Hades and in so doing fated the world to a cycle of spring and winter. Osiris's annual death and revitalization guaranteed the cycle of the Nile's waters and the fertility of the fields. From ancient times, humankind has incorporated the mysteries of the seasons into its culture. Furthermore, a perusal of literature from Aesop's fable of the ant and the grasshopper to the medieval books of hours to Kalidassa's six-season advice for lovers reveals that cultures have evolved guidelines as to the proper behavior in each season.

While the oscillations of the agricultural cycle may vary between years and may be modified by technical change, to a large degree this cycle is regular and known. Seasonal undulations are distinguished from the stochastic events that result in year-to-year fluctuations in agricultural output and productivity. Similarly, seasonal cycles differ from either longer-run climate cycles or business cycles, both of which are far less regular. Yet even outside agriculture, regular employment cycles do exist, as observed in construction, tourism, and various aspects of transportation and retailing. Given that the production process itself varies over the year, it is of interest to study how the modulations in the production cycle are matched by patterns in earnings and in consumption.

We will concentrate on agriculture and take the existence of a peak and a slack season as a given. Three broad issues are addressed. First, to what degree do wages adjust with shifting demand for labor? That is, how much more fluctuation is observed in earnings than in hours worked? Second, we look at technical and institutional arrangements to mitigate the seasonal patterns in agricultural production and employment. Third, we consider to what degree seasonal patterns of income translate into seasonal patterns of consumption. In this regard, the major issue is to examine opportunities and modalities of savings.

Seasonality of Income

The effect of seasonality in agriculture on the periodicity of earnings of rural households is determined by their sources of income. Specifically, one may distinguish among the major categories of rural income sources: net agricultural profits, including sales and home consumption; agricultural-related wage labor; nonfarm income from wages, salaries, and business profits; and nonearned income in the form of transfers, remittances, and so forth.

In turn, different types of households—that is, surplus producers, small deficit farmers, landless laborers—will earn varying shares of their income from these sources. Thus the periodicity of incomes is determined by the combination of the timing of these income sources.

In considering seasonal patterns of work, and thus earnings, the literature on farming systems, anthropology, and agricultural economics documents extensively the undulations in labor requirements. These are especially relevant to agriculture, where seasonality is most important. Agricultural calendars that detail the timing of work are frequently developed. They represent important information for formulating policies to address seasonal food insecurity in a given context. However, our purpose here is not to report on these studies, nor to recapitulate the myriad germane papers that establish that seasonality in work requirements exists in all agricultural systems, and that these patterns differ dramatically from one setting to another, depending on the resource endowment, technology, and policy context. Instead, we will concentrate in this section on the performance of labor markets from one season to the other, and the specific effects on seasonal wages. Although this discussion draws primarily on evidence from Asia, it is of increasing relevance to Africa as the share of income from wage labor increases. The question of wage formation in labor-surplus economies has generated a large literature focusing on whether surplus labor exists and, if so, why wages do not fall. Given the heterogeneity of agricultural employment and the diversity of environments, we will not attempt to determine which model of wage formation applies to which community. Rather, we will focus on the implications for seasonal earnings that the different models entail.

Models of Wage Formation

It is often convenient to present the alternative models of rural wages as either (1) models in which wages are institutionally and exogenously fixed or (2) models belonging to a neoclassical framework in which wages are determined by aggregate shifts in supply and demand, thereby reflecting the marginal product of the work performed. While such a dichotomy understates the range of alternatives, the neoclassical model does present a familiar base from which one can begin discussions or comparisons. In such a model, production and consumption decisions are independent maximizations of profits and household utility, respectively.[1] Producers equate marginal returns to the market wage rate. Institutional or cultural factors are not irrelevant, but come into the model principally through the utility function of the household. For example, the Moslem proscription against female participation in the labor market will shift the female labor supply curve inward, raising both female and male wages. Similarly, demand shifts coming from, say, off-farm alternative employment will affect wages. Thus the seasonal pattern in aggregate production will shift labor demand outward and entail a procyclical wage pattern.

One question of interest is the size of such a shift, which depends on the magnitude of the labor supply and demand elasticities.[2] Low elasticities of labor supply with respect to wages are reported for such diverse economies as Sri Lanka (Sahn and Alderman 1989), India (Rosenzweig 1984a), Malaysia (Barnum and Squire 1979), and Sierra Leone (Strauss 1982). Similarly, rural labor demand appears inelastic in India (Evenson and Binswanger 1984) as well as prewar Japan (Nghiep 1979). This inelasticity implies that seasonal shifts in labor demand (or supply) will have marked effects on wages. This leads to marked seasonal patterns in wages as well as sensitivity of wages to technological changes or the introduction of seasonally targeted public works in rural areas.

The neoclassical approach contrasts with the assumption of a constant subsistence wage rate found in a number of models of development, including the Lewis and Fei-Ranis models. Specifically, there are two main types of models that depart from the basic neoclassical model with market-clearing wages and no unemployment. These two types are both directly relevant to an analysis of seasonality and seasonal fluctuations in earnings among rural households. They can generally be classified as efficiency wage models and implicit labor contracts models.

The efficiency wage hypothesis, which purports to explain such a sticky rural wage in the context of apparently surplus labor, was first presented by Leibenstein (1957). The model posits that producers hire not labor time but units of work. The latter are assumed to be related to the former through a function that includes the calorie intake of the laborer. That is, labor power is functionally related to nutrient consumption, which

---

1. The model is discussed in detail and tested in Rosenzweig 1984a.
2. Assuming that there are no exogenous shifts in the supply of labor, the change in wages following a shift in the demand for labor can be calculated by the formula:

$$W_{t+1} = W_t\{1 + [D/(e_s - e_d)]\}$$

where W is wages, D is the percentage change in labor demand, and e_s and e_d are supply and demand elasticities.

is hypothesized to be increasing at an increasing rate at low levels of intake, thereafter diminishing in returns after an optimal level of consumption is reached. In this model, then, there is an optimal point below which reduction in wages leads to a sufficiently large reduction in work efficiency that net returns to the producer are reduced. This explains the rigidity in wages, since lowering them below the optimal point will jeopardize the nutritional status of the laborers and thus their ability to do work.

Three conditions are necessary but not individually sufficient for the relationship to be plausible. The first, widely but not universally verified, is that changes in earnings must lead to changes in consumption. The second condition is that institutions must exist that enable individual producers to capture the returns to their investment in the nutrition of their employees. The third is that improved nutrition must lead to increased work efficiency in the range of consumption observed for wage laborers. This latter condition, while intuitive, has proven difficult to test.[3] The recent work by Sahn and Alderman (1989), Strauss (1986), and Behrman and Deolalikar (chapter 7), however, provides some persuasive evidence of the nutrition-productivity link.

Of principal interest to the issue of seasonality in earnings is the prediction that efficiency concerns imply a floor level for wages, at least over low ranges of consumption. That floor corresponds to a level of consumption (and thus real income, to the extent that there is a fixed relation between income and intake) that provides for nutritional adequacy in all seasons when work is being done. Striving to achieve this floor should encourage in-kind payment in the form of food and on-site feeding of workers or similar mechanisms to discourage substitution of nonfoods for foods.[4] Note, however, that the theory does not rule out seasonal variation of wages when the demand curve for labor shifts outward during the peak season and wages rise above the floor. Like other minimum wages set by unions or by legislation, the efficiency wage implies higher earnings per employed laborer but fewer workers employed than the basic neoclassical framework. Whether these unemployed succumb to Malthusian forms of market clearing or eke out a subsistence livelihood in forage activities and small crafts is discussed in the literature but not resolved.

The efficiency wage theory also predicts that employers will prefer long-term contracts over daily hiring in order to capture the returns to their investment in nutrition. However, daily wages may be lower under

3. Supportive evidence is reviewed in Strauss 1986. See also Rogers 1975.
4. While the existence of canteens at the workplace has been pointed to in support of the biological determination of wages, an equally plausible explanation is that employers are simply trying to reduce the costs of workers taking long breaks in order to have their meal, coupled with the fact that it adds further structure and a sense of commitment to the workplace.

long-term arrangements because the nutritional floor wage depends on a long-term nutritional relationship. Also, since the efficiency of work, which determines the optimal wage rate, is a function of total consumption from all sources, wages would be lower for laborers with a small parcel of land or with more than one family member working.[5] Bardhan (1979), who found a negative coefficient for land cultivated when predicting wages, argued that nutritionally determined wages are not the cause. Rather, he suggests that it is more difficult to hire workers who have land to care for, and thus face risks and costs arising from delaying work on their own land. Employers face risks associated with a consequently unpredictable labor supply from smallholders and will therefore pay the landless laborer, who is more available, a wage premium.

While the floor wage is defined in terms of nutrition, it is only the nutrition of the worker and not the family that is of concern to the employer. Hence employers should show a preference for laborers with few dependents.[6] Rogers (1975) found this to be the case in Bihar, as did Bardhan (1979) for casual workers in West Bengal, although just the opposite was true for regular farm laborers in West Bengal. Also, if the probability of employment can be influenced by a worker signaling his nutritional status, workers could undercut other workers by increasing their share of family intake, especially in the off-season. The theory, then, implies interdependence of labor participation decisions by various members; and to the extent that there is a seasonal pattern to household members working, the intrahousehold distribution of food resources will be affected. Little research, however, has been done in this area.

Nutritionally determined wages may create conditions under which wages and employment are tied to other contractual arrangements, including credit. However, other conditions may also account for such arrangements (Binswanger and Rosenzweig 1984, Bell and Srinivasan 1985b, and Bardhan 1980). These fall into the domain of what may be referred to as implicit labor contracts theory, where labor services are not auctioned in the market, but are exchanged for an implicit commitment for work and wages on the part of the firm for a "reasonable" amount of time, with terms agreed upon between employer and employee in advance (Azariadis 1975).

Interlinked markets that tie employment or tenancy with credit or marketing activities may reflect the moral hazard of consumption loans

---

5. Bliss and Stern (1978) discuss the conditions under which this prediction should hold.

6. Another plausible explanation for preferring workers with few dependents is that they are less likely to be absent because of caring for other household members. A converse argument, however, is that laborers with few dependents require higher wages because of the constraints on intrahousehold diversification of income to limit risks.

given to individuals with little collateral, or may reflect various types of obstacles to (1) obtaining information on the capacity of workers and (2) monitoring the performance of laborers. Also, the risk-averse laborer who confronts great uncertainties in terms of the flow of income may trade his services for an insurance contract, whereby there is a tacit understanding that the firm will guarantee to maintain wage rates, hours worked, and so forth, despite the seasonal fluctuations in the production cycle. Thus wages no longer reflect the marginal revenue product of labor because the contract plays a dual role—compensating the worker for services and providing insurance against variations in income (Holmström 1983). Similarly, an employer's interest in reducing search or training costs or in influencing village-level politics by holding out the possibility of selective privileges encourages employers to enter into implicit labor contracts (ILCs).[7]

Bardhan (1979) contends that ILCs are not determined by customs or subsistence needs, which are invariant to changes in the demand for labor. However, with ILCs, the amplitude of seasonal wages is expected to be lower than would be predicted on the basis of marginal products, since the employer enters into ILCs during the slack season to save on peak-season recruitment costs. The implication is that the tighter the labor market and the lower the unemployment, the more difficult peak-season recruitment will be, leading to more tied labor contracts and lower seasonal variability. Accordingly, it is the level of employment that conditions the nature of labor relations and wages, not vice versa. If this is the case, the correlations between unemployment and tied labor contracts would be expected to be negative. Thus this model would posit that attached labor, which smooths out seasonality in wages, is not a manifestation of a feudal or stagnant agriculture, but quite the contrary.

In a similar vein, the model that suggests that high recruitment costs during the peak season lead to tied labor arrangements and higher wages has important implications for the effect of technological innovation on the seasonality of food security. It can be argued on the basis of ILC theory that technology that increases peaks in seasonal labor demand will not only drive up wages but will also encourage employers to enter into contracts, which will smooth the flow of income to poor households. Accordingly, introducing labor-saving technology during the peak periods might have especially deleterious consequences in terms of wages and seasonal food insecurity for the landless poor. This is in contrast to the argument that care should be taken not to introduce technologies that exaggerate labor demand peaks, because this only serves to raise the variability in seasonal wages.

7. See Hart 1986. Hart reviews evidence from Indonesia, the Philippines, Thailand, and Chile as well as many of the South Asian examples discussed in this chapter.

Despite the ongoing debate on the causes of such arrangements, there is widespread evidence that interlinked factor markets are prevalent at least in Asia (Bardhan and Rudra 1978; Bardhan and Rudra 1981; Bell and Srinivasan 1985b). Furthermore, while some evidence exists that such arrangements occasionally reflect extraeconomic coercion, the majority of examples appear to be voluntary agreements under which one or both parties benefit relative to untied arrangements, and neither loses. This is especially important since "implicit" contracts are difficult to enforce. From the worker's perspective, for example, it might be more advisable to reject occasional offers of higher wages so that he may get preferential contracts for "reliable" and seasonally stable contracts in the future. Similarly, the good reputation of the employer in upholding ILCs may condition the quality and cost of labor available to the firm in the future (Azariadis and Stiglitz 1983).

Although long-term rural labor contracts are not necessarily tied to other factor markets, they are often linked to the provision of credit. Whether or not they are tied with credit, such long-term agreements may reduce risk for laborers and search costs for employers. They may serve somewhat as a labor futures market operating between seasons. It is not, however, universally reported that such contracts reduce the daily or total wage paid to the worker; that is, there may not be any risk premium at all. In any case, such arrangements have the potential to smooth the flow of labor earnings over the seasonal production cycle. They are also likely to result in downwardly rigid wages, although, as in the efficiency wage model, mobility in labor markets will tend to make upward movement in wages possible. Also, like the efficiency wage model, the ILC theory raises the possibility that there may be less employment than a neoclassical equilibrium. However, the employer may not retain a full work force over the slack season in order to maintain his reputation and reduce the cost of finding reliable and inexpensive labor in the future.

When such contracts are linked with credit, the household acquires an additional means to control the flow of consumption from one season to the next through its savings behavior, which is distinct from the effect of assured employment. The net effect of tied credit arrangements, which of course may also be granted to casual laborers, depends in part on the conditions that motivate the link. On the one hand, there is the possibility of bonded labor in which the laborer's debts are used to obtain a serflike obligation to the lender. On the other hand, in models such as Hart's (1986), the credit would lead to a net addition to the household's consumption because it is used to ensure political loyalty, not to extract economic concessions from the laborer.

To date, there is insufficient evidence to assess the costs of such credit to the laborer. Bardhan and Rudra's (1978, 1981) data, which document the fact that employees frequently provide credit to laborers, also document the fact that wages are often reduced during repayment. Their studies, however, do not indicate whether such wage reductions imply interest rates above or below market contract rates. But as institutional credit is generally unavailable or rationed for smallholders and landless households, and frequently is also difficult to obtain in the informal sector (Bell and Srinivasan 1985a), the face value of interest rates is not a clear indicator of the value of credit to the household. When the demand for credit exceeds the supply, the value to the household of the loan component of the tied wage agreement is at least as great as, or greater than, the interest rate implied in the wage reduction. Under conditions in which tying labor and credit relationships provides a means of circumventing the laborers' lack of collateral, tied transactions tend to lower peak and raise off-season consumption by the laboring household.

Often, land as well as credit is rationed to individuals. Indeed some models of interlocked markets are predicated upon the assumption of rationing of land.[8] While the full welfare implications of such models are ambiguous without some restrictive assumptions, it is clear that the existence of such contracts should again influence the relative heights of seasonal peaks of earnings for laborers. Since labor provides income prior to harvests, supplying land as part of a labor contract provides for a different seasonal flow of earnings than either a pure tenancy or pure labor arrangement.

It is clear that the empirical evidence on the welfare implications of linked tenancy, labor, and credit relations is too scarce and localized for generalizations. Nevertheless, an understanding of intertemporal contracts is essential for understanding patterns of seasonal earnings and consumption in rural areas.

Specific Seasonal Effects

The theoretical underpinnings of seasonal wage formation discussed above and the implications for patterns of earnings are the subject of a limited number of studies that explicitly look at the seasonal dimensions of the problem. While a good number of studies do indeed describe seasonal variability in wages and are useful in establishing its extent, the data are limited in that the underlying processes responsible for shifts in wages from one season to the next are rarely discussed. For example, Norman et

---

8. Bardhan (1980) reviews his own and other models, pointing out the rigidity of assumptions used. Braverman and Srinivasan (1984) present a model with rationed tenancy, although labor is not directly tied. They show that if the landlord has reduced the tenant to his reservation utility level (by limiting the size of the plot), it is then in the landlord's interest that his tenant get credit as cheaply as possible, that is, from himself rather than from more expensive alternative sources of capital, such as other local moneylenders.

al. (1976) found that 44 percent of the total annual labor input occurred during June to August in Nigeria. This peak season of agricultural work was not characterized by a notable increase in wages, suggesting that some factors were limiting the flexibility in pricing. Similarly, the evidence on wages among the Masai in Kenya in crop year 1962–63 showed little seasonal variation, despite considerable amounts of labor being hired for periods of weeding and harvesting (Heyer 1981). Ryan (1982) estimated a very small coefficient for a dummy variable for peak seasonal wages when estimating wage equations. Bardhan (1979) also found that seasonal wages did not rise as expected given the basis of supply and demand shifts, and posits a number of explanations, which are discussed below.

Similarly, Chaudhury (1980) observed only a slight variation in seasonal wage rates, despite the fact that 40 percent of the demand for labor occurs during the period April to May. However, he notes that this rise is offset by the high rice prices that also occur during the peak seasons of labor. Thus real wages do not increase. This indicates that examining the movement of wages without regard to prices may lead to misleading conclusions about seasonal stress and wage formation.

In Java, Indonesia, despite the fact that there was a 52 percent decline in hours worked in wage labor from the peak to the slack season for households controlling less than 0.2 hectare of land, the percentage change in returns to labor per hour actually increased marginally from the peak to the slack season (Hart 1980). However, the higher wages and fewer hours worked seem to be the consequence not of the idiosyncrasies of fixed-wage employment models, but of the fact that during the slack season, workers are engaged in fishing rather than cultivation. This suggests the need to examine patterns of employment commensurate with data on seasonal wages.

In Egypt, Hansen[9] observed that wages are not truncated at some subsistence level, and that there is a close association between employment and wage fluctuations. He argues that there is a shift in the demand curve for labor, which brings about a highly seasonal pattern of wages. This—coupled with the fact that there are no wages reported during the slack season, presumably because there is no hired labor, both of which contradict an institutionally determined wage rate—is further evidence that wages reflect their marginal product. A similar conclusion is reached by Rosenzweig (1984a).

A final dimension to the seasonality of wages is the potential that seasonal patterns of morbidity have for affecting labor supply, thus affecting the price of labor. Specifically, the prevalence of diseases during certain seasons has been identified as a major factor in limiting the supply of la-

---

9. See B. Hansen 1969, 1971. These conclusions are challenged by J. Hanson 1971.

bor. The evidence as to whether illness has its most important effect during the busy season or slack season is somewhat conflicting (Cleave 1974; Chambers, Longhurst, and Pacey 1981). This contradictory evidence is partially explained by the fact that while the onset of disease may indeed be higher in the peak season, there is considerable latitude for postponing treatment or overcoming morbidity through determination. Thus both the timing of the agricultural calendar versus the seasonal pattern of morbidity and the ability of the workers to compensate for disease will determine the magnitude and implications of lost work time and its subsequent effect on labor supply and wages.

Smoothing Out the Stream of Seasonal Earnings

The decline of the marginal productivity of labor for on-farm work during the slack season often leads to a tradeoff between paid employment, either on another farm or in nonfarm activities, and working on one's own land.[10] For example, Norman's (1969) study on the seasonal pattern of work in Zaria in northern Nigeria indicates that despite marked seasonal fluctuations in the pattern of on-farm and off-farm work, there is little seasonal fluctuation in total days worked owing to the substitution between farm and other work. Similarly, as labor demand declines during troughs in agricultural activity, one can expect the laborer to seek alternative areas of more productive employment, assuming that wages fall commensurately with the marginal value product of labor.

The opportunity to exploit on-farm and nonfarm employment possibilities results in landowning families having a greater degree of freedom in reducing seasonality of work and income. Therefore, while labor inputs and earnings of agricultural households from own-farm activities may be highly seasonal, the household may be able to hire out its labor during the slack season and thereby minimize the impact of the production cycles in agriculture.[11]

The converse of this is that the agricultural household is likely to hire in labor during the peak season. The residual between required labor inputs and family labor thus presents work opportunities for landless laborers. Hence the opportunities for landless laborers to smooth out seasonality are fewer.

Evidence from India that there is more seasonal variability in labor

10. Also, see Cleave's (1974) discussion of the tradeoffs in Uganda among farm, nonfarm, and leisure activities.

11. Squire (1981) points out that in Nigeria, the coefficient of variation of total family labor input was 0.2 for total employment and 0.5 for own-farm activities; the corresponding figures for Malaysia are 0.20 and 0.35, respectively.

per hectare on smaller farms also suggests that the poor may be at greater seasonal risk (Ryan and Ghodake 1984). Lipton's (1983) extensive review shows that the poor, especially the casual laborers with few nonfarm assets and little or no land, are more severely affected by fluctuations in participation and wage rates.

One result of this seasonal variability is the relatively large allocation of time for job searches during the slack season. Similarly, a greater amount of time might be spent traveling to and from work during the slack season. Both of these phenomena were observed, for example, by Hart (1980) in a Javanese village. Undoubtedly, as these search or travel costs increase, potential workers are discouraged from seeking work. Observed differences in fluctuations in search and travel time for different population groups may reflect differences in reservation wages.

Technical Change

There are numerous strategies for reducing seasonal variability in production, such as irrigation and the more efficient use of rainwater. Better water use practices also include the collection and recycling of runoff, conservation of soil moisture through contouring and grading, and mulching. Perhaps more important is the availability of crops and the choice of crop ratios to reduce bottlenecks in labor requirements. All of these strategies increase the likelihood that a second or third crop can be cultivated.

In addition, a variety of management practices that more efficiently exploit the agroclimatic environment can help reduce seasonality of work and food availability. Dry seeding, which eliminates the loss of time between the rains and planting of crops, will also increase the likelihood of a second crop. Plant spacing, thinning, intensive weed management, intercropping, sequential or relay cropping, and splitting of input doses may also smooth out the demand for labor and increase the number of crops harvested per year (Jodha and Mascarenhas 1983). Technological advances, such as developing germ plasm that matures more quickly and is photoperiod-insensitive, will also increase the likelihood of multiple cropping, thereby smoothing out income flows.

These production strategies, which are intended to diminish variations in labor inputs and streams of seasonal income, are conceptually sound. The dilemma arises when strategies to smooth out seasonal peaks in production, and hence labor demand and income, are implemented at the expense of increases in output. For example, a change from shifting cultivation to semipermanent and permanent cultivation systems may aggravate seasonal labor bottlenecks. The reasons are that there is (1) less time spent clearing land, a traditional dry-season activity, and (2) increased labor required for weeding during the rainy season. Similarly, introducing new seeds and fertilizers and changing the power source may shift the bottlenecks from weeding, preparing land, and planting to harvesting (Norman et al. 1976).

This type of dilemma may also be manifested in selecting technologies to augment production. For example, Day and Singh (1977) found that although total demand for labor increased little with the introduction of high-yielding varieties and selective mechanization in the Punjab, the distribution of that labor was altered considerably. On the other hand, other studies have found that new technologies, especially fertilizer, have altered considerably the amount of labor required but have not changed the seasonal pattern of this demand. Thus tradeoffs may exist between technology that maximizes annualized farm earnings and production, and technology that minimizes fluctuations in the seasonal pattern of earnings and production. Potential conflicts arise between the desire to raise production and income, on the one hand, and promote stability in consumption and food security, on the other. These conflicts can be mitigated by improved capital markets, which will offset the negative consequences of more lumpy income or concentrated requirements for credit to purchase agricultural inputs, and a better market infrastructure, which will reduce seasonal price increases.

On-farm strategies to diversify seasonal income sources may also include shifts to cash crops and livestock production. For example, W. Jones (1972) reports that farmers in Mali plant cash crops after food crops. This behavior not only may involve producers diversifying their production, but may reduce the seasonality of employment for workers.

In his extensive review of the relationship between cash crop and food crop production in Africa, Cleave (1974) notes that sometimes there is, and in other circumstances there is not, a conflict between the seasonal labor requirements for the two sets of activities. For example, cotton planted in the Sokoto province of Nigeria appears to even out labor use, while groundnut production in the same region conflicts with production of millet. In Sudan, cotton production is shown to come into direct conflict with labor demands for production of cereal staples, while in Genieri, Gambia, both rice and groundnut production have coincident peaks in labor use. But more interesting is the nature of adjustments that farmers employ to address these potential seasonal labor conflicts. These revolve around changing the timing, intensity, and nature of operations; and adjusting cropping patterns (see, e.g., Heyer 1981; Cleave 1974).

Another production strategy that deserves consideration is the mixed-farm concept, in which livestock and cropping are integrated to smooth out seasonal labor demand and food availability. Once again, the evidence is conflicting and the outcomes contextual. In Uttar Pradesh, workers tend cattle during the slack season, which represents not only an instrument of savings but a source of work and income during a period of little activity in the fields (Lipton 1983). Likewise, livestock and dairy sales are noted to be a source of wet-season income in Nigeria (Kumar 1985). In contrast, the peak month of weeding for the Fulani in Burkina Faso is concurrent with the period in which the herds require the greatest attention, because their proximity to the fields requires supervision to prevent damage to the food crops. In this case, the integration of livestock and cropping only aggravates seasonal labor bottlenecks (Delgado 1979).

While there is little question that the seasonality of agricultural production and the marginal productivity of labor are largely determined by patterns of rainfall, the choice of crop, and technique, there remains another dimension to the relationship between seasonality and technical change. Specifically, not only does technology affect seasonal patterns of output, labor demand, and earnings, but technical change is guided by the relative prices (i.e., scarcity) of factors of production and by seasonal shortages and constraints in the availability of these factors.

Migration

Seasonal migration occurs for several different purposes, including search for work opportunities or food. This may involve migration from one rural area to another where cropping patterns and the agricultural calendar differ, or from the home region to plantations, mines, or cities. Other reasons for seasonal migration may include engaging in trade and marketing activities, cultivation of secondary landholdings, and pasturing cattle.

Seasonal migration is a widespread phenomenon that must be distinguished from transmigration or relocation. It may involve laborers migrating temporarily in search of employment from rain-fed regions of Bihar to irrigated tracts in the Punjab, from Lesotho to South Africa, and from truck farms in New Jersey to orange groves in Florida; traders in search of markets, such as the hill people of Nepal bartering Tibetan salt and woolen materials for foodgrains in the Terai of Nepal and the towns of India, and as far off as Burma, Thailand, and Singapore; and semisettled agriculturalists in West Africa in search of fodder during the dry season, or the seasonal migration of farmers and their cattle from the hills of Nepal when pasture is covered by snow to the Terai, or to the high plateau of Tibet with sheep, goats, or yaks (New ERA 1981).

Dry-season migration from rural to urban areas in the villages of Senegal may involve 30 or 40 percent of the active population. This enables them to earn cash needed for the next planting (Waterbury 1983). A study from a district in the hills of Nepal found that 30 percent of the households had one or more members migrating seasonally in search of employment. Similarly, in Matlab, Bangladesh, Chaudhury (1980) shows that during the lean season, especially during the winter months of January and February, laborers migrate to rural works projects and to other regions where crop conditions are more favorable, because of either environmental factors or the adoption of new technologies that alter the agricultural calendar and overall demand for labor. Inasmuch as demand for labor is determined by cropping patterns, technological differences between regions encourage and perpetuate seasonal migration in Bangladesh. Similarly, to the extent that ecological conditions differ in neighboring regions, encouraging varying patterns of work, migration will be prevalent. As agricultural technology becomes less differentiated, the opportunities for rural-rural migration will be reduced. Workers will be forced to travel greater distances and incur higher search costs for off-seasonal employment.

Seasonal migration may also involve workers heading to towns, plantations, and mines for work. Workers migrate to sugar factories from dry agricultural zones, as is witnessed in Maharashtra and Gujarat, India (Breman 1984). Similarly, women in Java, Indonesia, find off-seasonal sources of employment on sugar plantations (Hart 1986), men in Liberia migrate to rubber plantations in search of seasonal work, and farmers in the north of Ghana migrate after the harvest to work on the perennial cash crops in the south (Connell et al. 1977). This reinforces the potential role of cash crop production in smoothing out labor demand. However, there is a risk that seasonal migration will withdraw otherwise productive labor from food crop production if there are seasonal conflicts in labor demand.

Although the evidence on this potential conflict is limited, a linear programming model of seasonal migration in Ghana from the north, where subsistence agriculture predominates, to the south, where cocoa is grown and other food crops are cultivated, indicates that optimal solutions involve patterns of extensive migration similar to those observed in reality. The contention that temporary migration is inefficient or caused by factors such as imperfect labor markets is shown to be wrong (Beals and Menezes 1970). As pointed out by Berg (1965), for West Africa in general it is natural for laborers to migrate seasonally from the savanna zones to the forest regions where cocoa and coffee are grown. Without this type of seasonal migration, the inelastic labor supply in the forest zone would have restrained the growth of the export markets.

The prevalence of migration to areas with greater labor demand is not surprising. What is perhaps surprising is that marked regional wage differentials are observed in many areas. One possible reason for this is that lack of credit facilities and poor information impede interregional mobility of seasonal workers. Seasonal patterns are likely governed by the same factors as relocation—the magnitude of the wage differential; the cost of the job search, as well as the probability of success; and characteristics of the family such as age of the laborers, their dependents, and their landholding or tenancy rights. However, seasonal migration also differs from relocation because it is a regularly recurring event that provides opportunities for information gathering and—particularly when special skills are involved—for intertemporal contracting.

To understand fully the impact of such migration patterns on rural households, one must investigate the relationship between such migration and credit markets. For example, is credit extended to migrants for investment in travel costs as it might be for local farm employees, or can a family borrow against a migrant's expected earnings? There is some evidence from India that migrants working as sugarcane cutters receive credit from the broker who hires them (Breman 1984). Also of interest are the channels of remittances available to the migrant during his period of separation from the family. If secure channels for remittances are available, migration will have an impact on seasonal patterns of household consumption different from what would happen if the family must wait for the migrant's return.

Similarly, the impact for household consumption depends on whether the migrant draws upon household food stocks for his subsistence during the migration or obtains food from other sources, presumably outside the village market channels. Of course, there are circumstances such as those in Nepal where the whole family migrates seasonally, with the women working at paddy husking and the men obtaining employment in other agricultural activities and construction projects (New ERA 1981). The intrafamily distribution of the impacts of temporary migration—and its nuances, such as who migrates, for how long, and the patterns of remittances—remains relatively unexplored.

Under competitive assumptions, it is a relatively simple exercise to model the impact of such shifts of agricultural labor supply on the wages of laborers in the region receiving migrants as well as on the wages of those who remain behind. But once again, the shift in supply caused by alternative employment opportunities can be modeled according to either the neoclassical model or the fixed wage model. For example, Chaudhury's (1980) analysis that migrants are willing to accept a wage below that normally observed in the market, which in turn depresses wages at the destination, corresponds to the neoclassical model. Furthermore, he argues that seasonal migrants are "less troublesome" and more diligent since they are away from their families.

If, on the other hand, wages at the destination are nutritionally determined, migrants would not affect the wage rate appreciably. Similarly, at the region of outward migration, nutritionally determined wage levels will not be affected unless the withdrawal of labor is sufficiently prevalent to take the market beyond the region of low calorie consumption for which the efficiency wage model holds. Or, if remittances from migrants raise the consumption of family members remaining at home, migration may influence wages by lowering the cost of efficient wages for producers. Note, then, that in such a model, migration may have little effect on the off-seasonal patterns of consumption for most of the household. As discussed above, a similar difference between the nutritional and neoclassical models of wages exists in the case of off-farm employment.

What begins as seasonal migration may eventually lead to permanent migration or even constant wandering and homelessness. Concerning the former, cyclical migrants over time learn of opportunities for employment and availability of land; they gather other information that may precipitate a permanent move toward a more favorable economic environment. Concerning the latter, there is the prospect that the search for work will then often involve the entire family roaming in search of work, rather than individuals migrating seasonally to specific destinations from their home base. These mobile households, such as the Voddas of southern India (Epstein, 1973), are perhaps the most vulnerable and deprived, in part because of their "invisibility" and the difficulty of reaching them in any intervention programs.

Food-for-Work and Employment Schemes

One type of off-farm employment that is explicitly designed to deal with problems of seasonal employment are employment guarantee schemes such as that existing in Maharashtra state in India (Dandekar and Sathe 1980); this is also true for many food-for-work programs. While in many places legislation has provided a floor wage, the Employment Guarantee Scheme is designed to provide unlimited employment at such a floor. Although evidence on the Maharashtra scheme points to employment of 160 days per worker, there is currently little evidence on how much such a program is used as a seasonal floor rather than as a major source of employment for the otherwise marginally employed.

Food-for-work and other rural public work programs, on the other hand, are generally seasonal, providing in the neighborhood of one to two months of employment per worker employed (Wijga 1983; Ahmed and Kumar 1985). While the efficacy of such short-duration programs to fulfill nutritional objectives is not proven, they have the potential to provide a substantial portion of a laborer's off-season earnings, as well as a measurable increment to annual earnings.[12] However, unless funding is sufficient to employ all labor supplied at the wage offered, there is the need for some sort of rationing. While wages are often deliberately set below the prevailing market wage to avoid competing with private employment and to target

---

12. See Ahmed and Kumar 1985. The long-run impacts of the assets created are beyond the scope of this chapter.

the most needy families, there are few studies of the administrative guidelines that determine number of days and number of people employed and the tradeoffs between them when local supply is excessive. With the exception of a few regions, Bangladesh and Maharashtra, for example, few food-for-work programs employ a major share of a regional labor force, although they may do so on a district or village level.[13]

A final consideration in implementing any sort of employment generation or food aid scheme to reduce hardship in a slack season is the possibility that such efforts may change the employer's optimum wage. That is, if in fact the efficiency wage theory holds, increasing income during the lean season through employment programs may encourage the employer to reduce payment to workers, since it is now possible to gain the same level of efficiency for a lower cost. The implications of this are that if employment is provided in the slack season, the way to avoid a decline in wages is to generate enough work to absorb the entire excess labor supply (Rogers 1975). It is thus clear that the model of wage formation in a given context will affect the expected impact of seasonally targeted employment generation and food-for-work schemes. Besides employment generation schemes, temporal targeting of other food-related programs may counterbalance seasonal earnings. Specifically, if the poor do not have access to financial intermediaries or on-farm storage facilities to ensure access to food during the lean period, direct transfer payments may be appropriate. Similarly, if there is a clearly identifiable high-risk season when consumption declines, it may prove cost-effective to temporarily target food-related transfer programs.

Relationship between Income and Consumption Expenditures

In principle, wide fluctuations in earnings need not imply similar swings in the pattern of expenditures. Even if a household has income during only one season, it has the potential to keep consumption fairly constant through savings and dissavings.

The basic theory of intertemporal utility under certainty presumes that a household maximizes utility over time subject to prices in each period, earnings in each period, and the interest rates that determine the potential to increase future consumption by forgoing current consumption or to increase current consumption by reducing the stream of future in-

---

13. Wijga reports data on World Food Program (WFP) projects for 1976–79 and projects funded by U.S. voluntary agencies in 1980. Bangladesh and Lesotho were the major recipients of aid from the United States in per capita terms, while Botswana and Cyprus were principal recipients from the WFP. Note, however, that some projects, such as Maharashtra's, do not rely primarily on foreign aid.

come. There are, of course, myriad variations and practical extensions on the basic textbook framework, some of which are discussed below.

One variation that is a conceptually distinct subset of the larger set of models of intertemporal utility is the permanent income hypothesis, first introduced by Friedman. There is still much debate on how appropriate this hypothesis is for developing countries (Bhalla 1979, 1980; Musgrove 1980) or, for that matter, developed countries.[14] To the degree that seasonal patterns are regular, they are predicted and hence can be considered part of permanent income. Weather shocks, however, may be transitory (Wolpin 1982). Moreover, the permanent income concept may also be peripheral to the seasonality concern in that it is not generalizable to subsistence households or to households with low levels of consumption. Friedman explicitly ignored the possibility that a household is unable to reduce current consumption significantly, to plan for future consumption, without suffering extreme consequences such as malnutrition.

Nevertheless, tests of the hypothesis throw light on the process of adjusting consumption across seasons. For example, Bhalla's (1979) results provide evidence that the poor are prevented by liquidity constraints from realizing planned levels of consumption. That is, shocks are "absorbed to an unusual degree by alterations of planned consumption."

It is generally observed that savings out of transitory income are higher than those out of other income, even in developing countries, though not necessarily of the magnitude predicted by the permanent income hypothesis. This has implications for short-term welfare measures. Savings, however, do vary by income groups. Bhalla, for example, observes a marginal propensity to save out of transitory income for the poor of only 0.22, compared with 0.55 for well-off rural families.

The central question is whether the poor can use savings or credit to smooth the peaks and troughs of consumption. If the income elasticity for food is less than 1, then one would see a dampening of oscillations of food consumption relative to general earnings, even if no intertemporal planning occurred. Moreover, if families predict seasonal cycles and can plan in a manner similar to that postulated by the permanent income hypothesis, one would expect cycles in food consumption to reflect mainly price movements, regardless of income streams. For example, one recent study, albeit for the United States, hypothesized that consumption of food reflected permanent income in order then to calculate the transitory component of income (Dynarski and Sheffrin 1985). Such a view, however, is hard to reconcile with the often observed phenomenon of major fluctuations in total expenditures or in food expenditures in developing countries.

---

14. A recent test using household data in the United States and focusing on food consumption is that of Hall and Mishkin (1982).

Low correlations of consumption, as measured in terms of nutrients or actual quantities of foods for the same household between survey rounds, cannot be attributed to seasonal fluctuations.[15] While such observations may reflect large random variations in measured expenditures, they may also reflect more systematic effects, a view that is supported by movements in group means as well as low correlations. The low correlations, however, imply that different households respond in different manners to seasonal effects. Currently, there are only a few data bases one can use to explain rigorously the nature of fluctuations of expenditures and to study movement of the parameters of the curve relating food consumption to income and prices by using variance components or related techniques designed to combine time-series and cross-sectional data.

As mentioned above, models of intertemporal utility generally assume access to credit. If the costs of borrowing differ among income groups of landownership classes, one would quite naturally expect intertemporal patterns to vary among groups. Furthermore, if credit is unavailable or the supply is quantity-constrained at a given interest rate, the appropriate interest rate for the household exceeds the nominal, or observed, rate. Evidence that credit market imperfections are more likely to be faced by the poor has already been discussed. In fact, this problem of limited credit is also likely to be most severe in the preharvest season, when there is a need to purchase inputs and cash is most limited. Similarly, financing for processing, transporting, and storing crops after the harvest is also required by both farmers and traders. These two patterns combine to smooth the demand for credit, although there will be periods of low credit demand, such as just prior to harvest once inputs have been used and stocks are low.[16]

Note also that most models of intertemporal budget allocation apply the simplification that returns to investments are symmetrical with the costs of borrowing. Clearly, if there are fixed costs to lending money or to obtaining information, the returns to investing will be lower for those households whose lack of collateral also makes the costs of borrowing higher than they are for the general community. Of importance, then, for the study of seasonal consumption patterns are the prevalence and nature of financial and asset markets in low-income communities. For example, are postal savings or credit unions available and trusted institutions for savings? How liquid is investment in such assets as livestock, land, or gold ornaments? Such assets are of course important as a hedge against uncertainty, which is a major determinant of intertemporal allocation of expen-

---

15. C. Scott (1980) discusses such correlations and their implications for the concept of income and expenditures.
16. See Mears 1981 for an illustration of the seasonal demand for credit.

ditures. However, even when we simplify our discussion from this important topic to focus on earning and price cycles that are known with certainty, the suitability of these forms of investment for dealing with seasonal patterns must be addressed. For example, while cattle may be profitable investments in some environments, it may make little sense to purchase cattle after the harvest for resale four or five months later.

A variety of other mechanisms in terms of the timing of purchases are available to the poor to smooth out consumption. For example, the purchase of durables and semidurables can be timed to coincide with periods of greater income. While many nonfood purchases are not discretionary (e.g., shelter, bus fare), others are amenable to reductions in accordance with the timing of income (e.g., clothing, implements).[17]

There is also considerable latitude for shifting alternative food sources in accordance with seasonal availability. This simple price response may take the form of, for example, consuming corn in the period before the rice harvest, or increasing reliance on wild roots and tubers during the preharvest lean season. Thus one would expect substitution of consumption away from the more expensive commodity, both within and between seasons. This substituting behavior must be distinguished from the use of grain as a potential vehicle for investment. The returns of such an enterprise should be at least equal to the opportunity cost of capital and the physical diminishing of stock. These storage concerns are discussed elsewhere and may of course differ by income groups. But it seems that the poor, who derive a smaller share of their income from nonlabor sources (e.g., home production), are less likely to have grain reserves and the capital to make such purchases during the postharvest season.

Another issue that may differ by income groups is the preference for current versus future consumption in regard to actual preference ordering as opposed to the costs of capital. This may reflect risk avoidance but may occur even in fairly stable environments. For example, expenditure from monthly wages or transfers may not be distributed evenly over the month despite the regularity of income (Madden and Yoder 1972). This may represent a "feast and famine" preference ordering, a type of nonlinearity of intertemporal utility under which, say, two units of consumption in period 1 and one unit in period 2 may provide more utility than one and a half in each, even after accounting for price and interest rate effects. While such preferences have not been formally studied, they may reflect the needs of subsistence-level consumers.

---

17. See Simmons 1976b for an example of adjustments in expenditures by season in Zaria, Nigeria.

The evidence on the extent of seasonal fluctuations in income and consumption expenditures is extensive and quite divergent.[18] While these studies give some insights into the extent of variability for a population, they generally do not explain the causes of the observed seasonality,[19] and fail to distinguish between how different household types or income groups are affected. On the one hand, it would be expected that richer farm households have more liquid assets, greater access to credit, larger food stocks, and better storage facilities, which would tend to smooth out their consumption levels from good to bad seasons. On the other hand, the poor who are near the subsistence level will make more strenuous efforts to limit the variation in intake from one month to the next through mechanisms such as being more price-responsive in terms of substituting between goods.

Therefore, while it is possible to recount the myriad studies that document seasonal variability in consumption, or lack thereof, doing so is considered of relatively little value. This is true for a variety of reasons, including the questionable reliability of much of the data, the difficulty of drawing inferences about regular patterns of seasonal variability on the basis of one-year data, and problems in interpreting the implications of seasonal consumption patterns without knowledge of shifts in energy expenditures. In addition, in order to identify policy options that make it easier for households to smooth out consumption, there is a need for further research in order to examine more explicitly the causes of any observed seasonal variability, rather than simply describing these interesting patterns.

Consumption Patterns and Allocation of Time

We are mainly concerned with physical consumption of food, although generating income through work versus choosing leisure may also be a process of intertemporal decision making. Specifically, the literature on consumer choice is usually presented in terms of a household facing a given budget constraint. In reality, however, there is a joint determination of the consumption of goods and of leisure. With leisure L, providing utility, the consumer seeks to maximize the utility of goods and leisure subject to the constraint

$$pq + wL = Y_o + wT \tag{6.1}$$

---

18. See, for example, review articles by Schofield (1974); Longhurst and Payne (1979); and Valverde et al. (1985). Also, see Kennedy and Cogill 1988 and Kyereme 1984.

19. An exception is the chapter by Pinstrup-Andersen and Jaramillo in this volume.

where Y_o is nonwage income, w is the wage rate, q is the vector of quantities of commodities purchased, and p is the corresponding price vector. The right-hand side of this equality is termed *full income*.[20]

The most general application of this model is the relationship between the price of leisure and consumption, as influenced by labor supply. This has traditionally involved a determination of the extent to which there is a backward-bending labor supply curve. But of interest in the domain of seasonality is the complementarity between the price of leisure and food— the effect of wage rates on food consumption. If the price of leisure (wages) changes, one will observe the conventional income effects and the substitution effects as predicted by the Slutsky equation, as well as a change in the value of the time endowment that is in the same direction as the change in wages. This latter effect influences earnings and offsets the conventional real income effect (Deaton and Muellbauer 1980).

The implication for seasonality of this relationship between the wage rate and consumption revolves around the changing values of market time from one period of the year to another. To the extent that alternative sources of productive employment cannot be identified and the value of time declines during the lean season, there will be a shift to either leisure, household-level activities or social obligations that do not increase earnings. Cleave presents evidence from Uganda, Gambia, Nigeria, and Sierra Leone that ceremonial activities and social events are concentrated in the slack season. This is attributed to a combination of a lower opportunity cost of time, plentiful food after the harvest, and reduced fatigue associated with working.

Similarly, during the busy season the price of leisure or household-related productive activities (i.e., child care, food preparation) becomes so high that women and children who are not normally in the labor market are mobilized into the formal work force. For example, women in Kahangi and Kyarusozi, Uganda, replace leisure time with agricultural work (Cleave 1974). Interestingly, no substitution away from the nonagricultural domestic tasks in which they are engaged was observed. Similarly, in Java, although the percentage change in return to labor from peak to slack season is the same for females of different classes, the percentage of women who withdraw from labor force activities is considerably higher in better-off households (Hart 1986). This suggests that the labor supply of women from poorer households is less sensitive to changes in wages, and that they are less likely to substitute toward other commodities when the price of leisure declines.

In Egypt, female and child labor acts as a buffer, entering the labor

---

20. For a brief discussion of the use of a full-income budget constraint, see Deaton and Muellbauer 1980.

force in seasonal peaks corresponding to the wheat, maize, and cotton harvests. This is reflected in the fact that hours worked by women and children display more seasonal variability than hours worked by men. Interestingly, the greater seasonality in work of women is also the case with wages, which results in men's and women's wages being nearly equal during the peak season of the cotton harvest (Hansen 1969). The question remains, however, of the extent to which the low participation rates during the slack season are attributable to the preferences for leisure, the higher productivity of other household activities (e.g., child care, education) and household members, or the lack of demand for labor.

In a similar analysis, Spencer (1976) found that an agricultural development project in Senegal exacerbated seasonal peaks in labor demand, and one response mechanism used by the households was the mobilization of women and children into the labor force. The consequences of such adaptation for consumption and nutrition are open to conjecture. To the extent that other important tasks such as schooling and child care are neglected, such a process may be detrimental. For example, studies indicate that women's time constraints during the peak season interfere with breastfeeding (Chen, Chowdhury, and Huffman 1979); and children tend to be weaned at the onset of the rainy season (Carloni 1984). It is further suggested that during the peak season when women must work, infants are left in the care of siblings; meals are cooked only once a day; and pulses are consumed less because of their lengthy cooking time (Carloni 1984). Other effects of the increased demand on women's time during the peak agricultural season are reduced water and fuel collection and less gathering of green leafy vegetables (Schofield 1974). The converse is that the higher incomes that accompany such seasonal work by women may provide major benefits by increasing the household's ability to purchase food, receive health care, improve the local environment, and perform other similar initiatives that may improve nutritional status. Thus caution must be used in ascribing negative connotations to responses to seasonal labor bottlenecks, especially those arising from agricultural development.

It is also commonly observed that leisure activities are consumed differently when food is generally plentiful than otherwise (Immink and Viteri 1981; Beaton 1983). To the degree that a household can reallocate food from one period to the next in order to support the energy requirements of required activities, these adjustments reflect intertemporal utility decisions in a manner similar to physical consumption of goods. For example, Simmons (1981) found that Nigerian households increase their consumption to meet higher labor demand. Although this increase in consumption was not sufficient to meet the increment in energy needs, it illustrates the ability of households to adjust their intake through seasonal storage. This finding is not unique to Nigeria, as similar results were obtained for Burkina Faso (then Upper Volta; Blanc 1969) and Senegal (Boutillier et al. 1962).

Household to Individual Consumption

The above discussion focuses on the evidence and issues concerning seasonal variability in consumption at the household level. There is another dimension that thus far has remained unaddressed: the link between household consumption variability and that of the individual. Just as there is considerable latitude for the household to make adjustments intertemporally in its consumption, so too in theory can the household adjust the intrahousehold allocation of resources to maximize its objective function, such as nutritional well-being. There is the opportunity for the household to reallocate household food resources so that the proportion received by individuals differs from one season to the next. There is only a modest literature on this subject, owing mainly to methodological difficulties of measuring and modeling intrahousehold decision making.

In their review of the literature, Longhurst and Payne (1979) found no data that support seasonal discrimination involving children in intrahousehold food distribution. In a later publication, however, Longhurst (1984) examined some data from Nigeria on male and female energy intake that indicated that women generally consume less than men, and that "female intake is never satisfactory and it does get relatively worse compared to male intake in the pre-harvest period." However, the evidence from Nigeria contradicts the findings of Svedberg's (1988) extensive review of 50 data sets from sub-Saharan Africa, from which he concludes that women are not at a nutritional disadvantage.

In another study from the Philippines, the ratio of the calorie adequacy of the child to the household's calorie adequacy was regressed on a number of variables, including seasonal dummy variables, two of which were from the slack season. One of these slack-season dummies was during a year in which the overall economy was healthy. The coefficient was not significant, meaning that the share of household calories received by the child was not significantly different from the amount received in the harvest season, despite the fact that household intake declined approximately 2.5 percent.

The other slack dummy was from a year in which a major economic downturn occurred. Although the overall household calorie intake was the same as in the slack season when the economy was showing greater utility, the share of calories received by the child was lower during the slack season in a period of poor economic performance than that observed during either the harvest season or the slack season before the overall economic decline (Garcia and Pinstrup-Andersen 1987).

While it is not possible to reach any generalizable conclusions concerning the seasonal effects of intrahousehold distribution, there are a number of hypotheses that warrant further investigation. For example, if there is a seasonal change in the proportion of the household's food basket being consumed by each individual, why? One hypothesis is that the share of income earned and controlled by each family member varies from one period to another. It may be that women are more likely to favor children's consumption over that of other members. Thus if more income is earned by women at a given period of the year, the share of the household food resource consumed by the child of lactating women might shift accordingly.

Another plausible explanation for a change in intrahousehold distribution could be that women reduce their involvement in the labor force during the slack season, thereby reducing calorie requirements, while men maintain their dietary requirements for energy-intensive employment and job searches. Thus the short-term impact of mothers and children reducing their share of the household's food consumption to improve the ability of the men to find a job may be compensated for by a medium-term increase in income if the head of household is successful. In such a case, a reallocation of food resources to the men during the slack season might not be attributable to the different marginal propensities to consume of men and women, as intimated above, but rather may reflect a sound nutritional strategy to aid the man in finding productive employment.

These types of hypothesized relationships need to be the subject for further research. Improved models of intrahousehold behavior, coupled with new data collection efforts, are required.

Need for Further Research

The workings of the labor market and the savings and consumption behavior of households remain two areas that have preoccupied development economists. The existence of unemployment and underemployment, especially in the traditional sector, has led to the assertion either that labor could be withdrawn to work in the modern sector, or that measures should be taken to promote agricultural development that exploits underutilized labor capacity. While technological change has been visibly successful in increasing yields and outputs, equal success in generating employment has not been manifest.

One dimension of this disappointment in creating new jobs revolves around the seasonal nature of agriculture and limitations of our knowledge concerning labor markets that are imposed by traditional methods of examining employment and time allocation data. There is a need to gain a better understanding of the seasonal patterns of time use and wage formation. This is especially needed in the rural sector in order to develop strategies of agricultural development that take into account seasonal pressures in labor requirements and to determine how markets deal with such seasonality in labor peaks in terms of the formation of wages. Similarly, the introduction of innovations, in the form of either new technology or interventions that promote new crops—for example, cash crops—must take into account the timing of those operations in terms of their effect on seasonal patterns of labor requirements.

Improved understanding of food acquisition behavior in terms of intertemporal consumption and expenditure decisions is also an area in need of further research. This involves not only gaining insight into preferences over time but, more important, learning more about those factors that enhance the latitude of households in deciding when to consume what, despite the lumpiness of earnings. The ability to allocate income to consumption expenditures in the future revolves around household savings behavior and the functioning of financial intermediaries and stocking functions. There remains much to be learned about why certain households in certain circumstances maintain a relatively stable level of intake while others do not. Improved modeling of savings and intertemporal consumption behavior is therefore another research priority.

7 Agricultural Wages in India: The Role of Health, Nutrition, and Seasonality

JERE R. BEHRMAN AND ANIL B. DEOLALIKAR

There are two persistent themes in the development literature. The first is that there are important seasonal variations in nutritional and health status. The second is that there is a technically determined link between nutritional status or health on the one hand and labor effort and productivity on the other.

The first theme has been emphasized by Chambers et al. (1981) and Chambers (1982), who contend that seasonal deprivation in nutrient and other health-related intakes and changes in health status are considerable, particularly during the wet season for women and children. Longhurst and Payne (1979) also emphasize that hungry or lean seasons are "an important, if not the most important determinant of nutrition in less developed countries." But as these papers emphasize, there are serious deficiencies in available empirical research on such seasonality.

The second theme is often summarized as the "wage efficiency hypothesis" and has been discussed by Leibenstein (1957), Mazumdar (1959), Stiglitz (1976), and Bliss and Stern (1978), and by Alderman and Shan in this book (chapter 6). However, it has been subjected to little systematic empirical testing for a number of reasons. First, the nutrition-health-productivity relationship cannot be established by mere correlations between variables, since a correlation could be picking up the effect of increased productivity, and thereby income, on nutrition or health, rather than vice versa. A more rigorous test involves regression of productivity indicators on the nutritional or health status of a worker, recognizing that nutrition and health are subject to choice and hence are endogenous variables. Second, the appropriate concept of productivity is marginal, not average, productivity, which is rarely observed directly by social scientists. The measurement of marginal productivity often requires the estimation of a technical production function or the acceptance of the assumption that wages equal marginal products for labor. Third, there may be substantial interpersonal variations in nutrition, health, and productivity, such as those due to seasonality.

No study that we are aware of has addressed all of these problems. There are many that have been oblivious to all of them,[1] though some recent studies by Deolalikar (1988), Strauss (1986), and Sahn and Alderman (forthcoming) do advance the state of the empirical art concerning nutrition-health-productivity relations by direct estimation of production relations for individuals or households within a framework in which the possible simultaneity of nutrition and health with productivity is controlled. All studies estimate agricultural production functions with hired labor and nonlabor inputs in addition to family labor inputs. All find evidence of some health and nutrition effects on labor productivity. All suffer from some deficiencies, including very limited or no exploration of seasonality.

In this chapter, we attempt to integrate these two themes in the literature by using a rural South Indian sample to explore how seasonal changes in nutrient intakes and health status affect labor market productivity, as reflected in market wage rates. Our exploration attempts to improve on the earlier literature by treating health and nutrition as simultaneously determined with wage rates and by taking account of seasonal variations in the health-nutrition-productivity relationship.

We investigate questions such as the following: Do the impacts of nutrition and health differ between the agricultural peak and slack labor seasons? Is there evidence of a differential impact of nutrient intake *flows* versus cumulative *stock* measures of health on labor productivity, as might be suggested by the adaptability hypothesis promoted by Payne and Cutler (1984), Sukhatme (1982), Srinivasan (1981), and others? Do the effects of health, nutrition, and seasonality differ for males versus females?

The first section outlines the model underlying our empirical specifications, and the second describes the data set. The final section presents and discusses our empirical results.

Model and Estimation

There is a long and established tradition of estimating wage equations for developed countries (Mincer 1974). Recently, a number of wage equation studies also have been undertaken for less developed countries (Psacharopoulos 1982, Birdsall and Sabot 1988). For our application to a traditional agricultural setting in this chapter, we extend the standard wage equation framework in three ways. First, we explore the impact, if any, of health and nutrition on individual wage rates. Second, we test whether the parameters of the wage equation vary between the peak and slack agricultural seasons. Third, we treat health and nutrition as endogenous variables.

---

1. Many of these studies are reviewed in Behrman and Deolalikar 1988b.

Estimated individual wage equations typically include as arguments personal characteristics of an individual such as schooling, labor force experience, and gender. In an agricultural setting in a less developed country, however, it is likely that the labor market also offers a wage premium to healthy and well-nourished workers, since many agricultural tasks tend to demand physical strength or substantial energy expenditures. We therefore include health and nutrition as arguments in the individual wage rate equations that we estimate below, and expect their coefficients to have positive signs.

Health and nutrition cannot be treated satisfactorily as exogenous variables influencing wages in the above analysis, since there is substantial evidence suggesting that health and nutrition are choice variables (Behrman and Deolalikar 1987). We therefore treat health and nutrition as endogenous variables in the wage and labor supply equations, and use agricultural consumption and product prices and farm assets as instruments for them (Strauss 1985).

Earlier studies, such as those by Ryan (1982) and Strauss (1986), have included measures either of health status or of nutrient intake as explanatory variables in their agricultural production functions or wage equations. However, our review of empirical health production studies in Behrman and Deolalikar (1988b) indicates that current nutrient intakes do not affect stock measures of health, particularly for adults. We argue that such results suggest that short-run changes in nutrient intakes are reflected in changes in energy expenditure rather than in changes in health status. Therefore, we have included measures of both health status and nutrient intake as explanatory variables in our wage equations to allow nutrient intakes to have an additional impact (over and above impact through health status) on labor productivity.

Another reason for including both nutrient intake and stock measures of health in the wage equation is that the two may fulfill qualitatively different needs in agricultural operations. Health is associated with innate strength or "horsepower," while current nutrient intakes are associated with energy expenditure. Agricultural tasks vary in their requirements of these two attributes.

We estimate forms of the wage equation that allow all coefficients to differ across the agricultural peak and slack seasons. Tasks performed in the two seasons may differ in their requirements of innate strength versus energy expenditure (as well as other attributes such as education), and this would be reflected in differential labor market valuation of these attributes across the two seasons.

Wage equations can be estimated only for those individuals participating in the casual daily labor market and hence reporting a wage. The nonrandomness of this sample may result in biased estimates for the wage equation, particularly if labor market participation is a choice variable for the individual (Heckman 1976; Olsen 1980). We have corrected for this potential selectivity bias in our estimates by using Olsen's (1980) least-squares selectivity correction procedure.

The wage equations are estimated using OLS-IV estimation methods, with consumption and farm goods prices and farm assets serving as instruments for health and nutrition.[2] The wage equation estimated is semilogarithmic, since this is the functional form that has been most commonly estimated in the literature (Mincer 1974; Psacharopoulos 1982; Birdsall and Sabot 1988).

Data

We used the ICRISAT VLS (International Crops Research Institute for the Semi-Arid Tropics Village Level Studies) data set to estimate individual wage equations. The ICRISAT VLS data are panel data that have been collected at regular intervals since mid-1975 on production, expenditure, time allocation, prices, and socioeconomic characteristics for 240 households in six carefully selected "typical" villages in three different agroclimatic zones in SAT India. Within each village, 10 households are randomly selected as representatives of agricultural labor and nonlandholding households, and another 30 are a stratified (by size of landholdings) random sample of cultivating households.

Data on daily wages received and days worked by participants in the casual agricultural labor market were collected every two to four weeks. Since the precise intervals between interviews vary, the wage data were smoothed and then aggregated into peak and slack seasons. Peak seasons were defined as those months when opportunity costs of labor, defined as the product of wages and the probability of involuntary unemployment, were at a peak. Slack seasons were the remaining months of the year. The peak and slack periods so defined were not congruent across villages, even in the same agroclimatic zone (Ryan, Ghodake, and Sarin 1980). Table 7.1 reports the village-specific peak-period months calculated by Ryan, Ghodake, and Sarin and used in this chapter.

For the 1976–77 and 1977–78 agricultural years, four rounds of a spe-

---

2. Deolalikar (1988) has estimated a wage equation in first differences (but not specific to seasons) with annual data from the same primary data source that is used here. He finds both calorie intake and weight-for-height to be significant determinants of wages. In the present study, we are unable to estimate a fixed-effects model, since this would reduce severely the size of the sample and would require dropping a large number of individuals who did not report wage, health, *and* nutrition data for all the periods (namely, slack 1976–77, peak 1976–77, and peak 1977–78).

cial nutrition survey were undertaken by ICRISAT to record individual nutrient intakes in the past 24 hours and anthropometric measures of health status. Since the dates on which each of the rounds was undertaken were available, we were able to aggregate the health and nutrition data into the peak and slack seasons defined above. The health and nutrition data were then merged with the wage data. We use the average seasonal value for each casual labor market participant as an observation, which gives us a total of 468 individual and seasonal observations.

We use average daily intake of calories as the relevant measure of nutrient intake because calories are widely recognized to be the most important nutrient. For health status, we use weight-for-height, an anthropometric measure that is widely assumed to reflect short-run health status. We use the measures in their original units (calories and kilograms/centimeter) as well as in age- and sex-standardized units (with the Indian recommended daily allowances in Gopalan, Sastry, and Balasubramanian [1971] serving as the standard for calorie consumption and the modified Harvard standard in Ryan et al. [1984] serving as the standard for weight-for-height).

Table 7.2 gives the means and standard deviations by season and sex for the wage rate and for health and nutrition variables. One fact that immediately stands out is that agricultural wage rates do not vary much across the peak and slack seasons. Female wage rates are only 6.5 percent higher and male wages only 2 percent higher in the peak than in the slack season. Neither of these differences is statistically significant. The relative constancy of wage rates across seasons is largely the result of geographical mobility of labor in rural South India. In the peak season, migrant labor from other regions prevents wage rates from increasing, while in the slack season, out-migration of local labor keeps wage rates from falling significantly.[3] The observed constancy of agricultural wages across seasons, however, may hide large interseasonal variations in the relationship between health and nutrition on the one hand and wage rates on the other.

Like wage rates, weight-for-height is also largely invariant across peak and slack seasons for both males and females, as is calorie intake for females. For males, however, daily calorie intake is higher (by about 12 percent) in the peak than in the slack season, which suggests that the additional calories consumed by males in the peak season may go toward satisfying the greater energy requirements in physically demanding, peak-season male tasks, such as ploughing.

---

3. However, as noted above with reference to the definitions of the seasons, employment probabilities are higher in the peak than in the slack season, so expected wages are higher in the peak season.

TABLE 7.1 Peak and slack labor periods for adults in six SAT villages of peninsular India, 1975–76

District and Village	Category	Peak Period		Slack Period
		Months	Major Operations	Months
Mahbubnager				
Aurepalle	Males and females	Dec–Jan	Harvesting and threshing sorghum, pearl millet, castor	Feb–Apr
Dokur	Males and females	Nov–Jan	Harvesting and threshing sorghum, nursery bed preparation, paddy transplanting	Feb–June
Sholapur				
Shirapur	Males	Apr–May	Preparatory tillage, ploughing	Dec, Feb–Mar
		July–Aug	Sowing pearl millet, mesta, mung bean	
	Females	Sep	Harvesting and threshing pearl millet, mesta, mung bean	Apr–Aug
		Dec–Feb	Sowing and harvesting wheat, sorghum, chickpea, safflower	
Kalman	Males	Jan–Mar	Harvesting and threshing wheat, sorghum, chickpea, safflower	Aug–Oct
		May	Preparatory tillage, ploughing	
	Females	Mar–Apr	Harvesting and threshing wheat, sorghum, chickpea, safflower	Oct–Dec, Jan
		Nov	Harvesting and threshing pearl millet, mesta	

Akola				
Kanzara	Males	Mar	Harvesting cotton; harvesting and threshing pigeon pea	Apr–Sep
		Aug–Sep	Preparatory tillage, sowing wheat, chickpea; harvesting sorghum, groundnut	
	Females	Oct–Dec	Harvesting and threshing sorghum, groundnut, cotton	Apr–June
		Mar	Harvesting cotton	
Kinkheda	Males	Apr	Preparatory tillage	Aug–Oct, May
		June–July	Sowing, interculturing cotton, sorghum, pigeon pea, mung bean	
		Nov–Dec	Harvesting and threshing sorghum, groundnut; sowing wheat and chickpea	
	Females	May	Field cleaning	Feb–Mar
		Sep–Dec	Harvesting and threshing sorghum, groundnut; weeding cotton	

SOURCE: Adapted from Ryan, Chodake, and Sarin 1980, pp. 372–73.

TABLE 7.2 Means and standard deviations by season, SAT India, 1976–78

	Slack Season	Peak Season
Daily wage rate (rupees)		
Men	3.34 (1.08)	3.41 (0.94)
Women	2.01 (0.55)	2.14 (0.62)
Average daily calorie intake		
Men	2,123 (920)	2,373 (880)
Women	2,053 (850)	2,035 (784)
Weight-for-height (kg/cm)		
Men	0.246 (0.07)	0.248 (0.07)
Women	0.235 (0.06)	0.236 (0.06)
Years of schooling		
Men		2.5 (3.0)
Women		0.8 (2.2)
Years of experience[a]		
Men		19.8 (15.2)
Women		22.9 (17.1)

NOTE: Figures in parentheses are standard deviations.
[a] Experience is defined as age minus years of schooling minus 8. See Ryan and Wallace 1986.

Empirical Results

Table 7.3 presents the estimates for the wage equation estimated over all market participants as well as separately for male and female market participants. Since the peak-season slope dummies were uniformly insignificant at the 10 percent level in all the equations for all right-side variables other than calories and weight-for-height, we allow only the intercept terms and the coefficients on calories and weight-for-height to differ across seasons in the results reported in this table. In addition, since the use of unstandardized versus standardized values of calories and weight-for-height made virtually no difference in our results, we report only the estimates using unstandardized calories and weight-for-height. Finally Olsen's sample selectivity correction procedure did not indicate sample selectivity as a problem in any of the three wage equations. Therefore, we present only the estimates in which no correction for selectivity is made in table 7.3.

The OLS-IV estimates for all labor force participants in table 7.3 (column 2) indicate significantly (at the 10 percent level) higher wage rates (by about 5 percent) in the peak than in the slack season, holding all other wage-determining factors constant. In the slack season, calorie intake is not a significant determinant of wages, although weight-for-height is (with an elasticity at the sample means of 0.67). However, calorie intake has a

TABLE 7.3 Agricultural semilogarithmic wage equations, SAT India, 1976-78

Independent Variables	All Participants		Male Participants		Female Participants	
	OLS (1)	OLS-IV (2)	OLS (3)	OLS-IV (4)	OLS (5)	OLS-IV (6)
Intercept	0.038	−0.269	−0.126	−0.181	0.480	0.623
	(0.20)	(1.00)	(0.04)	(0.50)	(2.50)	(2.40)
Peak season[a]	0.001	0.054	−0.254	0.007	−0.056	0.060
	(0.00)	(1.80)	(1.00)	(0.10)	(0.30)	(1.60)
Calories[b,c]	0.037	−0.249	−0.240	−0.084	0.512	0.519
	(0.20)	(0.60)	(0.80)	(1.50)	(2.00)	(0.80)
Calories × peak season[b,c]	0.452	1.487	0.518	0.853	0.207	1.058
	(1.50)	(2.40)	(1.10)	(1.50)	(0.60)	(1.20)
Weight-for-height[b]	1.558	2.789	2.952	2.427	−0.216	−0.710
	(3.70)	(3.00)	(4.70)	(2.10)	(0.40)	(0.80)
Weight-for-height × peak season	−0.208	−1.330	0.314	−1.005	0.500	−0.397
	(0.40)	(2.30)	(0.40)	(1.80)	(0.80)	(0.50)
Schooling	0.015	0.020	0.029	0.037	0.003	0.003
	(3.10)	(3.90)	(4.50)	(5.70)	(0.30)	(0.40)
Experience	0.006	0.011	0.015	0.026	0.000	0.002
	(2.70)	(4.90)	(3.60)	(6.30)	(0.10)	(0.90)
Experience-squared[c]	−1.098	−1.643	−1.974	−3.283	−0.377	−0.583
	(2.70)	(4.10)	(2.30)	(3.80)	(1.00)	(1.40)
Male[b]	0.447	0.474	—	—	—	—
	(17.00)	(17.80)				
R^2	0.584	0.556	0.441	0.353	0.132	0.136
F-ratio	49.110	43.660	11.830	8.180	3.300	3.430
No. of observations	468	468	193	193	275	275
Residual sum of squares	25.450	28.080	9.490	10.230	11.030	11.980

NOTE: Figures in parentheses are absolute t-ratios. A set of four caste dummies also were included in each of the above equations. Since none of the caste coefficients was significant, they have not been reported here.
[a]Dichotomous variable with value of one in indicated state and zero otherwise.
[b]Endogenous variable. Instruments used in the instrumental variable estimates (OLS-IV) were the prices of milk, sorghum, rice, pulses, and sugar; farm size (in acres); and the percentage of cultivated area under superior soil.
[c]All coefficients in this row have been multiplied by 10^4.

significantly larger effect on individual wage rates in the peak than in the slack season (with an elasticity of 0.27 in the peak season), while weight-for-height has a significantly smaller effect on wages in the peak season (with an elasticity of 0.35 in the peak season). Thus the roles of nutrient intake and health status seem to reverse between the two seasons: while calorie intake is much more important in determining wages (and therefore, we assume, marginal productivity) in the peak than in the slack season, weight-for-height appears to be more important in the slack than in the peak season. (None of the four effects is observed to be significant and negative.)

One possible explanation for this finding may lie in the nature of the tasks performed in the two seasons. Tasks normally performed in the peak season, such as harvesting (see table 7.1), may require greater sustained human energy expenditure than slack-season tasks, but may not require innate strength (which is associated with greater weight-for-height) to the same extent as in the slack season. In fact small size may be a distinct advantage in certain peak-season tasks, such as harvesting and transplanting.

Interestingly, both schooling and experience are significant (at the 5 percent level) determinants of individual wage rates, even in the context of agricultural labor activities. The returns to schooling implied by our estimates are low (about 2 percent), however, compared with most estimates obtained in other studies of urban areas and nonagricultural activities (Psacharopoulos 1982; Birdsall and Sabot 1988). Our estimate indicates diminishing returns to experience, with a return of less than 1 percent for each additional year of experience at the sample mean. None of the dichotomous variables for caste is significant, while gender is by far the most significant determinant of wages (with males earning about 45 percent higher wage rates than females, holding other factors constant).

For male participants, the wage equation results remain broadly similar to those discussed above, though calories in the peak season have a significant impact only at the 20 percent level, and the returns to schooling and to experience are higher than in the relation for all participants. For female participants, however, the wage equation results are quite different. With the exception of the peak-season intercept dummy, which is significantly positive at the 10 percent level, none of the explanatory variables in the wage equation is significant. Schooling, experience, calorie intake, and weight-for-height do not appear to matter at all in female wage determination in either the slack or the peak season.

An explanation for this result lies in the segmentation between male and female tasks observed by Ryan and Ghodake (1984) for the same sample used here. They found that, of a total of 16 agricultural operations, only 2—harvesting and threshing—used considerable amounts of both male and female labor. Five tasks—nursery bed raising, transplanting, planting, weeding, and thinning—were almost exclusively performed by female labor, while male labor was almost exclusively used for the remaining nine operations, four of them involving bullock power. If male labor generally performs more physically demanding tasks than female labor, it is easy to understand why it receives a premium in wages as a result of better health and nutrition and why female labor does not receive such a premium.

Our results thus indicate significant support for the wage efficiency hypothesis for males within the rural SAT Indian context of this study, with significant differences between the peak and slack seasons. Food consumption, at least of adult males, has an immediate productivity impact through energy availabilities and a possible longer-run impact through health status. This implies that policies that improved males' health and nutrition would result in productivity improvements within this context. It also means that households might rationally allocate somewhat more food relative to requirements to adult males, especially during the peak season. There is the further implication that male wages should be treated as endogenous in an analysis of the determinants of other outcomes—for example, investment in children or in physical assets—for these households. With the current labor force division by gender and the extent of nutrition and health variation experienced in the sample, however, such results do not carry over to women.

8 Seasonal Food Insecurity and Vulnerability in Drought-Affected Regions of Burkina Faso

THOMAS REARDON AND PETER MATLON

Strict seasonal limits to cropping in the semiarid tropics of West Africa mean that farmers must manage or supplement household cereal stocks during the extended period between annual harvests. Following a poor harvest, farmers make a series of decisions to allocate their consumption, sales, and purchases over time with the goal of satisfying calorie intake requirements through the subsequent cropping season, when labor demands again peak. Failure can have substantial effects on welfare as well as productivity. If calorie deficits are sufficiently severe and sustained, and particularly if they occur during the cropping season, households may fall into a poverty cycle not easily resolved by most market-oriented policy instruments (Behrman and Deolalikar, chapter 7 of this volume; Chambers, Longhurst, and Pacey 1981).

This chapter explores the seasonal incidence and determinants of food insecurity as experienced by farm households in two regions of Burkina Faso (Sahel savanna and Sudano-Sahel) during the recent Sahelian drought period of September 1984–December 1985. Our objectives are to (1) describe the degree and nature of household consumption deficits in a seasonal framework; (2) determine how such deficits make households vulnerable to unpredictable factors in the market and in access to relief assistance; (3) determine whether these problems affect households equally, or rather are more narrowly concentrated in certain groups with particular wealth and demographic characteristics; and (4) highlight important policy implications of these cross-sectional and intertemporal patterns.

Defining Seasonal Insecurity and Vulnerability

We define *food insecurity* in a farm household as the consumption of less than 80 percent of what the World Health Organization (WHO) considers to be an average required daily caloric intake of 2,850 kilocalories (kcals) for a moderately active adult equivalent (FAO-WHO-UNU 1985). This then includes households that consume less than 2,280 kcals per adult equivalent (AE) per day. We define a household to have chronic food insecurity when consumption during two or more seasons is inadequate, particularly if consumption is deficient during the cropping season. Households that are chronically food-insecure constitute the highest-risk group and for policy purposes might be considered a primary target group for aid.

Households are defined as being *production-deficient* if their food stocks immediately following the 1984 harvest were below levels required to meet 80 percent of the WHO average caloric requirement until the 1985 harvest. A household can be production-deficient, and either food-secure or food-insecure in any given season, depending on (1) the resources available to buy food, (2) the receipt of transfers, (3) the intertemporal distribution of consumption that it chooses, and (4) the availability of food in the market or community.

Vulnerability denotes dependence on sources outside of domestic production to provide food for the household. These sources include the market (focusing here on purchases), transfers (external food aid and interhousehold gifts), and hunting-gathering activities which are generally done on common property lands. Under Sahelian and Sudano-Sahelian conditions, dependence on these sources can be highly precarious, for at least three reasons.

First, over time and across space, Sahelian markets exhibit wide and often unpredictable changes in prices and the availability of goods.[1] Even seasonal migrant labor markets, which can be a crucial component in the strategies of food-deficit households, are unstable.[2]

Second, because of high covariance of production, nonmarket transfers between households tend to be a positive function of the overall production sufficiency in any given year (Toulmin 1986). Thus years in which seasonal problems attain their grimmest aspect coincide with periods during which neighbor's surpluses would be least available. Dependence on external aid, where transport and soft infrastructure are poor and political winds capricious, is also a risky strategy.

Third, common property lands in many subregions are already seriously overgrazed and overhunted. While these problems restrict alternative income-generating activities, such as animal husbandry, periodic

---

1. A personal communication of preliminary research results from Chris Delgado.
2. For instance, during the 1985 hot season, politically inspired changes in urban rent statutes severely depressed the housing construction sector in Ouagadougou, upon which many Sahelian migrants depend for casual seasonal labor. Similar changes in regulations affecting interregional cereal trade substantially reduced cereal availability in grain-deficit areas during the 1983-84 season and widened interregional price margins.

drought and the expansion of farming have combined to decrease edible plant-gathering opportunities as well (Timberlake 1985).

Background and Methodology

Study Area

Two study villages were chosen to represent the Sahelian and Sudano-Sahelian agroclimatic zones following a review of secondary information and ground verification (ICRISAT/Burkina Faso 1981). The household sample in each village was chosen by a stratified sampling procedure, which, in an effort to obtain adequate numbers of households employing animal traction, overrepresents such households compared with the respective village populations.

Woure village, located near the regional market town of Djibo, was selected to represent the Sahel savanna. Low and variable levels of rainfall (long-term annual average of 570 millimeters [mm]) and soils that are very low in natural fertility and water retention result in farming systems with poor production potential and high risk of weather-induced crop failure. Our sample includes mainly the Rimaibe ethnic group, with some Fulani. Although both groups are sedentary agriculturalists, animal husbandry is very important to them, serving as the principal means to store wealth and as an insurance mechanism and source of liquidity.

Kolbila village, located near the regional marketing town of Yako, represents the Sudano-Sahel. As is typical of large portions of the densely populated Mossi plateau, rainfall is somewhat higher (with a secular average of 760 mm) and less variable than in the Sahel. Soils are heavier than in the Sahel, but shallow and with low natural fertility. The main ethnic group is the Mossi, whose central activity is farming. Livestock husbandry is relatively less important than in the Sahel, partly because of pressure on the land and partly because of ethnic tradition.

Rainfall in both zones is limited to the months of May to October, with up to 40 percent concentrated in August alone. Farm labor demands are thus highly seasonal and characterized by periods of acute labor bottlenecks. The rainfall problem was aggravated during 1984 because rainfall in the two zones' study villages was more than 40 percent below the secular average in each zone. Moreover, below-average rainfall in 1982 and 1983 had already reduced on-farm cereal stocks considerably as the 1984 harvest period was entered.

Both zones were considered to be in cereal deficit during the study period, and there is a more general concern as to the rapid deterioration of their agricultural and physical resource base because of the rising demographic pressure and recurrent poor rainfall.

Cropping systems in both zones are rain-fed and primarily subsis-

TABLE 8.1 Percentage of cultivated area sown to all crops, Sahel and Sudano-Sahel village samples, 1984

	White Sorghum	Red Sorghum	Millet	Maize	Peanuts	Other[a]	Total
Sahel	7	0	90	1	[b]	2	100
Sudano-Sahel	50	5	28	1	15	1	100

NOTES: The crops are either sole crops or principal crops in mixture. The Sahel village sample is from Woure and the Sudano-Sahel village sample is from Kolbila.
[a]The remaining crops include hungry rice in the Sahel village and rice in the Sudano-Sahel village.
[b]Less than 0.5 percent.

tence-oriented. Millet is the dominant crop and food staple in the Sahel, replaced by white sorghum as one proceeds south to the more diversified Sudano-Sahel (table 8.1). Insufficient rainfall and poor soil fertility limit maize plantings in both regions, though as an early-harvested crop maize can play an important role in meeting rainy-season food needs when carryover food stocks are low.

Livestock, along with stored crops, is among the main forms of wealth available to households to meet needs imposed by production shortfalls and periodic cash requirements. Routine sales of fowl and small ruminants are most common, with sales of cattle less frequent. Milk production, for domestic consumption, for sale, and in exchange for cereal, is of some importance during the rainy and harvest seasons in the Sahel, but not in the Sudano-Sahel.

Hunting and fishing normally provide little to meet seasonal dietary needs, although we will see that gathering activities can provide important dietary supplements for households facing the most acute seasonal consumption stress.

Seasons and Activity Patterns

To simplify our analysis, we employ four seasons defined on the basis of climatic and activity patterns, following discussions with the sample farmers. The distribution of activities across quarters can be summarized in order to highlight seasonal energy requirements in contrast to seasonal consumption patterns, which are reviewed in a later section.

1. *Harvest* (September–November): Depending on the zone's cropping profile (table 8.1), maize, red sorghum, and hungry rice are harvested in the first half of this season. These crops help deficit households meet consumption requirements until the more important white sorghum and millet harvests in the second half of the season.

2. *Cold* (December–February): Postharvest festivals and major family celebrations (e.g., funerals) begin at this time and require relatively large expenditures of crop stocks. Migration, artisanry, commerce, and husbandry are the main economic activities, and very little cropping occurs.
3. *Hot* (March–May): Activities similar to those in the cold season predominate. In 1985, the Moslem fasting period of Ramadan covered the last month of this season, shifting mealtimes for affected adults in practicing households to the evening and early morning. Land clearing is the only agricultural activity of importance. Household members engaging in seasonal migration usually return for the first planting in June.
4. *Rainy* (June–August): Farming activities are concentrated in this quarter, with labor bottlenecks in planting and weeding activities during June–July constituting a major constraint on production (Delgado and Ranade 1987). Because of available land and a generally equitable land distribution (and thus the absence of a landless labor class), labor hiring during this period is negligible. As a result, crop production is largely determined by the timing, duration, and quality of family labor, and major caloric deficits during this period can initiate the poverty cycle described earlier.

Survey Methodology and the Nature of the Data

The data used in this chapter were generated by a food intake survey administered in two International Crops Research Institute for the Semi-Arid Tropics (ICRISAT)/Burkina Faso study villages between October 1, 1984, and December 31, 1985. The survey was directed by the International Food Policy Research Institute (IFPRI) in close collaboration with ICRISAT. The survey instrument was administered in each household approximately twice a month to obtain (1) the types, quantities, and sources (produced, purchased, received as gift, gathered) of all foods consumed during the prior day by all members; (2) the number of adults and children (both members and visitors) eating each meal or snack; and (3) the types and amounts of food sent out of the household. In this analysis, we exclude all sauce ingredients (meat, vegetables) and include only cereals and seeds, pulses, nuts, and tubers, as well as milk and leaves when used as meal base ingredients.

Seasonal Food Insecurity and Vulnerability Related to Levels of Wealth and Demographics

This section examines the extent and nature of seasonal food insecurity and vulnerability among sample households and relates their incidence to household characteristics. By juxtaposing seasonal food intake levels with time-bound farm labor demands, we also suggest how nutritional factors may undermine productivity.

Sample Stratification

For an initial stratification of sample households, we sum the value of crops and animal stocks at the beginning of our survey period and divide by household size (in AEs) to form a measure of total liquefiable wealth per AE (henceforth referred to as *wealth*). Households are ranked by the resulting ratios and grouped into terciles (poor, middle, and rich) for each village.

This procedure assumes that seasonal food insecurity and vulnerability between the 1984 and 1985 harvests were influenced by the initial wealth endowment of the households. Initial stocks can be treated as exogenous determinants of the ensuing seasonal behavior pattern. In contrast, purchase patterns, the intertemporal distribution of consumption, and income-earning strategies to augment this resource base or reduce the number of consumers (through migration) are considered endogenous variables.

Table 8.2 presents characteristics of the sample households grouped by village and wealth strata.[3] Households in the Sudano-Sahel village are somewhat larger and have a higher proportion of children. Sample farmers in that village also entered the study period with less than one-third the Sahel sample's total value of livestock per AE, though with roughly the same crop (stocks plus harvest) endowment. Moreover, livestock are much more equally distributed across strata in the Sudano-Sahel sample, albeit at a lower level. The endowment of crops is nearly equally distributed in the two samples, with ratios approaching 3:1 for the respective extreme terciles.

Seasonal migration (during the cold and hot seasons) is substantially more important in the Sahel village sample and within that village shows a strong inverse relation to wealth. This can be explained in part by the significantly larger livestock holdings among the rich, which obviates the need to migrate.

Seasonal Food Insecurity

Seasonal average daily caloric intakes per AE for each stratum in the two-village sample are presented in table 8.3, and the percentages of AEs consuming less than 80 percent of the WHO average daily caloric requirement are shown in table 8.4.

---

3. A household is identified from the production perspective.

TABLE 8.2 Household characteristics, Sahel and Sudano-Sahel village samples

	Poor	Middle	Rich	Total
Sahel village sample				
(a) Household size	8	11	11	10
(b) Children/total (%)	46	35	52	45
(c) Animal traction (%)	0	38	25	21
(d) Crops/AE	5,956	10,625	9,349	8,957
(e) Stocks/AE	31	1,246	8,168	3,369
(f) Livestock/AE	2,224	19,731	131,182	54,448
(g) Migration (%)	24	14	3	13
(h) Land/AE	0.76	1.22	1.04	1.04
(i) Production sufficiency (%)	23	47	70	49
(j) Yield/ha	8,175	9,051	9,275	8,902
Sudano-Sahel village sample				
(a) Household size	14	14	11	13
(b) Children/total (%)	52	50	48	50
(c) Animal traction (%)	43	29	43	38
(d) Crops/AE	5,914	9,392	13,178	9,276
(e) Stocks/AE	135	197	6,881	2,120
(f) Livestock/AE	10,837	16,770	22,922	16,484
(g) Migration (%)	10	4	7	7
(h) Land/AE	0.60	0.59	0.68	0.62
(i) Production sufficiency (%)	25	39	80	46
(j) Yield/ha	10,758	15,733	20,727	15,445

NOTES: There are 24 households in the Sahel and 21 in the Sudano-Sahel village sample. All entries are rounded to the nearest integer. Variables are defined as follows:
(a) Average household size in unweighted members.
(b) Number children as a percentage of household size. Children are 15 years old or less.
(c) The percentage of sample households employing animal traction as opposed to manual cultivation.
(d) This is an adult equivalent (AE) weighted average of the value, evaluated by an average producer market price for October–November 1984, of the total cereal, pulse, and nut production from the harvest of 1984. Currency units are CFA.
(e) The value of the same crops as in (d) in storage immediately before that harvest, all deflated by the number of AEs at harvesttime 1984.
(f) This is an AE weighted average of the value, evaluated by an average price over age and sex of animal, of the market price at harvesttime 1984 of the total livestock holdings, deflated by the number of AEs at that time.
(g) The percentage of person-days in which household members were absent on seasonal migration during the study period. This was calculated as: [number of days of migration absence by household members in cold season (here Jan.–Feb.) and hot season 1985, multiplied by AE coefficients of migrant members], divided by [total household AEs multiplied by the number of days in those seasons (60 for cold plus 90 for hot)], weighted over sample by AEs.
(h) The total area of cultivated plots (thus excluding fallow) per household, deflated by household AE. The plot areas are unweighted by land quality coefficients. Households are weighted by AE. Coefficient of variation = 0.29 for Sudano-Sahel, 0.39 for Sahel village sample.
(i) The total quantity of harvested and stored crops (as in [e]), weighted by calorie coefficients, and deflated by AE times 365 days times 2,280 calories, to give the sufficiency of

TABLE 8.3 Average daily intake per adult equivalent by season and stratum, Sahel village sample (kilocalories)

	Harvest 1984	Cold 1985	Hot 1985	Rainy 1985	Harvest 1985	CV	Total
Sahel village sample							
Poor	2,595	2,335	2,269	2,380	2,604	0.25	2,506
Middle	3,926	3,165	2,793	2,912	3,049	0.28	3,234
Rich	3,163	2,903	2,542	2,864	3,242	0.22	2,990
Sudano-Sahel village sample							
Poor	2,036	2,203	2,251	2,137	2,875	0.37	2,312
Middle	2,069	1,985	1,972	1,860	2,260	0.36	2,038
Rich	2,631	2,548	2,802	2,395	3,100	0.34	2,704

NOTES: 75 percent of the Sahel coefficients of variation (CVs) of the seasons' average kilocalories per adult equivalent (kcal/AE) are at or below 25 percent; the rest are at or below 40 percent. 93 percent of the Sudano-Sahel CVs are at or below 40 percent; the rest are at or below 45 percent.

We used multiple comparison significance tests, Duncan procedure, at the 5 percent level, to test (1) over strata per season, and (2) over seasons per stratum, for average kcals/AE.

For the Sahel sample, for (1), all averages were significantly different, *except:* poor vs. rich in hot season, and middle vs. rich in rainy; the former is partly due to a very small number of observations for the poor in the hot season (because of migration). For (2), all were significantly different, *except:* (a) for poor, harv/84 vs. harv/85, and cold, hot, and rainy vs. each other (the latter is partly due to a small number of observations in the hot season); (b) for middle, cold and rainy vs. harv/85, and hot vs. rainy; (c) for rich, harv/84 vs. harv/85, and cold vs. rainy.

In the Sudano-Sahel, for (1), all averages were significantly different, *except:* poor vs. middle in harv/84. For (2), all were significantly different, *except:* (a) for poor, rainy vs. harv/84, cold, and hot, and cold vs. hot; (b) for middle, harv/84 vs. cold and hot, and cold vs. hot; (c) for rich, harv/84 vs. cold.

The CVs were computed without detrending data because of the short period.

Figures below the WHO average daily requirement for a moderately active AE of 2,850 are underlined.

---

extant stocks in meeting 80 percent of the average daily calorie requirement during the ensuing year, our study period.

(j) The value of crops produced per hectare of land cultivated. Average weighted over sample by AEs.

We used multiple comparison significance tests, Duncan procedure, at the 5 percent level, to test over strata per village for variables (d) through (j).

For the Sahel sample, all averages were significantly different except for variables (d) middle vs. rich; (e) poor vs. middle; (j) no two strata were significantly different.

For the Sudan sample, all averages were significantly different except for variables (e) poor vs. middle; (g) poor vs. rich; (h) poor vs. middle.

TABLE 8.4 Percentage of food-insecure adult equivalents per stratum per season, Sahel and Sudano-Sahel village samples

	Harvest 1984	Cold 1985	Hot 1985	Rainy 1985	Harvest 1985
Sahel village sample					
Poor	34	54	34	57	21
Middle	0	0	27	0	0
Rich	0	0	18	18	0
Sudano-Sahel village sample					
Poor	75	54	52	63	20
Middle	71	62	90	84	65
Rich	31	29	20	49	0

NOTE: These figures indicate the percentage of AEs per stratum whose average seasonal consumption fell below 80 percent of the WHO average daily required calorie intake of 2,850 calories.

Several features of these caloric intake patterns are important. First, consumption levels are substantially lower and more equally distributed among households in the Sudano-Sahel village sample. Nearly all strata are below the WHO average adequacy standard in most seasons, corresponding to a pattern of shared chronic insecurity. The Sahel case provides a striking contrast, because consumption inadequacy is largely concentrated among poor households.

Second, the degree of cross-seasonal variation of consumption (represented by the coefficient of variation [CV]) is also substantially greater in the Sudano-Sahel sample than in the Sahel case. This reflects the greater capacity of the Sahelian farmers to avoid large seasonal reductions in consumption by supplementing crop stocks through cereal purchases.

Third, the potential consequences of caloric deficits in the Sudano-Sahel sample are magnified by the fact that seasonal consumption and activity levels are *countercyclical*—caloric intake is lowest in each stratum during the cropping season, when demands for energy expenditure are greatest. Again, this contrasts with the Sahel case, where consumption is consistently lowest during the hot season, when labor demands are least. These comparisons suggest that the hungry-season hypothesis frequently found in the literature may hold only in situations of extreme deprivation when households are unable to ration consumption during the cold and hot seasons in a manner to meet rainy-season requirements.

Differences in the levels and distribution of wealth help to explain why the Sudano-Sahel case of "shared deprivation" contrasts with the more concentrated food insecurity observed in the Sahel village. Despite the somewhat higher crop endowment in the Sudano-Sahel village, the livestock endowment is inadequate to generate sufficient cash to close the caloric gap through large food purchases. Low livestock holdings place a similar constraint on poor Sahelian households, with nearly identical consequences.

Dependence on the Market and Consumption of Foods from Outside the Immediate Zone

For policy purposes, it is important to distinguish between *potential* vulnerability, as measured by the percentage of calories that would have to be obtained from purchases, aid, and gathering in order to achieve minimum caloric adequacy at any given level of production, and *actual* vulnerability, as measured by the percentage of total consumption that is in fact obtained through purchases, aid, and gathering. The difference between the two measures is a function of purchasing power, prices, physical availability of food, and consumer preferences.

When we measure potential vulnerability (table 8.5), it is evident that poor harvests forced households in both regions into positions of nearly equal overall dependence on sources outside of own production to meet caloric needs. However, a comparison of actual vulnerability across sites would suggest that market dependence is considerably lower among households in the Sudano-Sahel village and more equally distributed across strata compared with patterns in the Sahel village (tables 8.6 and 8.7). This difference is caused by the relative lack of purchasing power in the Sudano-Sahel sample and by the near-absence of food aid directed to farmers in that village.

TABLE 8.5 Average daily intake per adult equivalent by source, Sahel and Sudano-Sahel village samples, five-season period (kilocalories)

	Poor	Middle	Rich	Total
Sahel village sample				
Produced	1,278	1,617	1,884	1,670
	(45)	(57)	(66)	(59)
Purchased	802	1,358	837	1,014
Food aid, gift	401	259	269	298
Sudano-Sahel village sample				
Produced	1,410	1,264	2,163	1,528
	(49)	(44)	(76)	(54)
Purchased	879	754	514	741
Food aid, gift	46	20	27	23

NOTE: Numbers in parentheses indicate the amount produced expressed as a percentage of the WHO average requirement for a moderately active adult equivalent.

TABLE 8.6 Breakdown by food item and source of caloric intake per season, Sahel village sample (percent)

	Harvest 1984	Cold 1985	Hot 1985	Rainy 1985	Harvest 1985	Total
Poor						
White sorghum	12	16	41	12	20	16
Millet	62	21	5	13	46	39
Red sorghum	2	40	14	34	9	16
Maize	16	23	40	18	15	18
Wheat	0	0	0	18	2	5
Leaves	7	0	0	2	4	4
Produced	72	37	46	10	66	51
Purchased	25	17	54	69	14	32
Food aid/gift	1	45	0	21	20	16
Middle						
White sorghum	2	8	18	5	6	6
Millet	82	19	15	11	58	42
Red sorghum	2	19	28	58	17	25
Maize	8	54	38	10	15	20
Wheat	0	0	0	4	1	1
Produced	82	27	22	25	66	49
Purchased	16	50	78	70	21	42
Food aid/gift	2	23	0	4	13	8
Rich						
White sorghum	2	12	8	7	15	9
Millet	77	42	42	52	55	54
Red sorghum	4	15	12	27	12	14
Maize	7	27	36	9	11	17
Wheat	0	0	0	4	1	1
Milk	8	3	1	a	4	3
Produced	86	47	43	59	73	63
Purchased	10	33	57	34	19	28
Food aid/gift	4	20	0	6	9	9

NOTES: Items such as leaves may be included for one group but excluded for others when the numbers are inconsequential. Percentages are averages weighted by AEs.
[a] Less than 0.5 percent.

Our results also show consistent variation across strata within each region, which has potentially important equity effects. Households with smaller endowments of crops and livestock consume a higher percentage of purchased and donated foods and thus tend to be more vulnerable to market fluctuations and uncertainties in flows of relief assistance. Moreover, vulnerability for the poor is especially critical in the rainy season, when

TABLE 8.7 Breakdown by food item and source of caloric intake per season, Sudano-Sahel village sample (percent)

	Harvest 1984	Cold 1985	Hot 1985	Rainy 1985	Harvest 1985	Total
Poor						
White sorghum	77	49	61	97	52	68
Millet	10	23	7	0	28	14
Red sorghum	3	8	2	2	0	3
Maize	3	19	29	1	8	11
Peanuts	5	1	1	1	11	4
Produced	69	76	25	40	89	61
Purchased	29	23	74	58	10	38
Food aid/gift	2	1	1	2	1	2
Middle						
White sorghum	50	30	48	84	26	47
Millet	28	35	0	0	44	23
Red sorghum	13	12	21	0	4	9
Maize	0	17	29	13	11	14
Peanuts	7	3	1	2	10	5
Produced	86	72	38	24	88	62
Purchased	13	27	61	72	11	37
Food aid/gift	1	1	1	3	1	1
Rich						
White sorghum	48	68	81	89	50	67
Millet	37	18	2	0	32	18
Red sorghum	7	1	1	0	1	2
Maize	3	8	13	7	4	7
Peanuts	3	4	2	4	11	5
Produced	83	91	77	53	95	80
Purchased	17	10	22	43	4	19
Food aid/gift	1	0	1	4	1	1

their dependence on the market is highest, and when the negative effects on real income of preharvest price increases can be devastating in terms of both welfare and labor productivity. Between 60 and 70 percent of all calories consumed by poor and middle-strata households during the rainy season were in fact purchased.

These conclusions are reinforced by our finding that the diet of the poor households in the Sahel village sample is strongly skewed toward crops produced principally outside of the region. Although white sorghum accounts for only 4 percent of the cropping area, it provides roughly 25 percent of the calories consumed outside of the harvest seasons for poor households. The majority of this was transferred from surplus-producing areas of Burkina Faso. Even more striking is the consumption of red sorghum and maize, which account for only about 1 percent of the cultivated area but provide up to 60 percent of the diet of the poor during nonharvest seasons. Nearly all supplies of both commodities were provided through food relief.

Households in the Sudano-Sahel sample were also highly dependent on interregional food transfers and imports. More than half of all sorghum consumed during the cropping season was purchased, with the greatest volume derived from surplus-producing regions in Burkina Faso. Maize, though not significantly produced in the zone, was also an important item of purchase. Relatively cheap surpluses from the southwest of Burkina Faso and imports from Ghana served as the main sources. Some maize was also available as food aid in the 1985 rainy season and appears in the consumption patterns, albeit weakly.

Dependence on Food Aid

During the cold, rainy, and early-harvest seasons of 1985, donors poured food aid into the Sahel region, and important amounts were received by our sample households. In the cold season this included mainly red sorghum; in the rainy season this shifted toward wheat and in the early harvest period toward maize. The changing composition of the aid flow was clearly reflected in the diets of the poorest households, for whom aid served as a crucial source of calories. Dependence on aid peaked in the cold season, when almost *half* of the poor's diet was from aid, and held at about 20 percent of calories consumed during the rainy and harvest periods. These shares are roughly double those observed for households in the middle and rich strata, and reflect relatively efficient de facto targeting of aid among households.

Interregional targeting of relief assistance was far less efficient, if judged on the basis of need. While own production met 59 percent of the average daily caloric requirements in the Sahel village, where relief was widely available, own production met only 54 percent of requirements on average in the Sudano-Sahel village, where such transfers were negligible (table 8.5). In fact, food received per AE as relief assistance was more than 10 times larger in the Sahel village sample than in the Sudano-Sahel village sample (298 versus 23 kcals/AE). This reflected the perception by policymakers that punctual food relief was most needed where rainfall, yields, and market access were poorest. However, since it ignored interregional differences in purchasing power (compare absolute purchases between regions in table 8.5), the calculation of need was incomplete and ultimately in error.

Dependence on Common Property Resources

Caloric dependence on common property resources is represented here mainly by the consumption of gathered leaves and milk. Respondent farmers often asserted that gathered leaves and fruit were important as a dietary backup for the poor. While it is difficult to capture this kind of information in a survey (particularly with regard to foods eaten in the field while working), the seasonal importance of these foods to the poor is apparent in table 8.6. Although the leaf consumption shown relates only to leaves used as base products (not as sauce ingredients), that figure reaches 7 percent of the poor's total intake in the Sahel sample during the 1984 harvest. The percentages are substantially lower for the other groups (not shown).

The role of leaves is probably even more important than this figure suggests. Farmers claimed that the poor often use leaves to increase the bulk of a dish, thereby filling their stomachs while using a smaller amount of grain. Thus leaves appear to be linked to a strategy of rationing, in addition to their role as a source of energy and variety.

In sum, the quality of the common property resources and their usefulness as a source of gathered plants are marginal but nevertheless important seasonal factors for the poor in the Sahel. In this respect, famine food strategies of poorer households, despite their low livestock holdings, are increasingly vulnerable to the ongoing degradation and destruction of the tree and bush stands in the greater Sahel region.

Higher population pressure (hence low foraging potential) in the Sudano-Sahel village has already restricted gathered leaves to a minor stopgap role as a base ingredient during the pre- and early-harvest periods. For this reason, farmers in this village generally place greater emphasis on maize to meet early-harvest consumption needs when sorghum and millet stocks are at their seasonal lows. Because of poor 1984 maize yields, however, and the lack of milking herds, early crops such as peanuts and sweet red sorghum were relatively more important during the study period but still grossly inadequate in absolute volume.

Identification of the Chronically Insecure for the Purpose of Policy Intervention

Earlier we distinguished between two categories of seasonally insecure households: those insecure only in a single season (and presumably more able to recuperate in subsequent seasons), and those chronically insecure or nutritionally deficient during several seasons in the year (and as such probably less able to recuperate from the health and productivity effects of seasonal calorie deficits). The latter can be considered a priority target group for policy interventions to augment consumption. This section describes the salient characteristics of this risk group that could be used by policymakers to target interventions and to assess the likely impacts of various policy instruments on the seasonal and overall welfare of such households.

Table 8.8 compares a set of structural and behavioral characteristics for households that are, and are not, chronically insecure. When divided in this manner, only 17 percent of the Sahel village sample (four households) are included in the high-risk group, underlining once again the concentrated distribution of seasonal food insecurity in that village. In contrast, the majority of Sudano-Sahel sample households (13 households representing 62 percent of the sample) fall into the target group. The small size of the Sahel sample target group in particular cautions against generalizing too broadly from the inferences that follow. Furthermore, the CVs cal-

TABLE 8.8 Chronically insecure versus secure households, Sahel and Sudano-Sahel village samples, five-season period

	Sahel Village Sample		Sudano-Sahel Village Sample	
	Chronically Insecure	Secure	Chronically Insecure	Secure
Percent of sample	17	83	62	38
Household size	9	10	13	12
	(58)	(70)	(45)	(78)
Child./total	52	43	52	47
	(24)	(48)	(23)	(32)
Age of head	*35	*48	56	56
	(20)	(24)	(28)	(29)
Percent using anim. tract.	25	20	38	38
Crops/AE	*3,919	*9,870	*8,208	*11,136
	(83)	(69)	(50)	(55)
Stocks/AE	*765	*3,840	*1,023	*4,028
	(118)	(179)	(320)	(205)
Livestock/AE	43,416	56,446	*13,823	*21,117
	(113)	(125)	(48)	(52)
Land/AE	*0.91	*1.06	*0.59	*0.68
	(25)	(41)	(29)	(28)
Yield/ha	*5,502	*9,517	*14,292	*17,451
	(104)	(70)	(41)	(62)

NOTES: Numbers in parentheses indicate coefficient of variation, in percentages. Asterisk indicates significantly different at 0.05 percent level between insecure and secure groups. See note to table 8.2 for definition of variables.

culated on household characteristics are generally larger than the CVs calculated in the wealth-based stratification, which indicates that the criterion of chronic food insecurity aggregates households with diverse natures. Despite these caveats, the results do suggest several structural differences between target and nontarget groups.

Examining the Sahel village first, we observe that the average proportion of children in total household size is higher and household heads are significantly younger in the chronic-deficit group. This suggests that target households are in an earlier stage in a Chayanovian life cycle compared with those in the nontarget group. The hypothesized consequences of such life-cycle differences are reflected, though imperfectly, in the wealth and production characteristics. Households at an earlier developmental stage would be expected to have accumulated smaller livestock holdings. Our results suggest that this may be true, although the difference (30 percent) is not statistically significant. This apparently anomalous statistical result, however, is the artifact of an extremely skewed distribution of livestock within the target group. Among these four households, three owned essentially no livestock. The other possessed a herd valued at nearly 100,000 francs CFA but experienced a nearly complete crop failure.

Similarly, with a smaller proportion of workers (a consequence of the high ratio of children to total household members), a household's cultivated area per AE would also be expected to be smaller. The hypothesized sign of the relationship holds, but the magnitude of the difference is small (15 percent). Dominating the comparisons, however, is the large (73 percent) and significant difference in land productivity between target and nontarget groups. In the production environment of the Sahel village, which is characterized by relatively homogeneous technology and cropping patterns, productivity differences of this magnitude cannot be explained by life-cycle factors alone, but are more likely caused by variation in factor quality (including management) and stochastic events affecting yield variability (microclimate variation, household worker morbidity, etc.).

The simple comparison between target and nontarget households in the Sudano-Sahel village sample does not provide evidence that life-cycle relationships distinguish the two groups in that zone. However, several differences do emerge. Livestock holdings, though small, are significantly lower (53 percent) for the target group, suggesting that such farmers had less insurance to protect them from production shocks. Crop production variables are again significant. Cultivated area per AE is significantly lower for target households (15 percent), and there is a significant gap in land productivity as well (22 percent).

Comparing regions, we note finally that the extent of chronic seasonal food insecurity was substantially greater in the Sudano-Sahel village despite the fact that average land productivity was 75 percent higher than in the Sahel village. This reinforces the point that more limited land resources and lower livestock holdings in the more densely populated Sudano-Sahel village placed those farmers at greater risk when shocked by stochastic climatic reversals, and it underlines the inadequacy of employing yields or the magnitude of rainfall deficits as simple indicators of the potential impact of drought.

Conclusions and Policy Implications

These empirical results support several generalizations that have important policy implications.

First, it is essential to disaggregate by region, household wealth, demographic characteristics, and season in order to determine the magnitude, location, and consequences of food insecurity. Average measures can grossly underestimate the size of the food problem, depending upon the underlying distribution of consumption across households and seasons. When seasonal consumption patterns run countercyclical to energy demands of cropping activities, as observed among the poorest households in this study, the potential welfare and productivity consequences of a more general food shortage are greatly magnified. The use of chronic seasonal insecurity as a measure of need captures this dimension, but it is an empirically demanding criterion. The development of more cost-effective proxy measures that reflect the seasonal incidence and duration of food insecurity would be extremely valuable.

Second, following a poor harvest, dependence on the market and thus vulnerability to market inefficiencies were found to be extremely high, both across study sites and across wealth strata. A particularly important finding is the high dependence of the poorer strata in both zones on markets and the fact that the purchased share of their total consumption is largest during the preharvest seasons, when cereal prices normally peak. Improvements in market infrastructure to reduce interregional and interseasonal margins could have major benefits in efficiency and equity goals. Where purchasing power is lacking, seasonal generation of employment, such as through public works projects financed by food-for-work mechanisms, may be a highly effective complementary policy instrument.

Third, relief assistance was found to provide a crucial food supplement for a limited group of deficit producers in the Sahel sample, but relief largely bypassed the Sudano-Sahel site, where chronic food insecurity was as severe and more broadly distributed across households. This raises important policy questions about appropriate methods of targeting assistance both among and within regions.

In assessing the relative need for relief assistance among regions, gross measures such as yields or rainfall deficits reflect poorly the degree of food insecurity where there are important regional differences in the structure of production and in farm-level purchasing power. Because of regional variations in the land-to-person ratio and in livestock holdings and migration (and thus potential purchasing power), areas of greatest risk in the West Africa semiarid tropics are not necessarily in the low-rainfall Sahel, as is commonly held. This is due in part to farming systems in the Sahel having incorporated more risk-reducing strategy components. These include extensive farming systems, access to bush areas for gathering activities, large livestock holdings, and well-established seasonal employment channels in urban areas. When comparable shortfalls in precipitation occur, there may be greater risk of severe seasonal food shortages in the middle Sudano-Sahel belt. The efficiency of traditional strategies for risk reduction has been reduced over time in that region by a rising population, which limits cropped area as well as the potential for large livestock holdings and important gathering activities.

Fourth, within-region targeting of relief assistance and development interventions designed to benefit only households that are chronically short of food are difficult to achieve, since high-risk groups are not easily identifiable as structurally distinct household sets. Unfavorable household composition, low crop productivity, and small livestock holdings are major correlates of chronic food insecurity. A high ratio of children to adults occurs as a natural stage of family development and alone would not normally have permanent impoverishing effects except during consecutive poor years. Nevertheless, its association with severe seasonal deficits suggests that nutritional programs aimed at children may be an efficient means of directing seasonal food supplements to an important group of high-risk households, regardless of the subsequent distribution of intrahousehold consumption.

The number and complexity of factors causing low crop productivity in traditional cropping systems also suggest major problems in using this indicator as a targeting criterion. Rather, they underline the need to increase overall productivity (while reducing downside yield risks during drought) through the development of improved, broadly adaptable cropping techniques.

The practical as well as moral problems in evaluating land area and livestock holdings also present important obstacles to using these criteria for targeting food aid.

Finally, the sharp seasonality of food insecurity underlines the importance of timing in accurate targeting of interregional food transfers. In terms of policy, this argues for (1) the collection and broad dissemination of timely information on prices and on-farm stocks;[4] (2) where possible, the maintenance of reserve stocks in close proximity to high-risk groups, through such approaches as village cereal banks; and (3) the development of cost-effective famine early warning systems, which are necessary to ensure a large lead time to initiate requests for foreign assistance.

4. Although price is a robust and comprehensive measure of emerging food shortages, it should be recognized that price levels alone can be misleading in subregions with low purchasing power and thus little effective demand and thin markets.

9 From Seasonal Income to Daily Diet in a Partially Commercialized Rural Economy (Southern Cameroon)

JANE GUYER

All agricultural populations have to manage discrepant seasonalities, the most obvious one being that maximum energy expenditure in agricultural work generally precedes maximum food availability at harvest. A variety of strategies are used to augment the diet and adjust energy requirements in the time between the inevitable peaks and troughs of production. These include biological adaptations, storage of food, social distribution networks within and beyond the domestic context, hunting and gathering, and drawing on cultural knowledge about alternative dietary sources. Through living in a particular social and ecological environment over time, people develop a known repertoire of strategies that become part of customary knowledge. These strategies are socially maintained as a resource and mobilized when the need arises. The market now figures universally in the repertoire of constraints and resources through which uneven income rhythms can be transformed into a regular food supply.

Although it is certainly expanding in its influence, the food market in present-day Africa is still a long way from having taken over this function entirely. In the early 1970s, theories of economic change tended to assume incremental specialization of production and concomitant changes in consumption.[1] The emerging empirical picture suggests, to the contrary, that there is no clear direction in recourse to the market; it has been integrated as one—but only one—constraint and resource in production and con-

1. Writing in a neoclassical development mode, Johnston and Kilby (1975) imply that specialization of production can proceed by incremental steps. Writing from a Marxist perspective, Bernstein (1977) suggests that the logic of initial market involvement leads to the "simple reproduction squeeze," increasing indebtedness and increasing market dependence.

The fieldwork on which this paper is based was financed by a grant from the National Institute of Mental Health, and was carried out while the author was affiliated with the National Advanced School of Agriculture in Yaoundé.

sumption decisions.² Since "the market" can be incorporated into rural strategies in a variety of different ways, the question of whether resilience and the quality of the diet have been promoted or eroded can only be addressed by tracing the effect of the market on the total repertoire of consumption strategies. Recourse to the labor or product market has a direct effect on the amounts, kinds, prices, and sources of food, and an indirect effect on the rest of the strategies.

Recent policy directives aimed at increasing commodity production by African farmers have concentrated heavily on price incentives, tending to assume a cumulative and linear responsiveness. As Berry (1984) has pointed out at length, relative prices are important, but an approach through prices alone is narrow and weak as an interpretive and predictive framework. It is as difficult to predict the dynamics of production and consumption from macro-level interventions as it is to infer local dynamics from the macro data. For example, recourse to imported cereals has been rising in Africa, but the evidence can be interpreted in diametrically opposing ways—as the dangerous consequence of indigenous agricultural failure (Delgado and Miller 1985) or as the result of rising incomes, improved living standards, and the demand for more varied products (Morrison 1984). Recourse to the market could result from either process—absolute shortage or the planned use of cash income for food as an aspect of a general production plan. Both are probably true, in different regional populations and different social categories or classes, according to a broad range of social and economic variables. These include national agricultural policies, agricultural ecologies, type of market involvement (labor versus export crops versus food crops), relative poverty, and ethnic cultural differences.³

Von Braun and Kennedy (1986) take a much broader approach to the dynamics of commoditization, based on the assumption that home provisioning will remain one of the strategies of rural producers for a long time to come. The feedback model they use, however, is complex enough to pose methodological difficulties as potentially influential factors are added to the analysis.

What I want to suggest is that attention to seasonality may contribute not only substantive conclusions about the dynamics of poverty and vulnerability (see Chambers, Longhurst, and Pacey 1981), but also a methodological tool for analyzing certain aspects of commodity production. Sea-

---

2. Several of the papers in Chambers, Longhurst, and Pacey 1981 make this point, as do von Braun and Kennedy (1986). It is implicit in Hyden's (1980) notion of the "uncaptured peasantry" that farmers can vary their level of market involvement.

3. A study in Senegal (Josserand 1984) seems to be one of the few attempts to show the pattern of food purchases in the rural areas, relating it to commodity production and level of income.

sonal management is a separable dimension of farm decision making. Farm populations have to be concerned not only with the absolute level of income, whether through maximization of profit or minimization of risk, but with the ways in which various lump cash payments and incomes in kind can be transformed into a regular food supply. A change in crop prices, crop types, employment opportunities, tax impositions, school fees, and so on signals an effect not only on quantitative income levels but also on qualitative issues of synchronizing income and expenditure.

New cropping or labor patterns change the seasonal structure of income streams in several ways. They may alter the entire seasonal profile of income, smoothing over old troughs or provoking new peaks; they may substitute seasonal income in cash for seasonal income in kind; and they may add new seasonalities to old ones, as members of coresidential units take up personal activities on different rhythms. The vital transformation of income into consumption typically implicates a range of economic, technical, and social resources that people maintain and adjust at some cost, actively negotiating the terms of their availability and the degree of reliability.[4]

Thus if methods can be devised to separate out the element of time synchronization, this may help to define an important analytical method complementary to those concentrating narrowly on prices or broadly on the entire context of commoditization.

Case Study: Southern Cameroon

Many of the West African export-crop economies have been growing and changing for most of this century and are therefore well past the stage of transitional adjustment. It seems clear, however, that some populations have done much better nutritionally than others. An understanding of how they have managed this may suggest ways in which various different strategies of seasonal adjustment (ecological, social, economic, etc.) substitute for, complement, or undermine one another at the local level, and thereby provide a microdynamic through which data on consumption at the national level could be understood.

The case material used here is drawn from a historical study of the food economy in the hinterland of Yaoundé, the capital of Cameroon (Guyer 1984). The data on seasonality do not lend themselves to statistical

---

4. A series of papers in anthropology have made this point, particularly stressing the critical importance of reliability; people invest in ways of ensuring reliable support (Bloch 1973; Kuckertz 1985), and the degrees in their achievement of reliability rather than absolute level of income sources can be used as an indicator of differential vulnerability to impoverishment in deprived populations (Sharp and Spiegel 1985).

analysis; they are given as an example of what may be needed to explore the mitigation of seasonalities. Data are taken from two consumption surveys carried out in 1954 and 1964-65, and from the author's own 1975-76 budget study in a commercially active village, on a paved road 25 kilometers north of Yaoundé.

The Beti System

The Beti peoples of southern Cameroon enjoyed a rich agricultural hunting and gathering system in the period before colonial rule. Population densities in the forest and savanna-border environment were as low as five persons per square kilometer. Villages were mobile, gradually migrating into new territory or toward the coast and the trade sources. During the twentieth century, a number of interventions have altered production possibilities. Villages were permanently settled by government order; cocoa became the major commercial crop; and the food market for the urban center of Yaoundé opened up. Ecological management and the familial organization of resources and income have been made to adjust.

In spite of long-term, and at the outset forced, intervention to develop commercial growing of both food and cocoa, the Beti in the Yaoundé hinterland have not been the victims of serious inroads into nutritional standards. The National Nutrition Survey undertaken in 1977 showed this population, one of the pillars of Cameroon's peasant-based commercial production, to be closer to adequate nutritional status than any other in the country (RUC 1978). Level of chronic undernutrition were not as low as for the urban populations, but they were better than the levels in other regions, and not alarming by absolute standards.[5]

The main way in which consumption has been classified in surveys has been according to source—home production, other nonmonetary circuits of exchange, and the market. The SEDES study of 1964-65 found the following seasonal distribution of food, by source (table 9.1). The absolute, as well as relative, monetary value of the diet was highest in December to March, corresponding to the increased market and exchange activity following the cocoa harvest. In April to July, the season of greatest home provisioning, the absolute monetary value of the diet was at its lowest, at least 81 percent of the December to March value.

The nutritional quality of the diet, however, fluctuated somewhat differently. According to Masseyeff, Cambon, and Bergeret (1958), the seasonality in caloric and protein value was considerably less marked than the 1964-65 SEDES figures on monetary value, because of the kinds of food

---

5. Chronic undernutrition (under 90 percent height-for-age reference median) for children aged 3-59 months was 18.8 percent in the central south (excluding the cities of Yaoundé and Douala), by comparison with 22.1 percent for the country as a whole and 11.8 percent for Yaoundé and Douala (RUC 1978, p. 90).

TABLE 9.1 Economic origin of the diet, by value, according to season (percent)

	Home Provision	Market	Nonmonetary Exchange
August–November	60	20	20
December–March	54	21	25
April–July	62	19	19

SOURCE: Adapted from SEDES 1964–65, p. 60.

on which money was spent. The Masseyeff data are for a fairly isolated village, at a period when cocoa incomes were still in expansion rather than deeply established, and are therefore not to be taken too definitively. However, they show a very stable caloric intake, with a slight decline in December. Total protein intake rose slightly in September; the major fluctuation in protein consumption was in animal protein, which increased very significantly in September (83 percent) and December (55 percent) over a stable level in March and July. The Beti have high consumption levels of vegetable protein from groundnuts, so total protein intake is less affected by fluctuations in purchased animal protein than might be the case in other farm populations.

Calorie consumption is not given by seasons in the 1964–65 SEDES study, but the economic data quoted above are consistent with the Masseyeff study—the major items accounted for by purchase and nonmonetary exchange in all seasons, but especially after the cocoa harvest, were beef, fish, palm wine, and other drinks, rather than dietary staples. The content of the diet fluctuated seasonally, from what is considered an impoverished vegetarian diet when cash incomes were low to a more culturally adequate diet when cash incomes were high, but the technical effect on protein-calorie adequacy was less marked.[6] The basic stability and adequacy of the diet, in both protein and calorie terms, rested on the management of home-produced food, rather than cocoa income.

As can be inferred from table 9.2 on income sources and Masseyeff's (1958) nutrition data, the staple vegetable diet follows a seasonal rhythm similar to income from food cultivation, fluctuating very narrowly over the year. By contrast, the consumption of meat, fish, and drinks followed, though in a muted fashion, the unimodal pattern of cocoa income.

6. The author has used the term *technical* to distinguish biological effect from cultural perception. To many rural people, their diet has declined dramatically in value over this century because of the decline in availability of hunted meat and certain vegetable foods no longer grown in large quantities (yams and melon seed in particular). One man said, "We eat like goats," that is, a largely vegetarian diet. Given the regular consumption of green leafy vegetables, staples, groundnuts, and palm products, people can live above a technically adequate threshold while experiencing a real sense of deprivation. It is true of all food systems that people plan their diet to generate joy and commensality, not just survival.

TABLE 9.2 Indices of seasonal variation in cash income

	Aug.–Nov.	Dec.–March	April–July
Traditional agriculture	104	101	93
Export agriculture	152	137	13
Total monetary exchange	120	131	47

SOURCE: Adapted from SEDES 1964–65, p. 106.

Before moving on to the changes in seasonal management attendant on new sources of income from urban food sales, one needs to generalize this configuration and contrast it with others. In whatever way market crops have been integrated into local systems—by political pressures or by market incentives—these configurations around seasonal management may entail different social dynamics within domestic provisioning groups.[7]

The Beti pattern in food consumption is produced by the superimposition of the two separate seasonalities, one based on home production and the other on the cocoa market. The first is largely in the hands of women, and the second in the hands of men; the first is highly stable and the second is fluctuating. Superimposed seasonalities of income earning imply quite minimal mutual influence between the two components of income, home-produced in kind and earned in cash. As long as the home production system remains technically viable, shifts in conditions affecting the commodity crop will only affect items in the diet typically purchased with that income. The seasonal rhythm and the corresponding management mechanisms for home-produced food may be responsive to entirely different conditions. Within domestic or other provisioning groups, income management is generally specialized by sex or by age.

Control of the entire provisioning process, where it exists at all, depends on resource control defined outside the production-consumption nexus itself—through landownership by men, by male fulfillment of key tasks such as opening up new fields, or through bride wealth payment establishing generalized rights to a wife's produce.

There are three hypothetical alternatives to the Beti seasonal income profile, produced when a marketed crop is added to self-provisioning. The simplest involves increased marketing of a crop already produced, without entailing shifts in the amounts and kinds of other crops in the system. Already existing, regionally particularistic, seasonal patterns of income in kind are simply intensified and partially converted into cash. Claims on

---

7. The author has avoided referring to domestic groups as households, simply because the unit with respect to provisioning may not coincide exactly with the residence or administrative unit, or the unit pooling valued resources.

the cash may be a cause of renegotiation or dispute, but this may not affect the diet until the amounts involved begin to require changes in production patterns. Beyond this point, the reorganizational issues raised may be quite complex, resulting in shifts toward either the separation of spheres of responsibility or the integration of the labor and income of people eating from the same pot under a single manager. Many African systems, such as where upland rice is the staple among the Mende, or where an individually grown "minor" crop such as swamp rice in the same system (Richards 1985) and beniseed among the Tiv (Bohannan 1954), fall into this category.

If the seasonal rhythms of the various crops in the cultivation system, and in home-produced and market-produced income, are discrete but complement one another, one rising as the other falls, they yield an overall integration, a smoothing effect on total income in cash and kind. Integrated seasonalities are much more demanding in terms of constant mutual adjustment over the annual cycle and from year to year. Because no single source—crop, field, or person—accounts for food supply year-round, the means for mobilizing each seasonal strategy have to be maintained over the off-season, possibly involving both internal domestic negotiation and wider networks. Unless a strong customary regime of mutual expectations prevails and survives the other side effects of market penetration, control in a seasonally integrated production economy is likely to reach deep into the constantly shifting organization of work itself and day-to-day income management. Domestic pooling of income may be more important, implying either a single manager or a seasonally flexible framework for assigning responsibility for provisions that affect the entire domestic group. This is the kind of pattern described by Sabean (1978) for peasant Germany in the eighteenth and nineteenth centuries, when crops with different seasonal labor demands and different risks of failure were maintained "to tide the family over the absence of cash."

If home production becomes subordinated to cash cropping through substitution of one for the other, the seasonality of total income is dominated by the rhythm imposed by the cash crop. Substitution of cash-earning cultivation for home production logically implies a single manager and implies heavy dependence on effective storage, saving, or rationing institutions to smooth out food provision over the year, whether this is taken care of through the economic structure, as in a fully market economy, or assumed by local institutions such as savings associations or the extensive use of credit with retail traders.

The author is not suggesting these as static ideal types of domestic organization. It is simply a logical exercise to guide questions about the implications of changes in commoditization with different seasonal income profiles. Each configuration is associated with different ways in which season-specific income management is maintained—through separate spheres of responsibility, through locally specific distribution systems, through a single manager or seasonally shifting income transfers and responsibilities, and through rationing. Although all of them may involve some kind of "domination" of men over women, older over junior, they are qualitatively different in the points at which control is exercised and therefore in the locus of divergent interests and differential vulnerabilities in a situation of change. The question is how superimposition, intensification, integration, or substitution in income seasonality is produced, and what the ongoing internal dynamics are likely to be as income sources become yet more varied and pressures are exerted on the customary strategies for synchronizing the seasonalities.

As will be argued below, the gender division of labor and responsibility is central; all these processes are engendered, in the sense that the existing divisions of labor and responsibility between men and women in domestic groups affect the desirability, reliability, and even the possibility of the range of strategies taken. Since the remainder of the chapter explores the structuring power of gender relations, let it be said at the outset that a full analysis of seasonal management should encompass consideration of other social and ecological processes that may account for the emerging rhythmic structure of income, savings and storage, and expenditure.[8]

Income and Gender

In Africa, the superimposed pattern has been associated with economies in which one component of income is in the hands of women and the other in the hands of men. During the 1950s and 1960s in southern Cameroon, the classification of income by source could plausibly stand for income classification by sex. Women managed the home-produced, staple, and vegetable-based basic diet, and men managed the market-based, widely fluctuating animal protein consumption. The latter had relatively little impact on the technical adequacy of the diet, but in cultural terms it raised the dietary quality for both men and women very significantly.

By the 1970s, however, women were far more involved in market production, especially in the hinterland of major cities. This means that the old market-nonmarket categories no longer capture the male-female dynamics of seasonal management. To understand how market involvement affects the regularities of food supply, the gender difference in seasonal production and consumption rhythms has to be addressed directly. Each sex actually manages not a single seasonal rhythm, but a set of rhythms—

---

8. Donham (1981) and Lewis (1981), for example, show that extradomestic distribution of produce continues to be an important resource factored into the strategies of domestic groups.

in the timing of harvest, in the seasonal demand for their crops in the market, and in the seasonally changing demands they make on one another's incomes. Since women's sources of income cut across the old categories of self-provisioning-nonmonetary-market sources to a greater extent than men's, and since women are in charge of the diet, it is women's choices and decisions that provide insight into the new seasonal dynamics of consumption.

Women's income in kind from their fields has two different seasonal profiles. All the major staple foods—cassava, plantain, and cocoyam—produce on a virtually seasonless cycle. They are planted to be used as needed from the field, without storage. The other major food crops—groundnuts and maize—have two peak harvest seasons, in July and November, but their consumption patterns differ. Maize is not stored, except for seed. Maize consumption and sales both peak at the harvest. Groundnut consumption, by contrast, has been kept steady throughout the year through careful storage and rationing on the part of the women farmers, and by very limited market sales maintained in the face of high prices and a vigorous market in the city. Women have apparently resisted market penetration of the groundnut economy; they have neither reduced production to grow market-oriented crops nor sold at harvesttime to purchase for consumption later. A woman's success as a farmer and skill as a cook depend more heavily on the management of her groundnut crop than on anything else. It seems without a doubt that the resilience of nutritional standards reflected in the National Nutrition Survey can be attributed to women's protection of a staple-groundnut combination for family consumption, through a cultivation system and groundnut storage that provide almost constant year-round supply.

The profile of home consumption must not be seen, however, as a simple resistance to the market per se; it is carefully adjusted to keep the basic food supply regular in the context of market sales. Sales come from two sources—a planned absolute surplus of staples, and the equally planned surpluses of crops whose home consumption is also seasonal, such as the case of maize mentioned already. For the staples, greater areas are put under cultivation, yielding a constant availability of marketable crops. The absolute amounts involved depend on constraints on the expansion of field sizes and a relative vacuum in the storage or marketing institutions that would allow bulk sales.[9] Any expansion of marketed staples would be land-extensive.

---

9. Women's farming runs up against severe labor bottlenecks when expansion of production demands expansion in processing and storage. Processing is done entirely by women; both an expansion of processing by all women and the development of a division of labor among women are inhibited by the insistent rhythms of the farm work itself.

In addition to staples, women grow a variety of seasonal crops with short growing seasons, such as maize, leafy vegetables, market garden crops, and sweet potatoes, which give a sequential series of small seasonal peaks to income. Because of the nature of land use, any expansion on these lines would be land-intensive.

Women's sales and cash incomes from their own farms, therefore, contain two components—a regular income potential from staples, and a series of small harvest peaks as particular crops come to maturity. The overall effect might be expected to reproduce in cash the female-specific seasonality of home production, with women's income very stable over the year. In fact, however, there are now seasonal swings in women's cash incomes. They are not wide, but they do exist and raise important questions about the possible directions along which land use, income transfers, and food purchase are developing (table 9.3).

The cash controlled by women follows a marked seasonal pattern, with totals rising 53 percent in November over July. It should be noted again here that the harvest cycle of seasonal crops in the women's fields, being basically bimodal, in correspondence to the bimodal rainfall pattern, peaks in both months. The apparent shift in cash income toward the same seasonal peak as in the men's cocoa economy, has not been produced by cultivation shifts toward specialization, but by seasonal shifts in women's marketing strategies and in interpersonal cash transfers.

During November, men have greatly increased cash resources. If an integrated pattern of seasonal income management were developing, women's efforts to earn an income might concentrate in July when men's incomes are lower, and men's transfers to women might rise in November. The converse is the case; table 9.3 shows that, both proportionately and

TABLE 9.3 Women's cash income, by source, July and November 1976 (N = 13)

	July		November	
	Francs CFA	Percent	Francs CFA	Percent
Sales, own crops	23,885	38	41,600	43
Resale margins	3,615	6	4,210	4
Received				
Husband	24,840	39	34,770	36
Other men	3,900	6	—	—
Women	6,675	11	10,150	11
Cocoa	—	—	5,600	6
Total	62,915	100	96,330	100

NOTES: In 1976, 1,000 francs CFA was approximately U.S. $4. The values in this table are totals for the whole sample, not per person amounts.

absolutely, women's higher incomes in November are accounted for by increased economic activity on their own account rather than increased transfers from men.[10] The proportion of women's increased incomes coming from market sales of their own crops went up from 38 percent to 43 percent, while the proportion coming from men's direct cash contributions to the family economy actually decreased in this case from 45 percent to 36 percent of women's total income. Given that men have considerably less income in July, the proportions of that income that they contribute to women must be considerably higher to produce this effect. Put in another way, the reduced standard of living in July is characterized by proportionately greater routine income transfers from men to women; the higher standard of living in November is associated with heightened independence of income management with respect to food. In the women's own terms, they are taking advantage of increased commercial activity in the cocoa season to ensure a personal income that is much more dependable than the more or less voluntary contributions of their husbands and other menfolk.

Shifts in intrafamilial income transfer between men and women concerning routine food expenditures in a system where both sexes' incomes peak in the same season, therefore have a smoothing effect—men's contributions are higher in absolute terms, but proportionately lower, in relation to both men's and women's total cash incomes, during the time that cash income peaks. Small amounts of seasonal saving and overspending constitute another smoothing process to which women have recourse. Women's current income covered 81 percent of total expenditure in July, leaving a 19 percent deficit, whereas there was a 9 percent saving in November. As a result, food expenditures fluctuate less than women's cash incomes and, a fortiori, much less than men's; food expenditure was only 28 percent higher in November than in July, whereas women's cash incomes were 53 percent higher. Proportional to their income, women spent more on food in July (74 percent) than in November (62 percent), thereby keeping the absolute levels similar.

The internal distribution by item was also stable. Table 9.4 shows that the most important increase in seasonal food expenditure was on fish (74 percent of the difference between food expenditure in July and November).

When the meat and fish categories are looked at more closely, yet another aspect of the consumption smoothing process is revealed. Although higher amounts of expenditure are devoted to animal protein in Novem-

---

10. It is reiterated here that the author is referring to routine transfers for food and minor domestic needs such as soap and kerosene. After the cocoa harvest, often in December, many men give their wives a lump sum of cash for their own use.

TABLE 9.4 Cash expenditure on food, by item, July and November 1976 (N = 13)

	July		November	
	Francs CFA	Percent	Francs CFA	Percent
Rice and wheat	4,100	9	3,945	7
Other staples	450	1	1,850	3
Meat	14,475	31	18,550	31
Fish	19,060	41	28,905	48
Other (oil, salt, etc.)	8,660	18	6,740	11
Total	46,745	100	59,990	100

NOTE: The values in this table are totals for the whole sample, not per person amounts.

ber, the frequency of purchases is very similar in the two months—meat purchases averaged 1.8 per woman in July and 2.2 in November; fish purchases averaged 6.2 per woman in July and 6.7 in November. The difference in cash value in November is accounted for by larger purchases of more expensive kinds (e.g., fresh fish rather than dried fish). In July, women maintain the frequency of fish and meat consumption but reduce the quantity and quality.

In summary, the smoothing of expenditures over the seasons, in an economy heavily dominated by a highly seasonal major export crop, is achieved in two major ways: (1) self-provisioning in kind, mainly by women, using cultivation, storage, and marketing techniques and strategies that minimize seasonal peaks and troughs in the basic staples, and (2) women's provisioning through the cash nexus, devoting a high proportion of their income to food, augmented from men who increase their transfers to women (proportional to income) during the cash-income troughs. In addition, women do some seasonal saving and dissaving of cash income. The result is a diet of remarkable seasonal stability.

This dynamic is not deducible from income levels alone. It depends on the constraints and possibilities open to both men and women in relation to the production system, the market structure, and each other.

Conclusion: Long-Term Dynamics

The increase in women's income from food sales since the 1960s has entailed changes in seasonal management that were probably not yet played out in the mid-1970s. The historical trend seems to suggest the continued separation of male and female spheres of responsibility for the diet. Looked at over the long term, men's contributions—initially in labor only, later in certain labor inputs and in cash transfers—are fluctuating and discretional. At certain historical periods and in certain seasons, men's contributions to the food economy have been proportionately greater than at others, and women certainly work on increasing them. There are customarily accepted norms supporting a man's obligation to provide certain kinds of food to the family economy, while at the same time a man's discretionary power over his income is also recognized. A wife, by contrast, is expected to feed her family, and her husband has a right to eat from her kitchen whether he contributes directly or not. The terms of intrafamilial negotiation tend to concentrate around the demarcation of spheres of responsibility and the rights of spouses in one another's income.

Since the mid-1970s, two things have been happening simultaneously—first, in villages such as this with easy market access, some men have been trying to smooth out their incomes and to raise them at the same time by growing truck vegetables for the urban market; and second, in the somewhat more liberal political climate of the 1980s, women are increasingly developing credit associations in order to allow a portion of their income from food production to accumulate for investment in big-ticket items. Both of these developments have involved renegotiation, and the latter outright dispute, about the rights and mutual responsibilities of husbands and wives. These alterations in seasonal income patterns do not, however, involve mutual adjustments in the sequential integration either of tasks in production or of income sources over the year. The separation of spheres of responsibility seems to continue, with its debates about lines of demarcation as conditions shift.

One can now begin to assess the transitional dangers of major shifts to seasonal management patterns associated with other cultivation or income patterns. No major transition is sociologically or culturally simple, growing "naturally" out of the previous system. Substitution or specialization demands radical changes in institutions for saving and rationing income over the annual cycle. These exist on a very limited scale in the Beti economy, as in many other rural economies. Rotating credit associations exist, but their timing is specifically geared to producing a lump sum at a particular moment, not a regular small income. Only two men in this village, as far as the author knows, had bank accounts in Yaoundé. In all likelihood, the income from a monocrop would be in the hands of men whose expenditure obligations have been structured on a cycle totally different from daily dietary requirement in the past. In fact a voluntary transition to monocropping is impossible to imagine; it would have severe transitional nutritional effects, would probably undermine women's access to an independent income, and would entail a revolution in the framework of family relationships and in the social institutions available to ensure rationing of income over the year.

As noted above, the smoothing of income over the year as income sources increase through diversification seems to be developing on an individual rather than a familial basis. A shift to integrated seasonal management entails extensive redefinition of pooling and autonomy within the domestic group over the seasonal cycle. This is perhaps achievable only through fairly intense religious and political pressure, or through poverty. As Sabean (1978) suggests for peasant Germany, given the structure of domestic authority, women had little choice but to become increasingly active in the farm labor force as the seasonal cycle filled up with crops demanding assiduous attention. But evidence suggests that this strategy had costs in domestic welfare.

How threatening such processes might be to seasonal management of basic welfare levels cannot, of course, be predicted in abstraction from the other dimensions mentioned earlier: (1) the price structure and resource endowment that determine absolute levels of welfare, and (2) the broader institutional environment beyond the domestic sphere—such as the nature of savings, credit, and emergency relief organizations, and the conditions in the factor market—that may mitigate or intensify the dangers. Seasonal analysis is only one of a set of analytical frames needed to understand the dynamics of self-provisioning in a commercial world.

This chapter has suggested that understanding the dynamics of multiple and changing income sources in partially commercialized rural economies may be fruitfully addressed through seasonal analysis. It takes the management problem of how a set of individual production and consumption profiles can lead into the study of institutional arrangements and technical practices and shows their relevance to the issue and their connections to one another. Intrafamilial relationships can be seen as crucial mediators, but in a manner that avoids both extremes of modeling households—on the one hand, the oversimplification of the economic assumptions of a joint utility function as a uniform aspect of household strategic behavior across cultures and income levels, and on the other, the particularity of ethnographic analysis. The point is to ask questions that draw on different disciplines to reveal the dynamics producing characteristic patterns of income earning and consumption, which indicate the structures and struggles that account for them, and which suggest the trends already under way.

10 Seasonality in Food Systems: An Anthropological Perspective on Household Food Security

ELLEN MESSER

"Why seasons matter" (Chambers 1982)—in nutrients, foods, socioeconomic activities, and health—has long been a concern of anthropologists. Exploring seasonality from both biophysical (Western scientific) and sociocultural ("native") perspectives, they have documented how sociocultural cycles of time, superimposed on the annual rhythms of climate, fauna, and flora, influence human subsistence and the many dimensions of social, cultural, and economic life. From sociocultural perspectives, they have evaluated how people avoid seasonal food insecurity or adapt to seasonal shortages. From biophysical perspectives, they have described signs of adjustment to seasonal dearth. From both perspectives, they have questioned whether, as people pass from traditional local to purchased nonlocal diets, seasonal hunger disappears and, if so, whether it is only replaced by chronic undernutrition.

Seasonality enters into all dimensions of food systems—patterns of food acquisition, cultural preferences, rules for social distribution, and consumption—and their evolution. The sources and symptoms of seasonal food insecurity have been noted both in general ethnographic studies and in more specialized studies of food, nutrition, and health. While foragers (hunters and gatherers), herders (who move according to the seasonal availability of water and food), and agriculturalists (who plant and harvest according to the seasons) are touched most directly by the calendars of temperature and rains, those relying for their livelihood on crafts, trade, and industry may also suffer seasonal food insecurity due to periodicity in their incomes, in affordable food supplies, or in both.

Anthropological findings suggest that seasonal food insecurity at the household level is a problem for poor households in most parts of the world. Both its causes and its manifestations are complex; and no particular household structure or social organization provides a recipe for limiting seasonal risks. People throughout the world have developed diversified strategies for survival in good years and bad; but as natural resources, population, and market access change from area to area and over time, past sociocultural coping mechanisms may cease to be effective.

This chapter reviews the significance of seasonal food insecurity as observed by anthropological studies, drawing from them (1) ethnographic evidence of seasonal hunger and (2) concepts and methods that can be used to identify and address seasonal food problems. The concluding section includes an agenda for future research.

Seasonal Food Insecurity as a Dimension in the Anthropological Literature

Anthropologists studying food systems and their evolution have documented seasonality in climate and naturally available resources, in the ordinary and ritual activities that affect food provisioning, in the behaviors and emotions that affect social cooperation, and more generally in nutrition and health.

In archaeological and historical studies, seasonality of food sources and scheduling of human activities in relation to their availability are the two central concepts used to elucidate human settlement patterns, and also to show how diets and human social organization evolved (e.g., Flannery 1973). In sociocultural studies, seasonal cycles figure prominently in the plants and animals people manage in their agricultural, pastoral, and foraging strategies, and also in the crafts and wage labor that serve to supplement their incomes. Seasonal patterns of settlement and mobility are important aspects in the adaptations of gathering-hunting (foraging) and pastoral societies; anthropologists have sought to understand the logic of peoples' annual migratory rounds from both ecological and "native" perspectives (e.g., Evans-Pritchard 1940; Mauss 1979).

More generally, studies of the ritual cycle in its ecological context also document seasonal fluctuations in food supplies. The classic ethnographies from all parts of the world note the connections between food availability and sociability and also the importance of ritual hospitality and distribution in times of dearth. More specialized psychological anthropological studies have presented seasonal hunger as a motivation for behavior (DuBois 1941; Holmberg 1950) related to food choices and eating styles—including ordinary abstemiousness and ritual gorging (D. N. Shack 1969; W. Shack 1971).

In addition, socioeconomic studies have explored the linkages between seasonality and the sexual division of labor in agriculture and in child care arrangements (White, Burton, and Dow 1981; Burton and White 1984); seasonal diversifications and restrictions in occupational activities and diet at the household level (Rosemary Firth 1966); and the role of exchange and exchange rates (such as between pastoralists and agriculturalists) in determining seasonal food sufficiency for households engaging in seasonal or permanent occupational specializations (Bates and Lees 1977).

Ultimately, physical anthropologists have attempted to quantify seasonal changes in nutritional status among men, women, and children, and to assess in addition the effects of seasonal dietary changes on patterns of energy expenditure, illness, conception, and childbirth. Biocultural anthropologists then try to assess, from both "native" and scientific points of view, how well people are dealing with food supplies, given their nutritional and political-economic environment and their cultural beliefs and practices (Huss-Ashmore and Johnston 1985). With nutritionists, they also trace climatic, food, and other sociocultural factors involved in seasonal patterns of birth (Pasternak 1978; Condon and Scaglion 1982), breastfeeding (Serdula et al. 1986), and menarche (Chowdhury, Huffman, and Carlin 1978).

Ethnographic Documentation

Foragers

The remaining foraging societies provide models for how people have adapted socioculturally and physiologically to seasonal dearth by patterns of seasonal movement and settlement in relation to available food resources, including water; by consuming more diverse, and often less preferred, foods; by intra- and intergroup exchange and changing social group composition; and by loss of weight and other physiological adaptations. They provide examples of the circumstances under which people complain that they are hungry (they may be short of certain preferred foods, rather than of food in general) or use foods differently (Pagezy 1982, 1984). In addition, they provide examples of the impact of modern civilization on traditional resource use, settlements, and mobility.

In the Kalahari, for example, where during the lean season both water and foods may be scarce, people adjust to seasonal hunger by switching to less preferred foods and resting more (Marshall 1976; Lee 1979). They also may draw on extended social networks and move to zones of more plentiful resources (Wiessner 1981). Nutritional analyses indicate that under bush conditions, these people exhibit signs of seasonal infertility and weight loss as physical adaptations to food stress. These signs tend to disappear under settled conditions and more steady diets of mealie-meal, sugar, and milk: people are heavier, seem to show no seasonal weight loss, and have lost seasonal patterns in fertility (Wilmsen 1978, 1982).

In the tropical forest, by contrast, groups such as the Efe pygmies rely on symbiotic food exchange relationships with farming households (Bailey and Peacock 1984); it is questionable whether they or any other gatherer-hunter group could subsist in the tropical rain-forest at present densities without such relationships (Messer 1984a; Peterson 1978).

Although hunters and gatherers have been labeled "the original affluent society" (Sahlins 1972), biophysical evidence indicates that they suffer seasonal hunger and, along with it, reduced activity levels, weight loss, and periodic infertility. Given that they make up weight deficits, however, it is not clear that they suffer permanent health impairment. Moreover, it has been argued that foragers cope better than their settled neighbors with seasonal and permanent environmental stress (Flowers 1983; Cassidy 1980).

Pastoralists

Pastoral households also provide models for how people adapt socioculturally and physiologically to variable seasonal resources. As herders move transhumantly between sown and unsown zones, offering animals and animal products in exchange for grain, terms of exchange as well as ecological conditions dictate what production and consumption strategies are optimal and whether specialized herders can make a living in all seasons in all years (Bates and Lees 1977).

Many face a hungry season at the time when they are isolated from social relationships of exchange. Dimensions of hunger vary by society and region, and also change as traditional diversified economies and seasonal coping strategies move in the directions of greater economic specialization and cash market purchases.

Among the Senegalese Ferlo (Benefice, Chevassus-Agnes, and Barral 1984), for example, herdsmen face elevated nutritional deficits during the wet season from which they recuperate during the cool, dry season. The nutritional situation begins to deteriorate again in the hot, dry season. Yet they do not experience poor health or nutritional status, because they mobilize resources of subcutaneous fat and are not heavily subject to seasonal parasitic diseases. Their traditional ways of coping with seasonal hunger are to reduce energy expenditure, to increase reliance on gathered foods, and to engage in trade. Recently they have also begun to purchase rice and oil to compensate for decreasing forageable flora and game within their range, as well as for decreasing amounts of milk available from herds as they market more animals to meet increased demands for meat. In this region of predictable seasonal food crises, the traditional planned exchange through diversified agropastoralism, trade, and gathering of wild fruit is being replaced by expanded specialized production of meat and commercial trade networks.

It is not clear how much more vulnerable to seasonal hunger or sudden collapse this society will be under this more specialized economy. Both seasonal and permanent nutritional deterioration have been suggested for a Fulani group involved in a range development project to improve cattle production (Teitelbaum 1977). People were instructed to allow calves to consume all their mother cows' milk (to the nutritional detriment of human children). Replacing traditional exchange relations between cattle pastoralists and millet farmers, cattle were to be kept on cattle land; cattle manures were to be replaced by chemical fertilizers. In total, the plan proposed changing all human, land, and animal-plant, and social relations. The study infers that specialized land use and projects that develop animal commodities such as this one destroy traditional diverse sources of ecological and sociocultural adaptations to seasonal hunger, and before long cause biological and social deterioration of both pastoral and farming societies.

Seasonal Food Insecurity among Peasant Households

In contradiction to Miracle's (1961) denial of the existence of seasonal hunger in Africa, anthropological observations have consistently documented periodic food shortages there as well as in other parts of the world. Populations attempting to subsist on the production of a single post-rainy-season harvest have rarely covered their annual nutritional needs. Moreover, such populations have lacked slack-season sources of income sufficient to purchase adequate supplies of food.

British social anthropologists since the 1930s have reported how food supplies in many African societies dwindle from one harvest to the next, and have described the rationality of native cultivation, preservation, storage, and food distribution or rationing strategies (Fortes and Fortes 1936; Richards 1939; Sharman 1970).

For example, Fortes and Fortes (1936) alleged that native food production, distribution, and consumption practices among the Tallensi aggravated their precarious nutritional situation. Grain was short during the season of greatest energy needs—the planting season—which fostered underproduction; but they also interpreted the problem to be one of allocation of resources, since in the season immediately following the harvest, the Tallensi spent grain liberally on beer brewing and only later began rationing. In somewhat similar fashion, Sharman's (1970) comparison of rural Ugandan households suggested that individual household food sufficiency varied seasonally according to the management skills of the female household head. She suggested that it was not household food supply but the care and skill with which women rationed or distributed food that determined which households' children were seasonally or permanently malnourished.

Another theme in this ethnographic literature is the relationship of patterns of work, including male labor migration, to seasonal hunger; seasonal hunger is interpreted to be due to labor shortages, underproductive land, or both—and thus to be beyond the control of native managers. As Richards (1939) cogently demonstrated in the case of the Bemba, the hungry period coincided with the period of peak agricultural work and, therefore, energy demands, which meant that investment in agriculture was suboptimal. People failed to work hard (a primary concern for British mining and other economic interests), not from sloth but from undernutrition. In this society where men had been drawn away from their farming activities to the mines, women found it difficult to perform the heavy clearing tasks traditionally assumed by males in addition to their own cultivation and gathering roles. Analyzing the activities throughout the year and the social relations surrounding food, Richards found that food was in shortest supply during the period of the year when women most needed food energy for clearing and planting fields. Thus the Bemba were enmeshed in an ongoing cycle of underproduction and undernutrition.

Migration of male labor also caused hardship among Lesotho villagers, as women were left with major food cultivation and fuel-gathering tasks. Although food supply was not a problem in the years she observed, Huss-Ashmore (1984) saw that women and children suffered from lack of food and poor hygiene because women were too exhausted to cook and clean at times of peak agricultural work. In this case men had left agriculture for wage labor in the mines as a nutritional strategy to provide cash to buy food in bad years. However, this strategy did not seem to be insurance against either short- or long-term hunger.

Cash crops tended at the expense of subsistence crops have also been seen as responsible for some seasonal hunger. Surveys among the Genieri in Gambia in 1947-50 (Haswell 1953) concluded that both the absence of males and the tending of cash crops such as peanuts contributed to the wide variation in household food availability over the course of the year, depending on seasonal rains and individual household efforts and planning. Rice and other purchased grains were compensating in part for lower millet supplies due to lower male labor investment in food crops. People also attempted to stave off hunger by planting short-term crops such as early millet, *Digitaria exilis,* and maize—all of which, if they yielded at all, matured at the time when food needs were greatest. If drought followed the first heavy rains (the point at which early crops were laid in), however, these early crops failed and people suffered a hungry season. Or if the rains ended early, the main harvest failed, causing hunger early in the *next* season.

Illnesses of female adults at critical times in the production process also had an impact on labor efficiency and productivity. If cultivators were ill during planting or weeding, they might never catch up. Villagers, particularly women, tried to adjust their energy expenditures to reduced energy intakes, but many lost weight during the preharvest season. Also, in this case, wage labor was not making up the hunger gap. Restudies in 1974 indicated that 80 percent of the households would be short of food during the hungry season, although a few rich households, well favored in lands and human resources, had surpluses. The rich were further impoverishing the poor by selling them food at high prices and lending them money at high interest rates (Haswell 1981b). Haswell saw no solution, since technological gains such as improved roads and agricultural intensification appeared destined to benefit the rich, not the poor. She doubted "whether there are reserves of human energy either for more intensive physical activity or to summon the will to organize, plan, adapt, or innovate" (Haswell 1981b, p. 41).

Deterioration in the ecological conditions of production has also been seen as a cause of seasonal hunger among swidden cultivators. Both geographers and anthropologists working in Africa have noted a deterioration in the ecological conditions of production among swidden cultivators (Allan 1965). Under such conditions, even if cultivators are doing all they can—planting early varieties to avert seasonal shortages, attempting to retain soil fertility by diversified manuring practices, reducing the number of meals per day and the amount of cultivated food per meal—they are still doomed to periodic hunger. Among the Onicha Ibo of Nigeria, for example, Ogbu (1973) found that seasonal food shortages were due to insufficient farmland, low yields on farms, and high storage losses of staples. Even though the people worked very hard and manured their fields with animal dung, household garbage, and compost, after three years the land was completely exhausted and had to be rested; in the interim, yields were low. Moreover, the people suffered substantial postharvest losses as root crops rotted in the ground.

Among the Chakaka Poka of Malawi, farmland was sufficient but widely scattered, so people suffered substantial losses to predators. They also lacked sufficient labor to expand holdings because many males were away at other work. In this case, however, Ogbu (1973) also suggested that the people were somewhat careless cultivators—sloppy in tilling and ridging, particularly when working communally after consuming beer. Cash crops were also implicated, as people tended to devote insufficient time to food crops. Ogbu did not indicate whether more careful management of resources would offset deficits caused by socioeconomic and ecological factors beyond the people's control.

Among the Ntumba and Ngoni of central Malawi, by contrast, Nurse (1975) inferred that male labor migration did not influence production, since men did not work in local subsistence agriculture. The food shortages instead were blamed on inadequate storage facilities: wicker granaries allowed a large proportion of the grain to rot during the rainy season, and to fall prey to rats and mice during the dry season. People tended to sell their crops and then buy back grain, rather than suffer self-storage losses. They relied on cash earnings to attract grain supplies from other areas for consumption during the hungry season.

Although not focused on seasonal hunger, studies of poor Asian and Latin American peasants indicate that seasonal agricultural deficits and hunger have been common in their experience also. In addition, labor migration from areas of seasonal deficit to those of greater abundance has been a common means of making up seasonal shortfalls in food and income. In Asia, seasonal hunger occurs principally before the major rice harvests, when food supplies of land-poor households are exhausted, wage labor is scarce, and food prices peak (Hartmann and Boyce 1983).

Ethnographic studies from Mesoamerica from the 1930s onward have revealed that in this part of the world too, both farming and off-farm occupations vary seasonally and people suffer seasonal hunger. In Morelos, Mexico, peasants complained about *septi-hambre* (hungry September), the lean month when the maize from the previous harvest was exhausted and the new maize not yet harvested (Warman 1980). People sought to minimize suffering with seasonal crafts and other occupational diversification. In village Oaxaca, Mexico, by contrast, the hungry season continues to be May through July, the planting season, before early crops mature (Malina and Himes 1977).

More generally, in this part of the world insufficient home food production, combined with low seasonal incomes from crafts in the context of national policies seeking to assure cheap labor for plantations and haciendas, often pushed men or whole families into seasonal or longer-term migration in search of work (Wasserstrom 1983; Odell 1982). Seasonal shortfalls also could be aggravated by illness, especially the untimely incapacity of adult household earners, which interfered with planting and harvesting of crops, income, and longer-term food supplies (Scrimshaw and Cosminsky n.d.).

In summary, seasonal food insecurity and adaptations to it at the individual, household, and community levels have been extensively documented in the general ethnographic literature. We turn now to more specialized anthropological approaches to the topic, through studies of food strategies, food habits, and household socioeconomics.

Food Strategies and Coping with Hunger

The term *food strategies* describes the different routes by which households acquire food (DeWalt 1983, 1984) and extend food supply. These are (1) foraging, (2) home production, (3) income diversification, (4) gifts, (5) consumption behavior, and (6) adjustments in household composition. Their seasonal operations determine what kinds and quantities of foods reach individual households. In total, they determine whether households are seasonally deficient in food energy. The following sections consider the literature on each of the food strategies as well as their limitations as short- (seasonal) and long-term mechanisms for coping with hunger.

Dietary Diversification through Foraging

Dietary diversification, particularly seasonal reliance on foraged foods, has always been a major route to food security for populations faced with seasonal shortages in cultivated plant and domesticated animal foods as well as for those that rely wholly on foraging. Harvested seasonally in good years and year-round in bad, foraged plant and hunted animal resources have been shown to add essential calories and nutrients to the diets of farming populations in Africa (Brokensha and Riley 1980; Brokensha, Warren, and Werner 1980; Fleuret 1979a, 1979b), Europe (Forbes 1977), Asia, and Latin America (Messer 1977, 1978). They are also a means of weathering drought or other climatic or political disasters (Grivetti 1978).

Anthropological studies have concentrated on showing what species are seasonally available and how human behaviors favor or limit their persistence in particular environments (Messer 1984a, 1984b; Messer and Kuhnlein 1986). In most parts of the world, traditional diets selectively use a wider range of foods within local ecosystems than is common under modern cropping systems. These include both foods foraged in the bush and those that are the by-product of human cultivation practices. Related studies have calculated the nutritional values of indigenous foods (Robson and Elias 1978) and how human cultural knowledge, selection, and preparation affect their nutrient contributions to local diets (Messer and Kuhnlein 1986).

Although anthropologists are accustomed to lauding the nutritional contribution of indigenous foods and the cultural wisdom that preserves knowledge of the edibility, location, and preparation of these seasonal or starvation foods, they also note that dietary diversification through foraging is not a permanent solution to seasonal food shortages. Such foods become less and less available as foraging lands disappear into habitation, cultivation, and grazing sites; as herbicides eliminate edible herbs in field systems; and as people abandon their traditional relationships to the land and lose cultural knowledge of its resources.

Even where such species persist, younger generations may not know the plants, their location, or their manner of preparation as a result of their lack of contact with the land or their lack of immersion in traditional culture. Moreover, in certain regions such as Mexico, "wild" foods like uncultivated greens are not consumed because they are considered low in prestige. Therefore, these food sources are avoided by the upwardly mobile even if they lack good nutritional substitutes (Messer 1977; DeWalt, Kelly, and Pelto 1980). Although anthropologists and geographers urge the conservation of these native foods in any land development program (Messer and Kuhnlein 1986), the "let them eat weeds" fallback position is unrealistic and will not protect populations for very long from hunger and seasonal nutrient shortages. Nevertheless, it is possible that supplying traditional forage products to towns may become an occupational specialization for persons lacking other sources of cash and a limited source of food for those who cannot afford to purchase other fruits and vegetables (Messer and Kuhnlein 1986).

Home Production

Home food production provides the second food strategy in most rural and some urban areas. Contributions are limited mainly by inadequate land, labor, soils, and technological inputs, and by pests. Seasonal shortages in planting period, when the crops from the previous season have been exhausted and the new crops are not yet ripe, cause hungry seasons in most rural areas where these limitations prevail. These are overcome by occupational diversification to attract and pay for marketed goods, as discussed below.

Home food production has been declining as a food strategy for the poorest households in many rural areas of the world because of insufficient arable land close to home. Absolute quantities of land available to poor households decrease as plots are subdivided in inheritance from one generation to the next. This process is accelerated as poor households forfeit land to tide them over the lean season from one year to the next (E. Scott 1984; Chen, Chowdhury, and Huffman 1979; Warman 1980). Reduction of fallow land as a response to land shortage has caused concomitant declines in fertility and further reductions in productivity and home food supplies.

Labor is a second limiting factor. Although various observers over the years have attributed seasonal food insufficiency to native sloth or otherwise poor farming practices (Richards 1939), African studies since the 1930s have blamed overwork and exhaustion due to seasonal labor shortages. Sending men off to earn cash wages, a strategy meant to compensate for chronic underproduction, rarely seems to compensate for the lost contribution of males to home agricultural production (Richards 1939; Huss-Ashmore 1982, 1984). Illness or other weakness coincident with periods of labor demand also reduces human investment in agriculture and limits consequent harvests. The ill and those recently in childbirth plant less, weed less, and rarely catch up (Haswell 1953; Scudder 1962; Scrimshaw and Cosminsky n.d.).

The lack of labor may also affect choices of crops and consequent nutrient contributions of home-produced foods. Huss-Ashmore, among others, has reported how the seasonality of children's schooling has reduced their contribution to sorghum-millet cultivation and harvest. If there are no children to scare away birds at critical times, the less drought-tolerant maize becomes the preferred crop.

Lack of adequate storage further adds to pest losses in most parts of the world. Not surprisingly, in regions like West Africa, those zones with adequate storage facilities seem to suffer less severe seasonal undernutrition (Annegers 1973). As noted above, both seed crops and root crops suffer storage losses (Ogbu 1973). Some groups adopt a food security strategy of selling their crops rather than suffer storage losses, and then buy back again later in the season (Nurse 1975); cash from migrant male labor is said to attract grain from other areas during the hungry season. Expectations of losses due to pests may also limit efforts to increase home food production if people believe, however falsely, that such efforts would be subject to pilferage by vermin or people (DuBois 1941). It is not clear how widespread such anxieties are and how responsible they are for limiting agricultural efforts. However, it is likely that uncertainties about the rains and the functioning of markets are more prominent disincentives.

Other interference with production for home consumption includes commercial crops and low prices of food crops. Again, the shift of land and labor in all parts of the world from subsistence to cash crops has been mentioned as further reducing home food production. In some African examples, cash crops are specifically implicated in shortages of food for home consumption and seasonal hunger (Haswell 1953; Ogbu 1973). These studies suggest that food purchased with cotton or groundnut earnings does not compensate for food forgone in the shift away from home staples. In Latin America, where people have been offered or forced into cash cropping or herding "economic development" projects, evaluators have noted that the only families that appear to be better-off (i.e., their children appear to be well nourished) are those that maintain some of their own subsistence production along with the newly introduced cash enterprise (A. Brown 1978; Dewey 1980, 1981). Predicting the effect of cash cropping on seasonal food insecurity, however, is a complex and contextual exercise (von Braun and Kennedy 1986).

Complicating the evaluation of the food and nutritional benefits of cash crops, especially in Africa, is who controls the crop, and how often and in what form income accrues from the time and labor invested. Cash cropping schemes have been known to fail because they neglect to take into account the prevailing sexual division of labor and access to land and labor, among other household budgetary arrangements (C. Jones 1983). These same factors would affect the seasonal food and nutritional benefits of such cash crop operations.

More positively, traditional populations also engage in food production strategies that are supposed to minimize seasonal shortages, scatter risk among a number of diverse crops and crop zones should any particular zone or crop fail, and provide a diverse range of nutrients throughout the year. These include planting early-yielding crops to limit the length and severity of shortages of the main crop (Haswell 1953), diverse gardening canopies to provide a broad range of foods in good years and at least some food in bad, and multicropping strategies to provide calories, balanced vegetable protein, and some income over all the seasons. In addition, the introduction of new crops, such as the potato in Asia, is minimizing the traditional hungry season in some parts of the world.

As in the case of many of the foraging strategies itemized above, many of these seasonal survival strategies are atrophying in the face of the demands of commercial agriculture to rationalize production in terms of the highest return to single crops. They also suffer from the breakdown of traditional exchange networks between agriculturalists and pastoralists (Franke and Chasin 1980). Finally, in some cases like Ethiopia, they are victims of political destruction. Like many of the other strategies of survival indicated above and summarized below, in most cases they do not offer long-term solutions to the problems of seasonal or periodic hunger. Instead, people look to occupational diversification to offer longer-term solutions to periodic hunger.

Income Diversification to Purchase Foods

Anthropologists have looked into the cash income strategies of households in particular environments. Economic anthropological studies of land use, small-scale commodity production, wage labor, and trade all contribute information on how people earn money to purchase food. Among others, Ogbu (1973) has noted that those societies that have access to markets to sell and buy food crops seem not to suffer seasonal hunger. Thus opportunities to earn cash in food production and beyond are important determinants of household food security over the seasons. How people decide to allocate their land, labor, and other inputs among cash versus food crops to produce income in kind (food) or income to buy food determines their food sufficiency over the seasons and the longer run. What foods households choose to grow or are able to buy also determines their nutritional status.

SEASONAL CRAFTS. Historical and contemporary studies of societies practicing seasonal agriculture indicate that crafts, wage work, and marketing—singly or in combination—are usual activities to supplement household income, a large share of which is spent on food. However, local crafts production appears to have been deteriorating in many parts of the world, a decline that has become an important source of seasonal food insecurity. In Africa, for example, seasonal income from traditional village crafts has been interrupted by the introduction of industrially produced goods (Cooper 1982). In Latin America, decreasing access to the raw materials of manufacture has also been ruinous. Gudeman's (1978) exposition of socioeconomic change in an agricultural settlement in Panama noted that the disappearance of the reeds traditionally woven into hats ended this seasonal crafts employment. On a positive note, new trade in goods manufactured for tourist markets may make up income losses caused by the atrophy of traditional crafts.

Even where they persist, crafts, like agriculture, are often seasonal and contribute to seasonal shortfalls in income for food. In Bangladesh, woolen shawl production falls off in the hot season, causing hardship for those without alternative sources of income (Rizvi 1986). Indian studies also indicate seasonality in crafts production, although actual seasonal shortfalls in food budgets are hard to determine from household budgetary studies, which report income and expenditure in average annual figures. Crafts in Latin America, such as baskets to transport coffee, also sell seasonally, even when they are not threatened by competition from industrially manufactured goods. In particular, people with no access to land to produce food and no alternative sources of income suffer the most from seasonality in crafts production and sales (Odell 1982).

WAGE LABOR. To what extent do occupational patterns vary seasonally and regionally, and do they provide sufficient incomes to ensure adequate diets year-round? Ethnographers in Latin America, for example, have usually reported seasonal activities not only in crafts and marketing but in migrant labor to complement food acquisition activities in agriculture and foraging, and to contribute income to the food budget in the lean season (Wasserstrom 1983; Warman 1980). Unfortunately, opportunities for off-farm wage labor may tend to be low at precisely the time when the demand for seasonal agricultural labor at home and crafts sales disappears. In Aguacatan, Guatemala, for example, agricultural labor demand peaks during the last two weeks of August and the first two weeks of September, during which time one must harvest maize and seed garlic. Demand for baskets for the coffee harvest is also seasonal—from June through September. In October and afterward, people must seek scarce day labor or migrate to cities in search of work. Those who are land-poor do not have food stores to tide them over to the next harvest and work season, and suffer seasonal hunger in the absence of work (Odell 1982).

In Africa, labor migration of males has been viewed on the one hand as contributing to seasonal deficits in food production, as men are drawn away from forest clearing and field preparation to labor in the mines, and on the other as a response to seasonal shortfalls, through which men earn money to complement home food supplies through food purchases. However, as already noted, remittances often appear to be insufficient to ensure household survival, either because the men do not earn enough or send sufficient remittances home, or because what they earn is not put into food or sufficient quantities of the right kinds of foods (Huss-Ashmore 1982, 1984). Moreover, male labor migration may contribute to the deterioration of family and tribal ties. Under such circumstances, traditional hospitality mechanisms, which in the past mitigated the seasonal shortfalls of some households, may break down (Cliffe 1978). It has been suggested that the separate economies of male migrants and female agriculturalists who stay at home weaken the obligation felt by men to send remittances home, as their material and sentimental ties to the family grow more tenuous (Tully 1982). Thus migrant wage income as a source of income for food may not prove to be a reliable food strategy in the short run, and may have longer-term negative consequences if it hastens the breakdown of moral ties in the household and in larger social groups.

Seasonal labor migration from areas of seasonal and permanent deficit to those of greater abundance is also known in Asia. As these patterns change in response to changing labor demands associated with new agricultural technologies, it is not yet clear whether wages from such migrant laborers are overcoming seasonal food shortages in their home households, and whether such migrants working at lower wages are displacing labor from the prime agricultural areas (Messer forthcoming).

In Asia, where wage labor in agriculture has always been a principal means of supplementing household food production, such labor is seasonal and may come too late to offset the period of hunger prior to the harvest. Moreover, mechanical harvesting and hulling of rice is removing one of the principal sources of household income—earned by women—that in the past tided families over in times of hunger (Winikoff 1975; Hartmann and Boyce 1983). New cropping patterns, such as the replacement of (cool-season-processed) rice by (hot-season-processed) wheat in areas of Bangladesh, may also threaten the wage-earning potential of women workers (Lindenbaum 1987).

NONWAGE LABOR INCOME. Other income-earning strategies besides wage labor that seasonally help supplement home production and social network sources of food include sales of animals and other valuable possessions, retail sales of goods bought in bulk, and sales of home-processed foods such as beer. People begin to slaughter animals for food and to trade animals for grain when they perceive a hungry season coming on. Trading contacts both supply pastoralists with grain and also may provide them with refuge in times of disaster (Colson 1979). Because they rely on income from animal sales to tide them over bad seasons and years, pastoralists often try to keep varieties that provide a reasonable return under variable conditions, and stress quantity over quality. Sale of livestock, as mentioned above, is also the most common means of coping with food insecurity in certain mixed economies, such as South African villages. Alverson (1978), for example, was told by Tswana men in South Africa that when food supplies were insufficient to provide subsistence, their predominant recourse was to market animals.

Not all poor persons, however, can afford to keep livestock, and those who face the worst seasonal food shortages may find this mechanism unavailable. Likewise, as people experience hungry seasons they begin to sell off valuables, such as jewelry, purchased during good years to tide them over the bad (Colson 1979). In Asia and Latin America, people are more often forced to mortgage land or future crops to moneylenders or wealthier villagers. This measure helps tide the resource-poor over the slack season in a particularly bad year but may relentlessly impoverish them further (Chen, Chowdhury, and Huffman 1979; Kahn 1980). These strategies are limited also in that (1) they are soon exhausted, and (2) traditional valuables may not retain their worth in modern cash economies (Colson 1979).

In certain locations, those enterprising individuals who do not take the next steps of either migrating for seasonal wage labor or borrowing from relatives can also "break bulk," or throw beer parties for cash income. These cash operations replace traditional noncash social mechanisms of establishing credit. The socioeconomy has been transformed so that banking cash to draw on the future seasons of hunger is often thought to provide better security than social goodwill (Alverson 1978).

More generally, women supplement their own or household income through small-scale food processing (Simmons 1975; Basson 1981), small-stock-keeping (Alverson 1984; Nimis 1982), and trade (Tripp 1981). Operations such as food preparation may vary greatly with the seasons, as a response to seasonal home food shortages, holiday demands for cash, and holiday-season opportunities to prepare foods and other small goods in great demand by others (Messer 1982).

Women's work in trade and agriculture has sometimes been faulted as taking time away from child care and household food preparation, to the nutritional detriment of children (Popkin 1980). In addition, children suffer hunger, exposure, and illness while the mother works, and therefore are at greater seasonal nutritional risk than the children of the nonworking mother. The particular skill of the mother at managing work and child care, along with the cultural mechanisms for providing surrogate mothers of greater or lesser quality, however, has a significant impact on nutrition and health. There may also be significant intracultural variations in time and resource management. Skill in managing resources also applies to strategic giving of food as gifts, which carries with it some expectation of reciprocity.

Gifts

The final strategy for food provisioning is social: food flows acquired through feasting and food gifts. Even among foragers, exchange of food within and among groups counts as an important source of food security (Messer 1984a, 1984b). Reliance on exchange partners is also an important survival strategy for pastoralists when their stock runs low (Benefice, Chevassus-Agnes, and Barral 1984). In farming societies, disaster and seasonal shortages have always been met by gifts of food from better-off relations, neighbors, or both. Superimposed on regular or irregular exchanges of food among related households are cycles of feasting and fasting that redistribute food throughout the year.

Feasting may be important for redistributing food seasonally (Ford 1972). In Mexican societies, the institutionalized food sharing that goes on during feasting has been analyzed as taking place at times of the year when the population is most in need of protein and nutrients. The high-quality protein and vitamin-rich foods that are assembled and distributed during such festival cycles have been estimated to make significant contributions to the diets of poorer households, particularly in hungry seasons (Diskin 1978; Greenberg 1981).

It may also be the case that feasting in a season of plenty actually builds up credit for receiving food at some later time (Sharman 1970), but it remains to be demonstrated that such mechanisms are nutritionally effective, particularly under the modernizing conditions of economies based on cash (Alverson 1978). More generally, studies of reactions to crises demonstrate that mechanisms of food sharing, which redistribute food from those who have to those who do not, cease at points of severe food stress. We do not know, however, what level of resources triggers the decision not to share, nor is it clear who has the responsibility for deciding when to cease such sharing of resources.

As food shortages worsen, people may expand the socioeconomic networks on which they rely for aid, turning first to kin and trade partners in localities less seasonally affected, then to tribal or regional sources of supply. Migratory labor, particularly of males, has been noted as a strategy for developing contacts for employment, loans, and residence on a longer-term basis. It may be that such contacts are also seasonally important.

Finally, people rely on seasonal grain from state grain boards or relief efforts implemented nationally or internationally. As noted by Colson (1979), among the Tonga "coping methods are . . . now strongly affected by the existence of a relief system that removes some of the responsibility for long-term planning to ensure that they survive when their crops fail." Likewise, Scudder (1962), among others, has observed that such broader-based relief efforts have made it possible for people to survive in areas where seasonal risk was formerly too great. It may be questioned, however, whether reliance on government or more remote aid agencies for food will ultimately prove more adaptive than reliance on the weather; food aid may have its seasons also. Probably the most vulnerable societies are those in transition, whose traditional mechanisms for coping are gone and new mechanisms not yet established.

Household Consumption Behavior

Another way in which people deal seasonally or over the years with insufficient food or income is to eat less preferred staples. During periods of dearth, they rely more extensively on gathered over home-produced or purchased foods and more generally choose products that are cheaper, bulkier, and sustain a feeling of fullness longer. People may also change their manner of preparing grain and shift from pounding to grinding on a stone quern, which produces a coarser, bulkier, more filling product (Colson 1979). Examples are the shift from maize to sorghum in Africa and parts of Central America; from more expensive and tastier rice varieties to cheaper and less desirable rice in Asia; and from rice and other grains to manioc in many parts of the world.

As supplies dwindle, people also reduce the number of meals per day, the quantity of food per meal, and the types of foods consumed at each meal. In areas isolated from marketed sources of foods, rationing may begin shortly after the harvest in anticipation of shortfalls—a practice known in all parts of the world. It also appears that there is considerable interhousehold variation in the effectiveness of rationing limited resources (Sharman 1970).

Adjustment of Household Size and Composition

Land, exchanges, labor, and the social networks used by households to get by economically over seasons of uncertain resource availability may ultimately prove insufficient to feed all household members. In such cases the household itself may change. Size of the production and consumption units may expand to accommodate more labor and consumers or shrink to allow the household to survive on more limited resources. Adjustment of household size and composition may be seasonal or more permanent. One strategy is the fostering or sending out of children, which both spreads child labor around and secures children's nutrition. The practice of sending children to be raised by kinfolk is widespread in Africa. To the biological parents it is a way of reducing the number to be fed, and in some instances of responding to a crisis such as the death of one parent or the fracture of a marriage. To the recipient foster parent, the child represents an investment in future food security, particularly if the adult is childless (Goody 1982). Child fosterage is reported to be widespread in the Caribbean and Latin America, although the details of child care arrangements are quite variable. For instance, migrant parents may send money back for the keep of their children and thereby contribute income to the foster household.

Overall, eating units, food budget units, child-rearing units, and social networks may provide flexibility and options for meeting food needs as production and consumption units are adapted to current conditions of household food availability (Messer 1983).

Household food demands are also reduced by the seasonal exodus of migrant laborers or even daily wage laborers who consume food on the job, and may also bring home food in addition to remittances. Ultimately, scarcity of food engenders a deterioration in sociability and the social group itself disbands. Settings of food consumption change; occasions for food sharing diminish; and people retreat toward nuclear or individual units. Colson's (1979) description of famine applies to cycles of seasonal hunger: there is a "breakup into small family groups which comb the region" and a "refusal to share food with others." Theft increases. Unproductive members (the old and the very young) are sent away to where there may be more food. In days gone by, people would sell their children as slaves in return for grain, and in some cases children are still pawned for food (Alverson 1978).

Food Habits

What foods people choose year-round and in particular seasons is of course a key determinant of whether they are seasonally food-insecure. Dietary structure, content, and change can be investigated within materialist (ecological and economic) or cognitive (cultural or symbolic) frameworks, or both.

Ecological and Economic Dimensions

People choose foods according to their perceptions of time-energy-money costs, nutrient content, and cultural factors such as taste and prestige value. Such decisions can be evaluated in their own terms, and also contrasted with scientific assessments of the biological availability and of the energy and nutrient content of dietary items (e.g., Jochim 1981; Winterhalder and Smith 1981). Optimal foraging theory and related linear programming techniques are one method for setting up testable hypotheses on how people maximize return (measured in calories, protein, or some combination of elements) to effort (measured in energy, time, or some combination of effort-expenditure) in choosing what to eat and where and how to acquire it (Winterhalder and Smith 1981). Such methods might be used to evaluate optimal seasonal diets, departures from such hypothetically optimal diets under current conditions, and future nutritional improvements or losses under alternative cropping (herding, commercial) systems.

One weakness of such models is that people choose foods, not energy or other nutrients, in their dietary selections. Cultural data suggest that households allocate their time and energy (and money, in the case of market economies) to meet certain taste preferences, such as those for fat or protein, with foods that can garnish an otherwise monotonous diet. Thus optimality models that depict people as choosing nutrients, rather than food, cannot be used without attention to these and other cultural factors (Jochim 1981; Johnson 1980; Johnson and Behrens 1982).

Cultural and Symbolic Dimensions

Beyond ecological and economic availability, traditional knowledge and beliefs about the qualities of foods may also affect eating behavior in particular seasons. Certain foods mark the seasons, and their ceremonial as well as ordinary consumption marks membership in the social group. People also entertain beliefs about the healthfulness or harmfulness of some foods in particular seasons.

HOT-COLD CLASSIFICATIONS. Hot-cold classifications, which refer to an intrinsic quality rather than temperature or spiciness, interrelate the domains of food, health, and social relations in many cultural groups (Messer 1981). Ideally, hot and cold values are present within the human body in approximate balance. Overconsumption of either hot or cold substances, overexertion (overheating of the body), overexposure (to heat or chills), an illness agent classified as giving rise to heat or cold within the body, or a combination of these factors produces an imbalance that is believed to have negative health consequences.

Seasonal beliefs that either hot or cold foods are bad for children or adults may prevent the acceptance of a high-nutrient blend for weaning youngsters (as experienced by India's Project Poshak) or the consumption of high-quality protein, especially in particular seasons. Depending on the range of foods of variable quality available, and on how strictly dietary restrictions are applied, such rules would be especially threatening nutritionally for groups experiencing seasonal illness. They seem to be important in South and Southeast Asia and in Latin America, but unimportant in Africa.

HEALTH FACTORS. Other characteristics that affect cultural and individual food choices are cultural notions of the healthfulness of foods. These include concepts of "safe" or "harmless" foods (foods that do not make one sick) as well as foods that are believed to be positively good for health and well-being ("nutritious," "vitamin-rich," and "tonic" foods). Staples of the diet are often classified according to their "fillingness"— that is, their ability to produce and sustain feelings of satiety (Colson 1979). Such criteria affect food choices seasonally, and also under changing cultural circumstances as people move from home production to market food supplies.

ECONOMIC FACTORS. The costs of acquiring and processing food can be measured in terms of time and cash. Both flavor and these costs have been shown to take precedence over nutritiousness in people's food purchase decisions (DeWalt and Pelto 1977).

People with some cash often spend it in seemingly irrational ways in a period that follows the hungry season. Richards (1939), for example, reported that the Bemba would pay outrageous prices for fish after they had been living on a monotonous vegetarian diet for some time. In other instances, poor choices may be made because of a lack of information on better nutritional values. Still others may have to do with the prestige value of items like Coca-Cola or the negative prestige value of wild greens, the consumption of which, in certain cultural contexts, is considered to be a sign of poverty. Understanding how food decisions are made, particularly as household consumption shifts from home-produced foods to foods purchased for cash (Omawale 1980), is an important step toward evaluating whether seasonal insecurity for particular foods or nutrients increases or decreases under commercialization.

Changes in consumption of staples have been suggested as a measure of poverty (socioeconomic stratification); they also serve as a measure of seasonal shortfalls in favored foods. Similarly, changes in the types and quantities of more or less expensive relishes (sauces) and condiments (spices) that are eaten along with staple foods provide an additional measure of seasonal food management and seasonality in dietary structure.

Dietary Structure

Additional signs of seasonal hunger are changes in the dietary structure. The number of meals prepared and their timing during the day are culturally determined; and the number, timing, structure, and content of meals may vary with seasonal shortages. Rationing to extend limited resources is common in most cultures without unlimited food supplies. Many African horticulturalist societies adjust downward from two meals to one meal a day and rely on less palatable foods as harvest stores are depleted (Fortes and Fortes 1936; Ogbu 1973); similar cases can be cited from Asia (Mencher 1981) and Latin America (Messer 1982). The success of a household in circumscribing seasonal hunger may be closely related to how the household manager—usually the principal female—handles food supplies early in or midway through the agricultural season.

If firewood or other fuel supplies are seasonally short or expensive, and if the food provider-preparer has lengthy work responsibilities, the number of cooked meals may be similarly restricted. Employment and school schedules (seasons) also affect the number and content of meals consumed at home, and thus the nutritional adequacy among certain household members.

Other dimensions of food habits affected by seasonal supplies, which in turn affect seasonal nutrition, include cultural concepts of appropriate meal items, menus, food formats, and food cycles. Meal items are further classified as staples, relishes, or condiments in set proportion to each other. Household managers who can plan substitutions of approximately the same nutritional value as preferred food supplies dwindle are more likely to minimize the hungry period for their households.

In addition to how often one eats, the times of day that meals are served and the people who share food on these occasions are also indicators of seasonal dearth. In most societies, food sharing among relations is the rule that breaks down under conditions of dearth. When food is short, people lie about their food, saying they have nothing to offer; later, they eat alone (Richards 1939). In general, food obligations shrink; they apply to fewer people (Dirks 1978), and individuality prevails over social bonds (Turnbull 1972; Turton 1977). At the opposite extreme, festival sharing during times of plenty may help offset seasonal shortfalls and minimize longer-term consequences, as well as build future obligations for social sharing in less stressful times. We need more and better evidence, however, of the ongoing seasonal significance of such social transactions and their timing and frequency under conditions of plenty and dearth, and under changing social and economic contexts.

Dietary complexity provides a final way to conceptualize or measure seasonal changes in diet and nutrition. Ordinarily, those eating more of different kinds of foods are nutritionally better-off than those eating a monotonous diet, although nutrient intake and balance also depend on the types of food selected. Dietary complexity scores might be used seasonally as a way to screen households within communities known to be food-short.

Anthropological Studies of Time and Household Management

Anthropological and economic studies seem to indicate that among households within the same culture, suffering the same socioeconomic constraints, some fare better than others. Key issues in tracing which households within a culture will be seasonally short of food involve (1) how budgetary matters are managed and (2) how time is managed.[1] In particular, the time allocation of the principal household provisioner, usually the senior woman, appears to be important.

The time of income flows, the allocation of income to food over other goods, and the distribution of food within the household all affect the food and nutritional security of the household and its individual members. Details of how food purchase arrangements are worked out, of the circumstances under which male and female budgets are separated rather than integrated, and of expenditure arrangements, however, are complex and vary by culture in Africa, as in Asia and in Latin America.

Understanding the timing of income flows, particularly those of women, over the annual cycle is critical for understanding the circumstances under which households may suffer seasonal hunger. Any arrangement that decreases a woman's control over land, labor, or income, or that alters the seasonal timing of income flows, is likely to decrease the disposable income a woman has available for food and may increase the seasonal food insecurity of the household. It also follows that women's income must rise in proportion to seasonal food price rises.

Despite these observations, what we do not understand well in such cases are who sets the sexual division of labor, the women's bargaining power within the household for control of income, and the circumstances under which men do contribute to household nutrition and maintenance. We do not know whether such rules are socioculturally fixed or negotiated household by household, individual by individual, context by context.

Understanding time and cash budgets is also important in understanding which communities or households within them suffer seasonal deficits in food and nutrition. Time budgets—studied by time observations, opportunistic observations, spot observations, interviews, or diary methods (Messer forthcoming)—are one way to study seasonal work and eating patterns, to document seasonal food shortages at the household level, and to observe contexts of causation. Within communities, comparisons of the time budgets of those women who succeed in raising healthy, well-nourished children against the odds may indicate patterns of time and resource management distinct from those of less well nourished households, and may indicate whether such management has seasonal components.

---

1. These matters are discussed in greater detail by Guyer (chapter 9) and by Alderman and Sahn (chapter 6), and are thus dealt with only briefly here.

Summary, Conclusions, and Suggestions for Future Research

Seasonal food insecurity at the household level is complex, involving many different types of seasonality that must be interrelated in the resource strategies of individuals, households, and communities. Moreover, strategies to overcome seasonality and achieve food security undergo continual alteration as political, sociocultural, economic, and biological environments change simultaneously. The anthropological literature illuminates this complexity and offers concepts and methods for anticipating where seasonal food insecurity occurs now and might arise in the future. It also offers techniques for examining adaptations at the individual, household, and community levels.

A number of conclusions are therefore highlighted by the anthropological literature:

1. Seasonality is significant for understanding the evolution of diet, both in the past and in the present. Attention to seasonal factors allows one to understand how food flows to and through households, given ecological and sociocultural opportunities and constraints. Seasonality will continue to be an important dimension in evaluating how people construct adequate diets under conditions of variable food availability.
2. As a corollary, understanding seasonal adaptations to potential shortages can help us understand longer-term adaptations to more permanent dearth, and also help us predict which mechanisms of adaptation will be short-lived and which will be sustainable under programs of economic development.
3. The adaptive mechanisms of the past involved diverse and flexible techniques of exploiting ecosystem resources, occupational diversification to earn cash, reliance on food exchange both within and among social groups, and temporary or permanent movement of people from resource-poor to better-off households or areas. Under enough duress, the social fabric will ultimately unravel as these mechanisms are pushed beyond what they can sustain.
4. Intrinsic to short- and longer-term adaptations and the success of individual households in staving off seasonal hunger have been the structure and content of food exchange networks. The food passing through such networks—never recorded in marketplace transactions—has usually been critical for seasonal and longer-term survival. Traditional networks, like possibilities for diversified uses of ecosystem resources and seasonal migration, however, may be breaking down under the impact of socioeconomic development. It remains to be seen whether reliance on cash and seasonal grain from grain

boards and food aid are adequate substitutes. In Africa, it also remains to be seen whether such transfers are good for people in the long run, since food aid is distorting: such aid makes unsustainable environments and situations artificially viable, specifically allowing settlement of places unable to provide an adequate livelihood.
5. Many traditional occupations, such as food processing, crafts, and petty trade, that tided people over hungry seasons in the past seem to be atrophying with the onslaught of commercialization and industrialization. However, new adaptive mechanisms to cope with seasonal food insecurity are arising. These include new crops such as potatoes, which stave off seasonal hunger in Asia; new crafts opportunities, particularly in tourist industries; and increased opportunities to earn money from agriculture and wage labor. The impact of cash income on seasonal hunger depends on the timing of cash flows, how money is allocated for food, and how that food is distributed (see also point 2).
6. Nonfood factors, such as the availability and price of fuel, also affect seasonal food adequacy by affecting the time women allocate to fuel gathering, the family food budget, and the number of meals a woman cooks per day. These factors may also vary seasonally and affect diet.
7. Seasonal illness affects seasonal food security, since illness of principal adult earners during the height of agricultural activities can set back home food production for the year. Illness of adult earners may also cause exceptional hardship to families if they miss peak earning periods or forgo any earnings at all in a season. Seasonal illness of principal women in households may be particularly disruptive, since this interferes with food provision, food preparation, household maintenance, and child care.
8. Overall, greater involvement in the market may end seasonal hunger in areas traditionally isolated from adequate market sources, in that there was insufficient money to attract food. Yet the transition from the "moral economy of the peasant" to a more market-oriented economy may take time, and there is no guarantee that all will benefit adequately from such changes. Ultimately people may be trading seasonal food insecurity for chronic mild to moderate malnutrition.

Agenda for Future Research

To explore the general applicability of these conclusions and to test other hypotheses, we need more and better data. Certain specific studies need to be undertaken:

1. Psychological anthropological studies have suggested that seasonal hunger and food anxieties may be pervasive. They have also suggested that anxieties over seasonal hunger lead to gorging and abstaining behaviors. How widespread are such practices? What is their historical

basis? There is a need for further investigation of eating habits, their adaptive value for times of dearth, and how rapidly perceptions and behaviors change under changing environmental conditions.
2. Sociocultural studies of food habits suggest that it will be important to explore the utility of staple and supplementary food consumption as a measure of seasonal food insecurity and modes of adaptation.
3. Rationing for the purpose of extending limited resources from the post- to the preharvest period is common in most cultures, but the timing of such rationing and the determinants of saving behavior require more careful study.
4. We need much more information on social networks—those that in normal times provide flexibility and options for meeting food needs, and those to which individuals resort in seasons of food scarcity and dearth. Among the conceptual questions to be investigated in each case are the rules and practices for making demands on kinship relations, and altering household membership by out-migration or reorganizing eating and work groups. In cases of extreme food shortages, we also want to understand the circumstances under which hospitality rules break down, the point at which households refuse to accept additional consumers, and what signals an individual's refusal to share resources and tasks with others. Ultimately, what degree of food deprivation triggers out-migration?
5. On the subject of cash income, there is a need for studies detailing whether income from cash crops and activities that replace subsistence cropping adequately compensates for seasonal shortfalls in home production.
6. Further investigation is needed into the possible benefits of home gardens in avoiding seasonal food insecurity.
7. There is a general need to be more attentive to the interrelationships among the numerous seasonally variable factors, notably illness, income, household food supply, and nutrition.

PART IV

Grain Marketing and Price Variability

11 The Nature and Implications for Market Interventions of Seasonal Food Price Variability

DAVID E. SAHN AND CHRISTOPHER DELGADO

The purpose of this chapter is to raise the most important issues concerning the seasonality of food prices, review existing literature on the subject, and present data concerning the nature of seasonal food price variability and its relation to market interventions. The chapter's scope is limited to identifying major causes and consequences of seasonal price variability and identifying policy instruments for either reducing these price movements or mitigating their negative effects, or both. The chapter stops short of judgments about specific options, because adequate analyses of the overall costs and benefits of employing these instruments, even with respect to a specific goal, require in-depth treatment of specific cases beyond what can be attempted here.

Variation in seasonal food prices, both at the farmgate and in the retail markets, is a central factor in determining household food security. This is because food prices cut with a two-edged sword: they determine the flow of income received for food crops by agricultural households, and they affect the ability to purchase food of households that participate in retail markets as consumers.

Seasonal price increases may result in food insecurity for two reasons. The first is that as food prices increase after the harvest season, real wages of net food purchasers decline. If seasonal price patterns are known, intertemporal savings or the search for alternative income sources can help smooth out food consumption. However, in many areas the transaction costs of saving are high and opportunities limited. In those cases, seasonal price increases can limit access to food even if the magnitude of the price rise is predictable. Furthermore, a high internal rate of discount for the poor, in addition to the difficulties of gaining access to credit, discourages savings. This, coupled with the fact that the consumption of the poor is especially price-responsive because of the relatively higher impact of price changes on their real incomes, makes high preharvest market prices an especially serious problem for them. The seasonal price rise has the proportionately largest effect on the social group that is least able to deal with it.

The second reason that seasonal price variability may represent a threat to household food security arises out of the unpredictable pattern and level of the spread of seasonal prices. The potential for intertemporal savings to deal with seasonal increases in the price of important consumer goods, such as food, is determined by the accuracy with which seasonal price changes can be predicted. Therefore, the greater the interyear and interlocational stability of seasonal price increases, the more easily both food producing and consuming households can adjust their net sales and storage strategies to achieve food consumption objectives.

Therefore, after examining the magnitude and predictability of seasonal food price changes in the next section, this chapter will conclude by focusing on policies that reduce the magnitude of price increases relative to the costs of storage, and those that improve the predictability of seasonal price spreads from one year to the next.

The Magnitude and Predictability of Seasonal Food Price Variability

The Magnitude of Seasonal Price Effects

The effects of seasons on food price are of major importance in the third world, as shown by table 11.1. The entries in the table are based on monthly index numbers of prices for specific commodities. The indices express individual monthly prices as percentages of the centered 13-month moving average of the same price series. This standardizes the data for comparisons across countries and across years characterized by a trend inflation factor.

The results suggest a price index range (column 5) that is not obviously inconsistent with theoretical storage costs for most countries and commodities, although the variation across countries and commodities is large. However, a comparison of the highest and lowest annual values for the range in the next two columns indicates that seasonal price spreads are highly variable across years, an impression reinforced by the "high" coefficients of variation of the seasonal price spreads across the years reported in the following column. Furthermore, the variation of average long-term seasonal price spreads across countries and commodities is especially evident if annual spreads are divided by the number of months between annual highs and lows prior to being averaged across years, as shown in the last column.

The descriptive statistics in table 11.1 suggest that significant seasonal price spreads for food are a widespread phenomenon in the third world. The pattern of seasonality of price changes is best conveyed by specific cases, as for northern Nigeria in table 11.2. The data, from Hays

TABLE 11.1 Selected examples of seasonal range of agricultural prices in the third world (percent)

	Commodity	Wholesale/ Retail	No. Years	Mean[a]	Max[b]	Min[c]	CV[d]	Mean/[e] Month
Brazil	Millet	W	7	16.4	23.9	11.4	0.3	5.3
Ivory Coast	Millet	R	12	25.1	79.9	0.9	1.1	12.5
Niger	Millet	R	14	85.5	468.6	13.3	1.3	51.5
Niger	Sorghum	R	17	65.0	188.7	26.3	0.6	22.6
Brazil	Rice	W	7	19.2	48.1	9.0	0.7	6.2
Brazil	Rice	R	7	12.2	19.2	6.6	0.3	2.2
Costa Rica	Rice	R	7	6.0	24.1	1.2	1.3	4.8
Dominican Republic	Rice	R	6	12.9	24.4	4.4	0.6	5.2
Indonesia	Rice	R	8	24.1	40.6	17.5	0.4	4.7
Ivory Coast	Rice	R	11	41.8	89.0	11.2	0.6	15.1
Korea	Rice	W	5	17.6	29.3	3.9	0.5	4.7
Niger	Rice	R	19	36.9	81.4	10.4	0.6	11.7
Philippines	Rice	W	3	9.5	11.7	7.7	0.2	2.9
Philippines	Rice	R	5	16.3	56.9	3.4	1.3	3.6
Sierra Leone	Rice	R	15	46.6	96.0	11.8	0.6	12.2
Tunisia	Rice	R	8	4.8	6.9	0.5	0.6	1.8
Costa Rica	Maize	R	5	14.1	32.5	3.7	0.8	3.1
Dominican Republic	Maize	W	12	53.8	98.7	29.9	0.3	13.1
Indonesia	Maize	R	8	26.9	46.1	15.1	0.4	5.3
Ivory Coast	Maize	R	12	33.0	96.0	4.6	0.8	14.9
Bangladesh	Wheat	W	10	79.7	216.6	21.0	0.7	19.1
Brazil	Wheat flour	R	8	15.8	30.0	7.4	0.5	2.8
Ivory Coast	Wheat flour	R	11	7.3	27.8	0.3	1.1	4.5
Philippines	Wheat flour	W	14	15.5	48.9	0.1	1.0	7.4
Tunisia	Wheat flour	W	8	9.6	17.4	2.5	0.6	0.2
Tunisia	Wheat flour	R	5	8.3	10.7	5.8	0.3	5.9
Brazil	Sugar	R	13	27.1	111.9	4.3	1.3	10.3
Costa Rica	Sugar	R	7	10.1	27.0	—	1.2	6.0
Indonesia	Sugar	R	7	10.5	40.4	1.9	1.1	3.9
Ivory Coast	Sugar	R	11	9.4	34.1	0.1	1.3	7.3
Sudan	Sugar	R	10	14.6	40.4	—	0.9	3.4
Tunisia	Sugar	W	7	9.6	17.4	2.5	0.6	1.4

SOURCE: FAO food price data tape.

NOTE: A range-of-price index is defined as follows:

$$\left(\frac{\text{Max } [S_{ijk}] - \text{Min } [S_{ijk}]}{\text{Min } [S_{ijk}]} \right) \times 100$$

where the seasonal price index, S_{ijk} for commodity k, month i, year j, is a percentage of the corresponding centered 13-month moving average of prices.

[a] Average of the range over j years.
[b] Maxima of the range over j years.
[c] Minima of the range over j years.
[d] Coefficient of variation of the range over j years.
[e] Corresponds to "Mean," except that the mean spread is divided by the number of months between max (S_{ijk}) and min (S_{ijk}).

TABLE 11.2 Market town monthly wholesale prices, average of 15 northern Nigerian markets, 1958–65 to 1969–71 (to nearest percent of corresponding mean price in sample)

	Pearl Millet	White Sorghum	Both Crops
January	−4	−10	−7
February	−2	−9	−6
March	−3	−9	−6
April	0+	−5	−2
May	5	1	3
June	15	13	14
July	17	15	16
August	7	11	9
September	−16	2	−7
October	−12	4	−4
November	−5	−5	−5
December	−2	−8	−5
Adjusted R^2	0.26	0.23	0.20

SOURCE: Original data in the table are from Hays 1985.

NOTE: Entries are from generalized least-squares regressions of prices with means for each year and market removed, run against monthly dummy variables, with the estimated coefficients normalized by the appropriate samplewide means. See Delgado 1986 for the general approach.

(1985), pertain to monthly wholesale prices from Nigerian crop and weather reports for 15 town market centers, for the 1958–65 and 1969–71 periods (the intermediate period is missing because of the civil war). The data cover a broad spectrum of ecological areas spread over the northern half of the country (cereals belt).

The entries in the table, calculated across years and markets, correspond to average percentage deviations of monthly prices in each market from the average price in that market and year for that commodity. The underlying calculations are done within a generalized least-squares regression framework that weights the data in such a way as to allow for different variances of foodgrain prices in different markets, seasons, and years (Delgado 1986). The results show that pure seasonal effects (replicated on average across years) may amount to between one-quarter and one-third of mean food prices (that is, 17 percent + 16 percent for pearl millet). On the other hand, the low goodness-of-fit (R^2) suggests that, at most, constant *seasonal* price effects accounted for one-quarter of the price variation over the 1958–65 and 1969–71 periods. Inflation, intermarket variation in price spreads, and random variation accounted for the rest.

Seasonal Food Price Variability in Theory: Its Relation to Storage and Transaction Costs under Perfect Competition

In the immediate postharvest period, granaries are full and seasonal prices are typically at a low. While the stock of grain reaches its pinnacle after the harvest, it decreases steadily until it reaches a low prior to the next harvest. Prices tend to move in the opposite direction, increasing from one harvest to the next, a movement due in theory to the cost of storage. Savings allow consumers to maintain real purchasing power throughout the year even if consumers are assumed not to have mobility of employment across sectors, although the change in relative prices of different foods from one season to the next suggests seasonal variation in the composition of the food basket to mitigate the effects of higher prices for some foods. If the covariance of seasonal prices among food commodities is positive and employment opportunities are inflexible, intertemporal savings are even more necessary to prevent seasonal declines in calorie intake.

Under the assumptions of perfect competition, the rate of increase in the price of commodities is equal to the costs of storage from one period to the next, plus normal profits of traders. Storage costs include fixed costs (such as wages for permanent workers, rental value of go-downs and other privately financed infrastructure, and taxes) and variable costs (such as interest payments of borrowed capital, weight loss due to drying and infestation, insurance, variable labor costs, and variable material costs such as the depreciation of storage bags).[1] Since the assumption of perfect information almost never holds in practice, a marginal risk-aversion factor also needs to be taken into account. The entrepreneurial and managerial functions of traders are another cost that includes a return for the risk they incur in their stocking activities.

The major issue in this framework is therefore the extent to which observed fluctuations in seasonal prices correspond to the intertemporal price equilibrium predicted by the costs of storage plus a "reasonable" residual represented by normal profits. W. Jones (1972) estimates the minimum storage costs for six months of commodity storage under a variety of assumptions concerning interest rates and losses caused by pests and shrinkage. He does not include a variety of other costs mentioned above such as rent, a premium for risk, and labor costs, on the theory that these are of secondary importance compared with interest charges and storage losses. He finds that if interest rates are 3 percent per month, and 10 percent of the original amount purchased is lost to drying and insects, prices can be expected to rise 33 percent in six months, other things being equal.

---

1. It is reasonable to assume that the marginal costs of storage are approximately constant across the system of storers as a whole. The exception is the situation of "bin-busting" harvests, where additional units are improperly stored, leading to greater losses.

Similarly, if losses are on the order of 30 percent and monthly interest charges are 2 percent, the expected off-season price will be 61 percent higher than only six months previously at harvesttime. When the other storage costs and profits are added, even 100 percent increases in off-season prices are plausible in efficient markets under some conditions.

Casual inspection of table 11.1, however, suggests that both the magnitude of average seasonal price spreads and the range of such spreads across years vary greatly by commodity and country. The observed interyear instability in the pattern of seasonal prices could be accounted for in this framework by large differences across crops, regions, and years in production outcomes, in the context of markets characterized by high transaction costs across locations and years. Alternatively, storage and transaction costs across locations could vary by year and commodity.

Seasonal Food Price Variability and Trader Margins in Practice

Real storage and transactions costs may vary across locations but are presumably much less likely to vary for the same location and commodity across years. Therefore, it seems that the high variability across years in seasonal price spreads is an indication that they are independent of storage and transactions costs.

It is often contended that the dissonance between seasonal price spreads and costs of storage reflects a part of the larger problem of traders exploiting farmers by purchasing agricultural commodities directly at the harvest period when prices are low.[2] Households are purportedly vulnerable to such practices because they are in need of capital or do not have adequate on-farm storage capacity. Concurrently, similar analyses posit that oligopsonist traders drive commodity prices up unpredictably and erratically. Distress sales at low prices and precipitous price increases in the lean season are seen as part of a larger noncompetitive market structure that allows merchants engaged in temporal arbitrage to reap large profits.

A number of studies directly address the relationship between storage costs and seasonal price spreads. Table 11.3 shows annual wholesale price spreads for the major foodgrain in Nigeria and Bangladesh over a number of years. The Bangladesh example shows a relatively restricted range across years for seasonal price spreads of wholesale coarse-grade rice. Furthermore, there is no variation across 25 years in the lowest-price month, and relatively little in the highest-price month. This seems consistent with the view that seasonal price spreads are closely linked to storage costs in the Bangladesh case, a point reinforced by Ahmed and Bernard (1986).

The Nigerian case is based on fewer locations and concerns a rain-fed crop grown under much more variable climatic conditions. Even so, the

2. See, for example, Ellis 1982.

TABLE 11.3 Gaps between highest and lowest monthly wholesale price within years of major foodgrain in Nigeria and Bangladesh (percent)

	Sorghum in Nigeria 1958–71[a] (average of 15 markets, northern half of country)	Rice in Bangladesh 1960–84 (average of 500 markets, entire country)
Annual price spreads		
Mean across years	68	18
Maximum across years	104	21
Minimum across years	56	14
Grain price coefficients of variation[b] of highest and lowest price months		
Maximum month	33	18
Minimum month	67	0

SOURCES: The Nigerian data are from 165 price series for 15 market towns in northern Nigeria from crop and weather reports, published by Hays (1985). The Bangladesh data are from averages across markets reported by Ahmed and Barnard (1986).

NOTE: Despite the fact that northern Nigeria has only one major agricultural season and that Bangladesh has three, the average number of months between the lowest and highest price within a given year in Bangladesh is five, while in Nigeria it is six.

[a] Excludes 1966–68 because of missing data.
[b] Calculated by assigning months a number from 1 to 12 (January = 1, . . . December = 12). Excludes the 13–15 percent of the annual price series for sorghum in northern Nigeria that had (respectively) multiple maxima or minima.

variation across years in the seasonal price spread for sorghum is striking, as is the absolute magnitude of the average price spread. Furthermore, there is relatively high variation across years in the high- and, especially, low-price months. In the Nigerian case, consistency of seasonal price spreads with storage and transaction costs does not appear to be a reasonable hypothesis, in some years at least, a point forcefully made by Hays and McCoy (1984). Using the same Nigerian secondary data, they examined the extent to which reasonable storage costs in the northern Nigeria context corresponded to the seasonal rise in prices. They showed that the behavior of prices from month to month is such that prices do not increase in a smooth fashion from the post- to the preharvest period. In addition, the point is made that market arrivals are staged, often over a period of months; likewise, there is a continual turnover of stocks, since throughout the year there are purchasers in wholesale and retail markets because not all households are self-provisioning. Thus the intrayear timing of purchases and sales determines profits and losses of traders, not the total off-seasonal price rise. In most of the 15 markets Hays and McCoy analyzed, for both sorghum and millet, the average increase in prices over the entire season exceeded the estimated costs of storage. However, traders presumably took losses in some years while making greater-than-average profits in others. Even if grain storage is a profitable activity in most years, it is still possible that it can be highly unprofitable in some years (Southworth, Jones, and Pearson 1979).

Just as the Nigerian data suggest that interannual variation in seasonal price spreads is high in semiarid West Africa, analysis of monthly retail prices in Burkina Faso indicates that the variation of seasonal price spreads is also high across locations, and that this variability across locations varies by year. The four years of monthly data underlying the results in table 11.4 come from eight village markets that cover the three major agroecological areas of the country. Yet the areas are all connected by all-season roads, and none of the villages are more than 400 kilometers apart. Not only are seasonal price increases high (165 percent on average in 1982–83), but the high-priced north had a seasonal spread in 1984–85 of 24 percent, while the more variable south in the same year had a spread of 49 percent. Finally, it is noteworthy that the region with the largest seasonal spread tended to vary the most over the four years surveyed.

In a similar study, Lele (1971) examined the competitiveness of a sample of grain markets in India. The correspondence between the seasonal price increase and the costs of storage plus normal profits was determined. The results indicated a large variability in the gross margin on the investment of traders. In 4 of the 10 years, traders apparently took a loss in terms

TABLE 11.4 Variation in seasonal sorghum price spreads across locations in Burkina Faso, 1981–85 (percent)

	1981–82 (below long-term average)	1982–83 (bad year)	1983–84 (below long-term average)	1984–85 (worst year on record, esp. in north)
Spread between highest and lowest monthly price				
North (arid)	52	152	120	24
Middle (dry savanna)	31	85	95	47
South (subhumid savanna)	75	109	69	49
All three regions	84	165	139	105
Ratio max/min	184	265	239	205

SOURCES: Delgado and Matlon n.d. The three regions cover the major agroecological divisions of the country. The raw price data were collected weekly by the International Crops Research Institute for the Semi-Arid Tropics (ICRISAT). They were aggregated monthly and across markets within regions before calculating the spreads.

NOTE: Based on agricultural years from October to September.

of their variable costs, without even considering the fixed overhead of staying in business. In four years, however, potential profits were extremely high. There was an average 11.7 percent gross rate of return on investment, which would be lower if fixed costs of storage were taken into account. Actual returns to traders would depend on the volume of operations in each year.

Lele's analysis of the Indian data suggests that there was a problem of incorrect anticipation of supplies during many of the 10 years for which data are presented. Undoubtedly, traders would not have operated at a loss in four of those years if their knowledge had been better. Likewise, farmers would presumably have delayed their sales of grains if they had anticipated the steep price rises of 1958-59 and 1964-65. The fact that farmers sold and traders traded at a loss suggests that foresight is a scarcer commodity than hindsight, especially in grain trading.

The findings of Lele's work are further supported by the data on cowpea prices over a 15-year period in Nigeria (Ejiga 1977). Tremendous year-to-year variability in seasonal price increases was noted by Ejiga. He compared these increases with the costs of storage. The number of years in which there was a profit from storage only slightly exceeded the number in which there was a loss. These results are compatible with other studies in Asia and Africa (Mears 1981; Lele 1971; W. Jones 1984; and Eicher and Baker 1982).

The fact that farmers as well as traders confront great uncertainty in terms of seasonal price movements also serves as an explanation for heavy market arrivals after the harvest. While some have used such data as evidence of distress sales, collusion of merchants, and the limited on-farm storage capacity of farmers, the alternative explanation is that farmers' behavior, like traders', is attributable to avoiding the risks of monumental losses that holding grain portends in intermittent years (Southworth, Jones, and Pearson 1979). In countries such as Burkina Faso, private grain traders and public grain marketing account for only 5 percent of annual grain stores in an average year (Houghton 1986). Therefore, the seasonal movement of prices, which depends upon the seasonal availability of market offerings, cannot be understood without understanding the behavior of farmers as grain traders.

Explaining "Erratic" Seasonal Price Variation

Clearly, current sale and purchase behavior is in part determined by expectations regarding future shifts in the supply and demand schedules. Unlike the determination of a spatial equilibrium where farmers, traders, and consumers react to the costs of market activity in one location versus the other, it is less easy to foresee the total size of the present harvest and that of the future crop, or the amount of imports (arrivals at a given moment) and exports, or government intervention through open market operations. Although the economics of intertemporal price equilibrium is analogous to spatial equilibrium, the former is considerably more complex owing to the difficulty of predicting the future.

A theory of intertemporal price formation must take into account uncertainty as to what the future will bring. Such a framework was developed by Working (1958) and expanded upon by Goldman (1974) and Bouis (1983). Its major contribution is that seasonal fluctuations in prices are not solely attributable either to current supply and demand forces bringing about an equilibrium price, to which the storage costs are added over time, or to a single forecast of future conditions. Rather, expectations concerning supply and demand conditions in the future are constantly being revised during the year. Prices fluctuate from month to month in response to changing anticipations of supply and demand forces. Thus seasonal price spreads may differ markedly from year to year as anticipations differ.

There has been a limited amount of empirical research to support the theory of anticipatory price formation. This is largely explained by the dearth of good seasonal price series that stretch over the numerous years necessary to employ appropriate models of seasonal price formation and to draw meaningful inferences from the results.

The study by Bouis (1983) in the Philippines provides some of the best empirical evidence in support of the anticipatory price model and illustrates the effects of trade policy on seasonal price changes. He found that instability in seasonal price changes was attributable to the government's ineffective and erratic management of buffer stocks and imports, and not monopolistic behavior by traders. Government intervention in markets aggravated the problem of instability in seasonal price fluctuations by creating uncertainty about supplies of rice in the future. This made for more drastic and more frequent adjustments in expectations by traders, who had not only to form expectations for yields, but to anticipate the unpredictable level and timing of imports and market operations.[3]

Goldman (1974) tested the anticipatory price hypothesis for an Indonesian case, focusing on the uncertainty created by the highly variable yields of the dry-season rice crop. In contrast to Bouis's findings for the Philippines, Goldman found that production instability was the major determinant of the irregular pattern of seasonal rice prices from one year to the next, rather than uncertainty connected with imports, buffer stock operations, or marketplace inefficiencies.

Empirical work along price anticipation lines is virtually nonexistent for West Africa, because of the lack of sufficiently long time series of prices

---

3. For another similar analysis of seasonal price variability in the Philippines, see Unnevehr 1983.

and other necessary data. Yet the price anticipation model appears to fit the stylized facts. It is hard to explain otherwise why grain prices in some years begin falling in June even though the harvest does not begin until September.

It is precarious to draw conclusions about market efficiency and competitiveness on the basis of one year's data. Similarly, in some regions (such as West Africa), a grain trader can incur severe losses by speculating on seasonal price rises alone. The expected return necessary to induce intertemporal arbitrage would therefore presumably need to be that much larger. This is consistent with large average seasonal price spreads. Finally, the fact that month-to-month price movements are often erratic because of constantly changing anticipations implies that it is the timing of purchases and sales within a given agricultural cycle, rather than the entire seasonal price spread, that determines the profits of traders and of farmers who hold back stocks in anticipation of higher prices in the future. On the other hand, net purchasers of foodgrains are frequently in a difficult position to know when to stock up. The difficulties imposed for both producers and consumers by "erratic" seasonal price variation suggest the need for policy to reduce both the absolute magnitude of seasonal price spreads in some cases and their unpredictability in all cases.

The Scope and Role of Policy in Affecting Seasonal Price Movements

Governments frequently intervene to reduce seasonal variability in food prices. They may seek simply to reduce the level of uncertainty about seasonal price increases, without reducing the average price rise expected on the basis of storage costs and normal profits. Alternatively, they may choose to reduce average seasonal price increases across years. However, the key distinction concerns the way in which the objective chosen is pursued.

This may be through either direct or indirect government intervention. The former involves the government actively participating in markets as a buyer and seller of foodgrains. The indirect approach, on the other hand, recognizes that there are market imperfections and inefficiencies that can be remedied by promoting certain conditions that will have an effect on market performance, and thus prices, without direct involvement in the procurement and sale of commodities.

When markets are competitive, the average increase in off-season prices across a number of years will be commensurate with the real resource costs. Reducing average seasonal price spreads will then have true economic costs. Governments can either pay traders to reduce off-seasonal prices or directly subsidize consumers by entering the grain markets as a buyer during the low-price season and a seller in the time of scarcity. The implicit subsidy borne by the government is a result of reducing variation around the average annual price. Marginal costs of price stabilization will likely increase as one moves toward complete stabilization.[4] The more efficient storers are squeezed out of the market as the seasonal price spread becomes less and less, and it is likely that the cost per ton of smoothing out prices will rise at the margin.

As a result of producer prices being driven up by procurement operations, a marketed supply response can be expected. Given that the elasticity of marketed supply is positive, farmers will sell a larger share of their output than would be the case without a stabilization program. A higher marketed supply response implies a larger amount of grain that will have to be handled by the government. Thus seasonal price stabilization by direct government marketing operations will involve greater budgetary outlays than those anticipated solely on the basis of current levels of marketing, unless the elasticity of marketed surplus is zero, which is very unlikely.

If for any reason the government is not willing or able to maintain the floor (or ceiling) price that it has set, the question then becomes one of which farmers get closed out of the opportunity to receive the government price. It seems probable that small farmers are less likely to have access to grain procurement agencies. Since it is also their ability to store grain that is at issue, seasonal stabilization will likely have negative distributional impacts if rationing does indeed occur.

The budgetary costs of government intervention to smooth out seasonal price increases can be calculated as follows (Timmer 1974):

$$C_i = Q(N_i - S_i)P_h \qquad (11.1)$$

where:

C_i = total government subsidy when government stores for i months,
Q = quantity of grain procured,
N_i = normal storage costs, including interest, depreciation, etc., for i months,
S_i = actual seasonal price increase for i months, and
P_h = harvest-season price.

This formula has been applied to data from the Philippines and Indonesia to determine the costs of government stabilization programs.[5]

---

4. The government may have unlimited access to capital at constant costs. In such a case, the marginal costs of increasing storage operations would be relatively stable.
5. In equation 11.1, it is assumed that the marginal cost of supplying marketing services is constant. This will not be the case, if, for example, the cost of capital used in storage activities increases at the margin.

In the Philippines, Unnevehr (1985) analyzed the actual costs incurred by the government in defending a seasonal price spread. She found that the extent of government intervention was not sufficient to defend the official targets for the seasonal price spread. On the average, it would require nearly twice the size of the actual intervention to maintain the official price spread, and presumably at least twice the cost as that actually incurred.

In a similar study, Siamwalla (1984) has estimated the costs incurred in the relatively successful attempt to smooth out seasonal price fluctuations in Indonesia. The total cost of the storage subsidy over the period April 1979 to March 1980 was only 4.5 billion rupiahs (U.S. $7.2 million). However, procurement in Indonesia has expanded five- or sixfold in the years since 1980 in order to maintain price stability. Nevertheless, stabilization policies have exerted only a relatively small drain on the national budget.

Based on these and other studies, there are four important lessons that underscore market interventions to reduce seasonal price increases. First, policy objectives in market intervention need to be consistent with the resources available to achieve these objectives. Given resource constraints, policies can be designed that meet limited stabilization objectives, with limited cost. The precise policies that should be adopted in this context will depend upon the weight that the government attaches to price stability. Given a specific objective, optional policies using a mix of instruments can be devised to achieve partial stabilization (see Pinckney and Gotsch 1987).

Second, within this context, there are at least four prerequisites to an effective government stabilization policy: (1) market intelligence on prices available to the procurement agency must be accurate, (2) procurement operations must be well timed, (3) confidence in the government's performance must be high in order to establish credibility, and (4) a grain marketing agency must have adequate access to credit in order to carry out its mandate. In the absence of such conditions, government interference in markets will probably increase speculative behavior and exacerbate abrupt seasonal price changes.

A third lesson is that over time there is inevitably growth in the stores that the government must assume if the floor and ceiling prices are to be successfully maintained. This can be attributed to farmers learning how to maximize the subsidy received by selling sooner after harvest. When the government first enters the market, many farmers are accustomed to holding stocks for sale at a later point in the year. They initially are not confident that a flat seasonal price can be maintained. Farmers are reluctant to liquidate their annual holdings, especially when it will require identifying alternative investments. As time passes and credibility is attached to government storage activities, more and more grain will be sold earlier and earlier after the harvest.[6]

The fourth lesson is that a key ingredient in the success of government stabilization activities is targets set in terms of floor and ceiling prices, not quantities. This procedure is in contrast to that followed in many countries, where a fixed amount of money is allocated to import or purchase grain. As noted by Siamwalla (1984), it may prove less expensive to have a procurement system with an open-ended commitment than one where allocations are fixed. In the latter case, speculators can exploit the fact that there will be major movements away from price targets as budgetary resources near depletion.

In sum, a key aspect of direct government intervention in food prices is credibility. There will be justified uncertainty about public storage activities in environments characterized by rapid inflation, instability in annual production (as from regular droughts), severe budgetary and foreign exchange constraints, limited physical infrastructure, and insufficient managerial capacity. Where these conditions obtain, direct government intervention in grain markets to reduce seasonal price increases is probably not advisable.

Using Grain Imports for Direct Price Stabilization

Governments can (and frequently do) use control over grain imports, including food aid, as a means to control aggregate national grain stores and to stabilize prices across seasons. Even without this intent, direct government involvement in food imports implies that public institutions are inextricably linked to the storage function.

Either storage or trade can be employed to reduce the gap between seasonal supply and demand. The use of imports to reduce the seasonality of supply requires an institutional ability to respond rapidly to expected seasonal shortages. For reasons discussed earlier, the latter depend on producer and consumer expectations in addition to production outcomes.

The time of arrival of imports is obviously important to seasonal price stabilization. In Indonesia, the heaviest period of imports is from November through March, in keeping with the lean season before the harvest (Mears 1981). In Bangladesh, imports are planned for arrival in July through August, when prices are at their peak. In the Philippines, 68 percent of the imports arrived between June and October in seven of the nine years for which data are available (Bouis 1983). In Gambia, the months of June through September are high-price periods, because of the relative

---

6. This scenario was given as an explanation for the experience in Korea, where the percentage of rice production purchased by the government has steadily increased since seasonal stabilization policies were initiated. See Tolley, Thomas, and Wong 1982.

lack of availability of sorghum and millet prior to the fall harvest. The timing of rice imports generally corresponds to this lean season, helping to stabilize millet and sorghum prices (CRED 1977).

Nevertheless, the timing of domestic scarcities is not the only consideration in using imports for price stabilization. Rice prices on international markets also vary seasonally. The annual low price for exports from Bangkok, for example, occurs at the beginning of the year, following the harvest of the major rice exporters—Thailand, Burma, the Philippines, and South Korea (Mears 1981). Rice purchased at this time would arrive in Indonesia after the preharvest shortages. Thus the optimal timing of imports also depends on expectations concerning seasonal price changes in supplying countries.

A further issue is that if expectations for imports during certain periods of the year are created and then do not materialize, it could have graver consequences for seasonal price stability than no imports at all. This was demonstrated in the Philippines with regard to the anticipatory price model discussed above (Bouis 1983). Specifically, if imports arrive too early, the government must either pay the cost of storage or release them onto the market. The former has a cost to the government, and the latter will make it more difficult for traders to formulate reasonable expectations and will cause an erratic pattern of seasonal prices. Similarly, late market arrivals of imports will initiate precipitous seasonal price increases, at least until the supply is augmented by the distribution of the imported goods.

Indirect Government Intervention to Reduce Seasonal Price Increases

Indirect measures are designed to promote competitive markets and reduce the costs of storage by lowering the risks, constraints, and transaction costs faced by farmers and traders engaged in intertemporal arbitrage. This approach is juxtaposed with that of direct government intervention in procurement and sales. To quote W. Jones (1984, p. 133):

> There is a great difference between price stabilization that consists of buying and selling at fixed prices over a period of weeks, months, or years and price stabilization that attempts to reduce lags and restrain over-reaction to changing market conditions. The first transfers the risk of price change from farmers and merchants to government at the risk of progressive distortion of supplies from requirements. The second increases the reliability of prices as indicators of the relationship between supply and demand, and in this way it enhances their adjustment. The first destabilizes, the second equilibrates.

Indirect interventions to reduce the extent and instability of seasonal price increases may be divided into two categories: the first is concerned with increasing the competitiveness of markets and marketing opportunities for farmers; the second involves developing infrastructure and providing subsidies to reduce storage costs.

There are a variety of reasons why increasing the efficiency of markets will reduce the costs of intertemporal arbitrage. First, any storage operation requires access to credit. Competition among traders is reduced if access to credit is impeded. Poorly functioning capital markets increase the major cost to seasonal arbitrage beyond the opportunity cost of the credit itself, regardless of whether there is competition among traders. Conversely, government efforts to lower the costs of capital will facilitate lower storage costs, although whether or not this is desirable policy depends on the opportunity cost of the subsidy as well as its potential benefits.

A second indirect area for government intervention involves addressing infrastructural constraints that hamper the ability of the farmer to respond to market signals. This includes promoting and improving availability of on-farm or cooperative storage facilities, information facilities, and removing transportation bottlenecks that dissuade the farmer from bringing crops to wholesale markets where the best price can be received. Improving communications and reliable information about the forces that affect prices will go a long way toward reducing risk and speculative gains and losses.

Conclusions

Nine salient conclusions follow from the analysis above. First, seasonal food price effects are frequently large in the third world. Second, although storage and transactions costs might explain the average magnitude of seasonal spreads, they cannot explain the high variation of the price rise across years. Third, it is probable that traders who store do very well in some years and lose in others, although it is reasonable to assume that while professional traders presumably store in all years, they would not do so if they could foresee losses. This implies that there is a high element of risk to engaging in temporal arbitrage. Fourth, expectations of future, more than present, grain supply and demand conditions have major impact on storage behavior, and thus on seasonal price variation. Incorrect expectations lead to supply and demand imbalances in the future, which contribute to the erratic pattern of seasonal price increases.

Fifth, government intervention to eliminate the spread of seasonal prices will have large net costs in terms of limited budgetary and manpower resources; however, governments can intervene to reduce seasonal price increases in accordance with the resource constraints they face, employing a mix of policy instruments.

Sixth, the likely effect of direct marketing intervention by governments to reduce seasonal price variability is to increase the proportion of domestic grain production marketed by the public sector. This is because price stabilization within years will reduce the quantity of grain stored by the private sector, in addition to encouraging farmers to sell (to the government) earlier so as to avoid storage costs and to reap the higher price that prevails at harvesttime. Furthermore, the higher prices on average to farmers at the time they usually sell (even if the annual average is not higher) are likely to elicit a supply response on their part.

Seventh, the optimal timing of direct intervention, given that a decision to intervene has been made, is both critical and a skill-intensive function. It depends not only on domestic seasonality, but also on seasonality in the principal world suppliers of grain. Eighth, direct intervention in procurement of grain by governments has worked well only where the market information available to the intervening agency is very good and where the seemingly unlimited credit available to the agency helps give it credibility in its promised (or threatened) interventions.

Finally, indirect intervention for seasonal price stabilization seems preferable. Specific priority interventions are location-specific, but better transportation and communication infrastructures are likely to be central. Most important, however, is promoting increased efficiency in the functioning of capital markets so that there is improved access to capital, at lower prices.

12 Seasonality in Burkina Faso Grain Marketing: Farmer Strategies and Government Policy

LYNN ELLSWORTH AND KENNETH SHAPIRO

In the early years of the Sahelian drought, Burkina Faso (then Upper Volta) and Niger were encouraged by donor agencies to establish national cereal marketing organizations with broad and ambitious mandates to help both producers and consumers. Burkina Faso created the Office National des Cereales (OFNACER) in January 1971. The objectives and authority of OFNACER have varied over the past 15 years, but generally have included (1) stabilizing grain prices within and across years; (2) offering producers a high enough price to stimulate increased production; (3) offering consumers a "fair" price; (4) decreasing exploitation by traders, who were thought to be a major cause of price fluctuations, of low producer prices, and of high consumer prices; (5) distributing food aid; (6) maintaining a security stock; and (7) supplying deficit areas.

With the benefit of hindsight, many observers (including some of the aforementioned donors) believe that national marketing organizations are probably unable to be important, efficient actors in African grain economies. This is especially so when the organizations have very broad mandates with some conflicting objectives, as is the case of OFNACER. In addition, several research projects have cast doubt upon the assumption that African traders are seriously exploiting consumers and producers. Thus there is the general question of what role government should play in domestic staple food marketing, but more specifically of how governments can improve food security by dealing with seasonality in their domestic grain economy.

The objective of this chapter is to help answer that question for Burkina Faso by examining the marketing behavior of farmers surveyed in

---

The survey described in this chapter was undertaken by the University of Wisconsin under subcontract to the University of Michigan contract no. 686-0243-C-00-2063-00 with the U.S. Agency for International Development. This contract funded the research component of the Burkina Faso Grain Marketing Development Project.

the western part of that country in 1984. The first section presents background information on the sample villages and their grain economies. The second examines the sample's seasonal pattern of grain sales and purchases and concludes that the situation is not so extreme as generally thought. Nonmarket transfers are analyzed next and are shown to be quite important compared with grain market transfers. The fourth section disaggregates the sample according to the different marketing patterns that farmers are following and shows that most farmers avoid the potential losses that might result from sharp seasonal price swings. The final section discusses policy options.

The Survey Villages and Their Grain Economies

Two hundred and twenty farm households were sampled in five villages across three ecological zones. Between 42 and 51 farm households are in each village sample. The two northernmost villages studied, Mené and Bougouré, are in the Yatenga region in the northern reaches of the Sudanian savanna zone, bordering the drier Sahelian region farther north. Annual rainfall in this area is between 500 and 800 millimeters (mm), the lowest and least reliable of the areas studied. The rainy season is typically shorter than in the other villages. Yatenga has recently been a grain importer, and prices here are typically higher than in the central or southern areas of Burkina Faso, although they exhibit a similar seasonal pattern.

Two villages, Tissi and Dankui, are in the west central Volta Noire region, an area that often exports grain and has more fertile soil. While it is also in the Sudanian savanna, the area receives slightly more rainfall than the Yatenga villages to the northeast. In recent drought years, such as the survey year, this area has become less of a surplus exporting region.

The southernmost village, Baré, is in Hauts Bassins in the northern Guinean ecological zone, which is wetter and more humid than the Sudanian savanna where the other villages lie. Baré has often been a grain exporter.

Almost all households in all five villages purchased grain. In contrast, there were significant numbers of grain sellers in only two of the villages. For the sample as a whole, about 15 percent of the grain harvest was sold,[1] but the southern surplus village of Baré dominates these statistics. The Baré sample sold 22 percent of its harvest, while none of the other four village samples sold more than 10 percent. A contemporaneous survey by Purdue University and the Semi-Arid Food Grain Research and Development (SAFGRAD) program in relatively favorable environments found

---

1. Millet and sorghum are the most important grains, followed by maize. There is also a little rice.

sales equaling 15–25 percent of the harvest in four villages in 1984 (Pardy 1987). These data are presented in table 12.1.

In the five surplus villages (Baré plus the Purdue/SAFGRAD villages), the average selling household sold 480 kilograms (kg) of grain. In the four deficit villages, the average was only 95 kg.

Analyses by Szarleta (1987) and Pardy (1987) indicate that marketed surplus in these nine villages is not significantly price-responsive. (Their analysis was limited to seasonal and spatial price differentials within one year and is thus not a traditional test of supply responsiveness. However, similar approaches by K. Bardhan [1970] have shown significant responses to price.) The overwhelmingly most important determinant of sales was harvest. This of course has implications for the importance of price policy versus research and extension for increasing marketed surplus, an issue to which we return in the analysis of seasonality.

Seasonality in Grain Prices, Sales, and Purchases

In the survey area, the rainy season typically lasts between 134 days (in the most southerly village of Baré) and 100 days (in the most northerly villages of Bougouré and Mené). "Normal" annual rainfall in Baré is 1,100–1,200 mm and in Mené 500–800 mm. The rains begin in May or June and end in September. The main grain harvests begin in October or early November, with some maize harvested in September. Grain prices in markets serving the sample villages plummet at harvesttime and then in-

TABLE 12.1 Grain sales and harvest in nine Burkina Faso villages, 1983–84

	Households in Sample	Number of households that Sold Grain	Total Grain Sales (kg)	Total Grain Harvests by All Sample Households (kg)	Sales/ Harvest (percent)
Mené	46	7	1,175	37,319	3.1
Bougouré	42	6	75	8,017	0.9
Baré	50	47	30,001	136,003	22.1
Tissi	40	24	2,043	21,247	9.6
Dankui	42	3	506	18,620	2.7
Dissankuy	27	13	4,859	32,809	14.8
Nedogo	75	28	8,033	38,873	20.7
Poédogo		18	8,799	35,355	24.9
Diapangou		22	9,765	45,800	21.3
Total	322	168	65,256	374,043	17.4

SOURCE: Data for Dissankuy, Nedogo, Poédogo, and Diapangou are from Pardy 1987, and the remainder are from a survey performed by the University of Wisconsin.

TABLE 12.2 Grain sales and purchases by quarter in five Burkina Faso villages, 1984 (kilograms)

	Jan.-Mar.	Apr.-June	July-Oct.	Oct.-Dec.
Sales[a]	9,520	4,885	6,347	7,811
Purchases[b]	25,366	35,846	57,158	12,085

NOTE: Data are for Mené, Bougouré, Baré, Tissi, Dankui.
[a]Sorghum, millet, maize.
[b]Sorghum, millet, maize, rice, food aid, miscellaneous foods.

crease throughout the next 12 months, reaching a peak in October. Intrayear seasonal increases of 50 percent in consumer and producer prices are not uncommon.[2]

Perhaps the most common characterization of staple food marketing in Africa is that farmers sell in the postharvest, low-price season and buy in the preharvest, high-price season. Explanations for this supposed pattern include outdated notions of peasant irrationality, sales tied to paying off loans, pent-up demand for cash purchases, taxes, fear of having to share surplus grain as the hungry season approaches, and seasonal variations in access to buyers and sellers. In the five villages studied under this project, the conventional wisdom is borne out for purchases more strongly than for sales, but there are large amounts sold and bought in all seasons.

Grain prices start falling rapidly in October in the markets serving the surplus villages. Thus the postharvest "quarter" is defined as October 11 to December 28. The remaining quarters are January-March, April-June, and July-October 10. The seasonal transaction pattern for the total sample is shown in table 12.2.

The sales pattern appears more evenly distributed than is typically thought. The largest grain sales do not occur immediately after the harvest, but rather in the following quarter. No quarter has less than 17 percent of sales and none more than 33 percent. The purchase pattern follows expectations more closely, with most grain purchased in the preharvest period. The range is from a high of 43 percent preharvest to only 9 percent postharvest.

This picture of seasonal transactions is modified if one looks at individual household behavior. Table 12.3 shows the number of households that made the largest amount of their sales or purchases in the quarter

---

2. The cost of storage is unlikely to be this high. Common estimates of postharvest crop waste in the Sahel are less than 10 percent. In the villages, there seemed to be adequate physical storage capacity on most farms. Indeed, following several years of poor harvest, there is likely to be a loss of spare capacity. The opportunity cost of holding grain rather than converting it to money should probably be evaluated well below returns on capital invested in, for example, trading, since most farmers do not invest the proceeds from grain sales.

TABLE 12.3 Number of households with their largest volume of sales and purchases in each quarter in five Burkina Faso villages

	Jan.–Mar.	Apr.–June	July–Oct.	Oct.–Dec.
Sales[a]	21	5	8	35
Purchases[b]	45	61	64	7

NOTE: Data are for households selling more than 25 kg and buying more than 50 kg.
[a]Sorghum, millet, maize.
[b]Sorghum, millet, maize, rice, food aid, miscellaneous foods.

shown. Examination of individual household selling patterns reveals that the postharvest quarter is indeed the heaviest sales period for the largest number of households, and that the two postharvest quarters are the heaviest sales period for 81 percent of farmers selling more than 25 kg in the year. Purchase data at the household level show that many households do buy heavily before the high-price, preharvest season, but the preharvest season does show the greatest number.

In sum, the seasonal transaction patterns observed in the five sample villages provide some support for the conventional wisdom that African farmers sell low and buy high. However, the pattern is not extreme, and there is evidence that some farmers contradict it. For example, 22 of the 68 households that sold more than 25 kg made less than 15 percent of their sales in the postharvest quarter. That is, almost one-third of the sellers did not sell very much at the worst time with respect to price. This subset of households is investigated further below.

Disaggregating by village provides further insight into seasonal patterns. The sharpest differences are between Baré, a relatively wealthy, surplus village, and the others, which are poorer and either chronically or occasionally deficit in cereals. Nearly half of Baré households made their heaviest purchases between January and March rather than waiting until later in the season when prices were higher. In the other four villages, the following two quarters of progressively higher prices (i.e., April–June and July–October) saw an increasing percentage of households making their heaviest purchases. Comparative analysis of sales is difficult, because two-thirds of those who sold over 25 kg were in Baré. Comparing Baré with Tissi, the village with the next most sellers, reveals that proportionately more Baré farmers avoid making their heaviest sales in the postharvest quarter. Thus the relatively well-off, surplus village seems to suffer least from the strong seasonality in grain prices. The "sell low, buy high" characterization fits Baré less well than the other poorer, deficit-prone villages.

Marketing Compared with Other Grain and Cash Transfers

Income derived from grain sales by the sample was small compared with total cash receipts. Grain and other crop (e.g., cotton and peanuts) sales together amounted to only one-third of other cash receipts. In addition, sales (kg) in the four deficit villages were smaller than the amount of grain those villages exchanged in nonmarket transfers. In contrast, grain purchases in those villages far exceeded grain received from nonmarket sources. These data are summarized in table 12.4.

The seasonality of other transfers has elements similar to the seasonality of grain marketing. As shown in table 12.5, nonmarket grain transfers peak in the postharvest season, with the preharvest season being next in importance. This is consistent with Saul's (1987) contention that payments for agricultural work and "obligatory gifts" result in high transfers around the harvest, and that these are subsequently sold and swell the postharvest sales glut.

This point is reinforced by examining the reasons for nonmarket grain transfers in different seasons. For example, farmers most often cited the offering of a gift as the reason for transfers during the postharvest quarter. Emergency help is cited only 9 percent of the time as the reason for grain transfers in the preharvest quarter, although this frequency is even lower during other periods of the year. In the second quarter after the harvest (January–March), funerals are cited as the main reason for nonmarket transfers. In the following quarter, relatively little cereal is transferred and various reasons are given (e.g., "maintain relations," "miscellaneous payments," "other"). In sum, there is a substantial interhousehold nonmarket movement of grains, with the most commonly stated reasons varying through the year. In the preharvest period, the majority of grain transfers are for payment of work parties; for gifts in postharvest; funerals in January–March; and a variety of reasons in April–June, when the volume transferred is lowest.

TABLE 12.4 Market and nonmarket transfers of grain in four Burkina Faso villages (kilograms)

	Sales	Purchases	Nonmarket Transfers In	Nonmarket Transfers Out
Mené	1,175	30,340	1,817	476
Bougouré	75	18,420	1,894	596
Tissi	2,043	24,163	8,847	14,593
Dankui	506	15,576	591	1,913
Total	3,799	88,499	13,149	17,578

SOURCE: Adapted from Szarleta 1987.

TABLE 12.5 Nonmarket transfers of grain and money by quarter in five Burkina Faso villages

	Jan.-Mar.	Apr.-June	July-Oct.	Oct.-Dec.
Grain received (kg)	2,192	1,656	4,153	7,590
Grain given (kg)	5,601	801	5,867	11,458
Money received (1,000 francs CFA)	1,484	3,691	1,399	899
Money given (1,000 francs CFA)	1,228	2,294	1,187	570

NOTES: Data are for Mené, Tissi, Dankui, and Bougouré, with minimal data for Baré. Amounts given do not match amounts received because of inflows and outflows to and from the sample.

With regard to money transfers, perhaps the most interesting fact is their concentration between April and June. Some of this may represent gifts aimed at helping families buy cereal as the hungry season approaches. The data have not yet been analyzed to the point of allowing us to see the role of remittances from family members away from home. The reasons for money given in transfer do not show very clear seasonality.

Household Variations in Marketing Patterns

Some households appear to suffer from the "sell low, buy high" syndrome, while others avoid the worst effects of seasonality and still others seem to take advantage of seasonal price movements. The government of Burkina Faso would be able to target its interventions more efficiently if it could identify different marketing patterns and the types of households pursuing each one. This section presents one effort to do this.

Fifteen households had a sales pattern that took advantage of the seasonal price increase. They made more than 60 percent of their sales in the six months preceding the harvest, that is, during the higher-price periods. This group is wealthier and perhaps more market-oriented than the rest of the sample. While accounting for only about 7 percent of the 220 households in the sample, they own 25 percent of the samples' cattle and 10 percent of the value of durable goods, and they account for 38 percent of total sales by the sample. The average household in this group sold 944 kg of grain, compared with a sample average of 155 kg (among those who actually sold any grain, the average amount sold was 389 kg).

In contrast to this group are 27 households that seem to be taking losses because of the seasonal price movements. This group includes those that buy more than 60 percent of their year's purchases in the six months prior to harvest or make more than 60 percent of their yearly sales in the six months after the harvest.³ These households are extremely poor in cattle but are slightly above average in value of durable goods. They have a relatively small area under cultivation (6.33 hectares [ha] versus the sample mean of 7.62 ha) even though their family size is larger than average (7.89 consumer equivalents versus 6.86 consumer equivalents for the sample). Also, they have more of their land in cash crops (cotton and peanuts) than average (13 percent versus 10 percent). The average amount sold and purchased by these households is close to the sample mean.

There are two types of marketing patterns that seem to imply neither significant gains nor losses from seasonal price movements. A group of 25 households is fairly autarkic, making no purchases and selling less than 100 kg during the year. They cultivate more land (8.07 ha) than the sample average and have less of it in cash crops (7 percent versus 10 percent for the sample). These households have considerably fewer cattle than average (0.38 versus 1.81 head per household) but about the same value of durable goods. Another group, constituting slightly more than half the sample (112 households), managed to avoid taking losses caused by seasonal price lows or highs. They made either less than 15 percent of their total sales in the three months after the harvest, or less than 15 percent of their total purchases in the three months prior to the harvest. That is, they avoided the worst-price periods for their sales or purchases. However, they were not particularly adroit in concentrating their sales in the high-price period. In most respects, this group was similar to the sample mean, which is not surprising, since they are the majority of the sample.⁴

In terms of household size, the largest numbers of consumer equivalent units (CEUs) were found in the group that took advantage of price swings (7.89 CEUs) and the group that suffered most from them (7.29 CEUs). The other groups comprised between 6.4 and 6.8 CEUs.

Perhaps the major finding here is that 137 of the 220 households either do not interact significantly with the grain market or manage to avoid taking major losses from seasonal price movements. A very small group seems to benefit from those movements, and about 12 percent seem to be suffering from them. This latter group may be the appropriate target for government action related to seasonality.

Conclusions and Policy Implications

Government intervention in grain marketing may be motivated by four circumstances related to seasonality. These have to do with exploita-

---

3. The latter criterion also requires at least 100 kg of sales in the year.
4. Another group of households that closely resemble the sample average is best characterized by a pattern of buying heavily during the low-price season.

tion, inefficiency, welfare, and incentives. Our survey did not gather data to assess the extent to which traders may be exploiting producers and consumers by exaggerating seasonal price swings, nor do we have data to determine whether government efforts to overcome market inefficiencies would cost more or less than resultant benefits in the form of decreased costs of holding grain from one season to the next. Examination of farmers' marketing patterns and other transfers does offer information concerning the welfare and incentive issues.

Aggregate data show some concentration of sales in the low-price, postharvest periods and a concentration of purchases in the high-price, preharvest periods. However, the pattern is not extreme. Disaggregation by types of marketing patterns reveals that most farmers sampled were probably not significantly affected either positively or negatively by seasonal price movements in 1984. A small group of poor farmers seems to have suffered and could be the target of interventions motivated by welfare considerations.

Incentives arguments find little support in our analysis. The group of farmers that was able to take advantage of seasonal price increases was the group with the largest sales per household. Government efforts to increase prices in the postharvest period are unlikely to provide meaningful incentives to these large sellers, since they evidently have the ability to hold most of their grain until prices rise later in the year. For the sample as a whole, cash receipts from crop sales were small compared with other cash receipts, and hence grain marketing may not be a key economic activity that is likely to respond strongly to price incentives. In addition, the supply response analyses of Szarleta and Pardy do not show price to be a dominant factor. Finally, official producer grain prices in Burkina Faso have already been rising rapidly, both absolutely and relative to other crops. For example, from 1971-72 to 1984-85, the official producer price of millet rose from 12 francs CFA/kg to 80 francs CFA/kg, while the official cotton price rose from 32 francs CFA/kg to 80 francs CFA/kg (Moussie 1985). Thus there may not be much scope left for government to try to influence grain prices.

In conclusion, our research does not offer strong support for government action to affect seasonal aspects of the Burkina Faso grain economy. Many writers consider price swings a potential detriment to farmers, who may be forced to sell low or buy high. However, most farmers we studied avoided such losses. On the positive side, the largest grain sellers are able to take advantage of high preharvest prices. Finally, in many cases nonmarket grain transfers are far more important than market purchases and sales, and grain sales are a relatively minor source of income.

If the study reveals any need for government intervention, it may be on welfare grounds. There is a small group of farmers who do suffer from seasonal price swings. The wealth profile of this group is somewhat ambiguous. They are relatively poor in land and cattle, but above average in durable goods and in percentage of land in cash crops. If these farmers can indeed be categorized as "poor," then welfare interventions may be appropriate.

PART V

The Role of Technology and Policy

13 Indigenous versus Introduced Solutions to Food Stress in Africa

JON R. MORIS

The existence of marked seasonality in food production within the African tropics has long been recognized, and is evidenced by the prevalence of indigenous terms for the hunger period when food is insufficient to meet daily needs. It is also readily apparent from gross rainfall statistics, which show that a four-month dry season is typical over much of the continent and five or six months are not uncommon.[1] However, the purpose of this chapter is not to reiterate the evidence of extreme fluctuations in food availability at the local level, but rather to address how indigenous and introduced technologies have been employed by households to cope with this typical circumstance. While drought marked by a failure of an entire season has received considerable attention in the semiarid areas of Africa, the more routine hunger period has not.[2] How are typical fluctuations in food supply buffered by local economic and social systems so that producers survive in good health from one growing period to the next?

In examining the mechanisms for surviving extended periods of food deficit, we should note that the degree of stress experienced will be a function of several interrelated aspects:

1. The adequacy of food supplies at the start of each dry season, reflecting interyear instability in production during the usual growing period.
2. Available technologies for food storage at the local level, in combination with the storability and perishability of the crops farmers choose to grow.

---

1. For reviews of seasonality in African rainfall, see Walsh 1981, Katz and Glantz 1977 on the Sahel, and Tyson 1979 on southern Africa.
2. In the huge literature on African drought, sources relevant to this chapter include van Apeldoorn 1981, Patton 1971, Hitchcock 1979, Akong'a 1982, Wisner and Mbithi 1974, Wisner 1977, O'Leary 1980, 1984, Campbell 1978, Franke and Chasin 1980, J. Shepherd 1975, and the Botswana Society 1979.

3. The strategies at household and community levels for ensuring food security.
4. The length of time between harvest and when fresh food supplies become available—depending on the mix of enterprises as well as on climate.
5. The energy required for production at times of food scarcity, which is also a function of the mix and scale of production activities (an aspect explored by Bayliss-Smith 1981).
6. The positive and negative influences of trade and participation within a larger economy in regard to food security.

The discussion that follows draws mainly upon areas of East and southern Africa, but evidence will also be adduced from the Sahel (where seasonality is even more pronounced). The point of departure is the hunger period, first brought to academic attention in Audrey Richards's (1939) classic study of the Bemba, which retains its relevance today. Under agricultural, mixed, or purely pastoral subsistence strategies, the various traditional and modern mechanisms aimed at ensuring adequate food supplies during the prolonged time of food deficit have had limited success in reducing seasonal food insecurity.

This is so for basically two reasons. First, some of the earlier adjustments (reviewed below) are no longer available to all households. And second, imperfections in the national marketing system or other emergent economic constraints deny poor consumers or those in remote areas the possibility of relying on modern sector alternatives—a deteriorating supply situation made worse by the adverse economic trends African countries are experiencing. On the downside, the inability of poorer households to counteract food deficits at this time explains why they are often forced into highly inequitable short-run "solutions" that deepen their poverty—a point stressed by Chambers, Longhurst, and Pacey (1981). However, the deficiency in institutional arrangements also influences the up side; where risks are high and commercial food supplies insufficient, even progressive farmers may be unable to make a permanent transition into larger-scale, commercial production.[3]

Indigenous Solutions to Seasonality

A household might adopt a variety of strategies to counteract or ameliorate the hunger period. We can usefully distinguish six production options (some of which are subcategories of the others): (1) diversification;

---

3. See, for example, the situation in Zambia documented by van Donge (1982).

(2) root crops; (3) exploitation of vertisols; (4) livestock enterprises; (5) bush collecting; and (6) off-farm income.

Diversification

In traditional African farming, households often grew a wide array of crops, sometimes spread over two or more ecological zones. Obviously, crops differ in their moisture and soil nutrient requirements. By growing a range of crops with different response characteristics, farmers ensured some yield under a broad spectrum of seasonal conditions. Diversification could also take the form of staggered planting, so that portions of each crop were planted at dates that bracketed the onset of the growing season. Plots might be located along the catenary sequence, allowing a better match of crops to soil type and increasing the chances that a disaster at one site would not destroy the entire season's output. Farmers often adopted the same strategy even within a single field, practicing intercropping, relay cropping, crop succession, and sometimes the planting of different crops around the field border (Keswani and Ndunguru 1982).

These tactics, which have been documented by many observers in different parts of Africa, are usually explained as risk-minimizing devices. By the same token, they directly influence the household's food security through diversifying the sources of food and extending the period when it is available. They are evidence of a deliberate choice by farmers not to optimize production of the highest-yielding crop. Instead, they safeguard a minimum food supply and ensure that the labor force has access to some sources of energy at the critical food gap period.

For example, among the Meru of Kenya, agriculture was spread over eight ecological zones. Bernard (1972) identifies three crop complexes suited to the homestead, intermediate, and lowland areas, each made up of 11 or more traditional crops. In total, the Meru grew 16 food crops and several nonfood ones as well. The full array is seldom captured in farm-economic surveys aimed at computing farm budgets and annual income. Furthermore, Bernard's listing glosses over varietal distinctions, which were (and are) especially important for some root crops and legumes. A nutritional survey of the Shambaa in northern Tanzania found that people had 10 varieties of bananas, some used in cooking and others eaten fresh (Schlage 1969). Most homesteads in East Africa grow several types of beans and various spices, fruits, and semiwild vegetables as well.

One strongly suspects that this diversity is adjusted to increase food availability throughout the year. Lagemann's (1977) study of farming in eastern Nigeria graphs the main harvest and storage periods for some 13 staple and leaf vegetable crops. In most months at least four of these are in harvest, and there is only one month with only a single main crop (yams) ready for harvest.

As one becomes more familiar with African peasant farming systems, the apparently haphazard diversity can be seen to have an underlying structure, what Ruthenberg (1968) termed "hidden order in seeming chaos." The patterns he identified from a pioneering study (1968) of 10 Tanzanian farming systems include (1) adaptation of cropping to different soils within a catena; (2) adaptation of cropping in concentric belts; (3) adaptation of planting to different rainy seasons; (4) adaptation of cropping by changing the composition of mixed crops; (5) adaptation of cropping by staggered planting; and (6) adaptation of cropping by crop rotation.

In real life, there is tremendous variation in possible yields, information is poor, and risks are high. Farmers allocate their inputs under an intersecting matrix of constraints—soil moisture status, pest outbreaks, an unexpected sickness, lack of ready cash—that can change quite rapidly. The necessity of having enough food to survive the hunger period remains an ultimate objective. However, in the short run attention is concentrated on the varying mix of constraints and events, which can have quite different implications depending upon the stage of crop maturity. One can model this kind of day-by-day opportunistic planning as a hierarchical sequence of "yes-maybe-no" decisions subject to different constraints, but such analysis is only just now being applied to peasant farming.[4]

For example, in the 1960s and 1970s the Kenya Ministry of Agriculture was promoting high-yielding maize populations planted in pure stands. Farmers disagreed emphatically, claiming that the mean rainfall was seldom actually experienced and that an interplanted stand at lower populations had a far better chance of surviving in adverse seasons (Moris 1970). In the season observed, there was a dramatic shift in weather, with a premature onset of drying winds. Lowland farmers rushed into their fields to thin their crop stands, while others began planting sweet potatoes and other minor crops they said could still be established in the valley bottoms.

A more recent study in the same area by Franzel (1984) looked specifically at reasons given for and against adopting an early-maturing variety of maize. Franzel found that while farmers do grow some of the Katumani maize recommended by the Ministry of Agriculture, they do so for reasons related to food security, not yield. Two separate decisions are involved: (1) which maize variety to grow for its earliness properties, and (2) which to grow as the main crop. Both "progressive" (40 percent) and "poor" (60 percent) farmers plant about an acre to Katumani, despite their view that it yields less and stores more poorly than the longer-maturing local variety.

---

4. See Gladwin 1982.

The reasons they have in mind are different, however (Franzel 1984, p. 203):

> Low income farmers have more of a propensity to grow Katumani for early maize because they lack regular sources of income, and, thus, have the most to fear if they run out of food. . . .
>
> Planting Katumani also makes sense for high income farmers. It is likely that the cost savings from growing Katumani, in terms of not having to purchase maize during August, outweigh its yield disadvantage.

In addition to the sequential aspects of peasant farming, there is also an underlying spatial structure. Crops that are used continuously by the household, that are easily damaged by vermin, or whose harvest requires a large effort tend to be planted nearby, with others located at greater distance (Ruthenberg 1968). In Kenya, we found that farm surveys generally overlook the horticultural "women's" crops planted close to the house, which perhaps consist of a mixture of individual plants, each with a specific purpose.[5] Beyond the intermediate zone of arable field crops, one would find the grazing land for goats and calves, and still farther out the limit of the area grazed by cattle. This pattern of production within concentric belts, varying from an intensively used area around the house to lightly utilized bushland, is a well-known feature of African shifting cultivation found wherever land has not yet emerged as the limiting factor (Okigbo 1984).

Root Crops

Aggregate estimates suggest that root crops are more important in Africa than in any other continent, exceeding even Latin America's consumption (table 13.1). Food and Agriculture Organization (FAO) estimates for 1974 indicate that root crops provided about 40 percent of total calories in Zaire, Ghana, and Togo (Goering 1979); even in other countries usually thought of as grain producers, the actual contribution of root crops exceeds what outsiders might anticipate. A recent FAO/Tanzania Ministry of Agriculture study suggests that there are more root crops produced than cereals, 217 kilograms (kg) per capita versus 185 kg for cereals. The term *root crop* includes bananas as well as cassava, potatoes, sweet potatoes, and cocoyams. *Starchy staples* might be a better designation, if one considers the key role such crops play in African peasant agriculture. Of these, cassava is by far the best studied. Its popularity among smallholders as a "famine breaking" crop arises from its low labor requirements and the fact that it can be left alive in the ground until needed during the peak of

---

5. Beyer (1980, p. 37) quotes Noel Vietmeyer to the effect that when the National Academy of Sciences asked for a listing of underexploited tropical plants, over 2,000 species were suggested.

TABLE 13.1 Root crop production in the tropics (million metric tons)

	Cassava	Potato	Yam	Sweet Potato	Other Root Crops[a]	Total
Africa	45.1	1.7	18.7	5.2	3.9	74.6
Trop. America	36.9	8.6	0.2	3.9	0.8	50.4
Near East	1.1	3.1	—	0.5	—	4.7
Far East	21.1	6.9	0.1	7.7	0.2	36.0
Other	0.1	—	—	0.2	—	0.3
Total	104.3	20.3	19.0	17.5	4.9	166.0
Percent of total	62.8	12.2	11.4	10.5	3.0	100.0

SOURCE: Reprinted with permission from Goering 1979.
[a] Include taro, arrowroot, arracocha, yam beans, oca, olluca, and *Xanthosoma*.

the dry season. Cassava's high productivity, as measured in calories yielded per hectare in comparison with alternative crops, must also be a significant consideration. The FAO (1984) statistics show it returns per hectare (ha) 10 times the gross weight and 3 times the energy yield of maize.

Bananas are often also a very popular crop wherever rainfall is sufficient for their production, having a traditional role in the seasonal food budget much like that of cassava (both crops can be dried and stored for extended periods). Bernard (1972, pp. 52-53) describes the situation in traditional Meru life as follows: "Of the crops mentioned above, certainly the most important to the Meru was the banana. It helped bridge food shortages between seasons; it provided famine relief in dry years; and it was a social stabilizer and a measure of time. A Meru man was not ready to marry until his banana plantation *(rurigo)* had become established." As used locally in African communities, bananas were treated not as a single crop but instead as an array of types suited to quite distinct purposes.

A feature of root crop production in East Africa is the wide range in actual yields being realized at the farm level. Swanberg and Hogan (1981) estimate that the difference in yield between "traditional" and "improved" methods of cassava production is 1,400 kg/ha. Such major differences probably reflect the fact that most farmers plant these crops as insurance in case the dry season is prolonged or the early crop of cereals fails.

Exploitation of Vertisols

Vertisol areas (or hydromorphic soils) possibly constitute Africa's main underexploited land resource. The name *vertisol* derives from the deep vertical cracks shown when such soils are dry, accounting for their common designation as *cracking clays*. Because of this feature, vertisols incorporate organic matter from the surface—otherwise a rarity under hot, tropical conditions—and their high clay content causes slow infiltration but long retention of surface water. They form wherever the slope of the land leads to impeded drainage, sometimes on higher-elevation benches but more typically in valley bottom and depressional areas. They vary from being highly alkaline (salt pans common to the East African Rift valleys and the Kalahari) to highly acidic (the peat bogs of highland Rwanda or the coastal mangrove swamps of West Africa). Nevertheless, until recently such soils were ignored by most agricultural research and as a consequence do not figure in official extension recommendations.

While vertisols vary widely in regard to soil nutrient status, their key agronomic advantage derives from their superior moisture retention. Runoff is concentrated into depressional areas, where the deep cracks store and then retain water long after the surrounding porous soils have dried out. On some clays, this advantage is so pronounced that a single preirrigation over the land surface can provide enough moisture for establishing a crop.

Vertisols are found in all countries of sub-Sharan Africa, being a prominent feature of the ancient land surface. In Zambia, for example, it is estimated that 14 percent of the land is either swamps or seasonally flooded (Elling 1981). Such *dambos* are prized for their animal carrying capacity in the dry season (FAO 1979). Elsewhere, the names differ, of course: *mares* and *bas fonds* in Francophone countries, *fadamas* in Nigeria, *mbugas* in East Africa, and *vleis* in Zimbabwe and South Africa. If it seems that this point is being overemphasized, we might note that in the Zambian instance just cited, *dambo* lands cover 104,391 square kilometers, in area 74 percent as large as the country's entire area of cropped land (Elling 1981).

For West Africa, P. Richards (1985) describes how farmers have evolved their own wet-rice cultivation in these valley bottom lands, usually as a supplement to a main crop of rain-fed upland rice. This combination gives a flatter labor input profile across the year than either upland cultivation or "improved" practices for swamp cultivation. He notes that throughout West Africa, smallholders obtain their early-season crops (which are most significant for breaking the hunger period) from the toe-slope of a catenary succession. Soil conditions are often complex at the margins of a depression, with the silty soils just above the clays being better for early-season cultivation and the clays best for extending the growth regime later.

In recent years, intensification of farming within vertisols has occurred in response to population growth and technological innovation. Suitably equipped tractors can work the heavy clay soils, converting large areas to arable cultivation. In addition, farmers have learned how to construct shallow wells to draw upon ground water for small-scale dry-season irrigation. In northern Nigeria, it has been estimated that while this *fadama* cultivation covered about 120,000 ha in 1958, 20 years later it had grown to 800,000 ha, vastly outstripping the expansion of "official" irrigation in Nigeria's expensive schemes (Wright et al. 1982). The paradox is that farmers are responding to proximate opportunities with little governmental assistance, while governments continue to fund extremely expensive formal schemes that may even flood lands already under production.

The highlands of East Africa may also offer a prospect for more intensive use of seasonally wet swamps. Jones and Egli (1984, p. 23) describe how in Rwanda farmers are moving downward.

> to clear, develop, and farm the swamps. This process began not too long ago, and continues. Papyrus or other cover is cleared. Soil is worked into ridges and drainage ditches with hoes. Crops are planted in the dry season . . . when they cannot be grown elsewhere. . . . Beans and sweet potatoes are the principal marsh crops. While marsh cultivation is physically demanding, the results are impressive.

They note that because the alluvial soils capture nutrients deposited from the upper slopes and enjoy an extended moisture regime, the output from vertisols is "much greater than their proportion of cultivated lands would suggest." (And, from our perspective, what is crucial is the timing of this activity, yielding a cash and food return during the dry season.) For western Kenya, Toksoz (1981) notes that there are about 800,000 ha of land with impeded drainage, of which he thinks at a "conservative estimate" some 600,000 ha could be reclaimed by a combination of drainage and flood protection measures.

Livestock

When *fadama* lands come under irrigation, this use conflicts directly with the principal alternative employment of valley bottom lands: early dry-season grazing for livestock.[6] Throughout Africa's Sudanic and Sahel zones, mixed systems combining livestock raising with cultivation are common. Given the enormous differences in local systems (Jahnke 1982; Sandford 1983; Dahl and Hjort 1976; Simpson and Evangelou 1984), one must be cautious in formulating broad generalizations about African livestock production. Here we take as our focus seminomadic and transhumant pastoralism (like the Masai of East Africa or the Fulani of West Africa),

---

6. In their northern Nigeria study, Norman, Simmons, and Hays (1982) found that households concentrating on *fadama* utilization for crops tended to be those without much involvement in livestock. Earlier accounts drawn from this excellent study include Simmons 1981 and Norman 1973.

rather than mixed, agropastoral systems (like the Akamba of Kenya or the Sukuma of Tanzania).

As Swift (1981, p. 80) points out, "Seasonality is at the very heart of nomadic pastoralism." Dutch researchers (de Ridder et al. 1982) provide quantitative data from the Sahel showing the seasonal fluctuations in biomass under different ecological zones (figure 13.1). These underscore Swift's observation that nomadism represents a form of social organization designed to tap the huge but variable potential of rapid plant growth during the short rainy season, while using an array of coping strategies to survive the long dry season when plant productivity ceases.

For outside analysts, full awareness of the degree of success achieved under traditional systems was retarded by the fixation on measurements of weight gain *per animal.* Now that measurements of *total biomass* sustained per hectare are available, it is clear that under adverse conditions the indigenous strategies for livestock management equal or exceed the performance of "improved" methods (Penning de Vries 1983).

African livestock producers who must cope with extreme seasonalities in availability of forage and water prefer hardy, light animals that can survive with a minimum of grass and water but show a rapid production response when better conditions occur. M. Price (1981) suggests these same tendencies occur in natural systems under conditions of high seasonality. Behnke (1985) argues convincingly that much of the supposedly greater technical efficiency of commercial livestock production over indigenous African pastoral system is indeed an analytic artifact caused by employment of inappropriate per animal weight gains as *the* measure of productivity, and by a persistent undervaluation of livestock products used by households.

The question remains: How do livestock-oriented producers cope with extreme seasonality?

First, animals are *herded* rather than pastured. This gives a more even utilization of plant species and a shorter duration of pasture use. During the brief growing period, it obtains maximum plant productivity while protecting the more palatable species (because animals are grazed at a walk and are not as free to concentrate on certain species).

Second, *long-distance trekking* permits movement of herds to the areas receiving rain and extends the radius of pastures utilized. Location of household camps at a distance from water (to which stock are taken on alternate days) further extends the utilization radius. Seasonal movement into the lower-rainfall areas during the short growing season taps vegetation on lands that could not sustain permanent occupation, while also protecting forage reserves around dry-season camps.

Third, pastoralists use the animals themselves as a *store of energy,* capturing high-quality forage during the brief period when it is in surplus

FIGURE 13.1 Seasonal fluctuations in biomass under different ecological zones

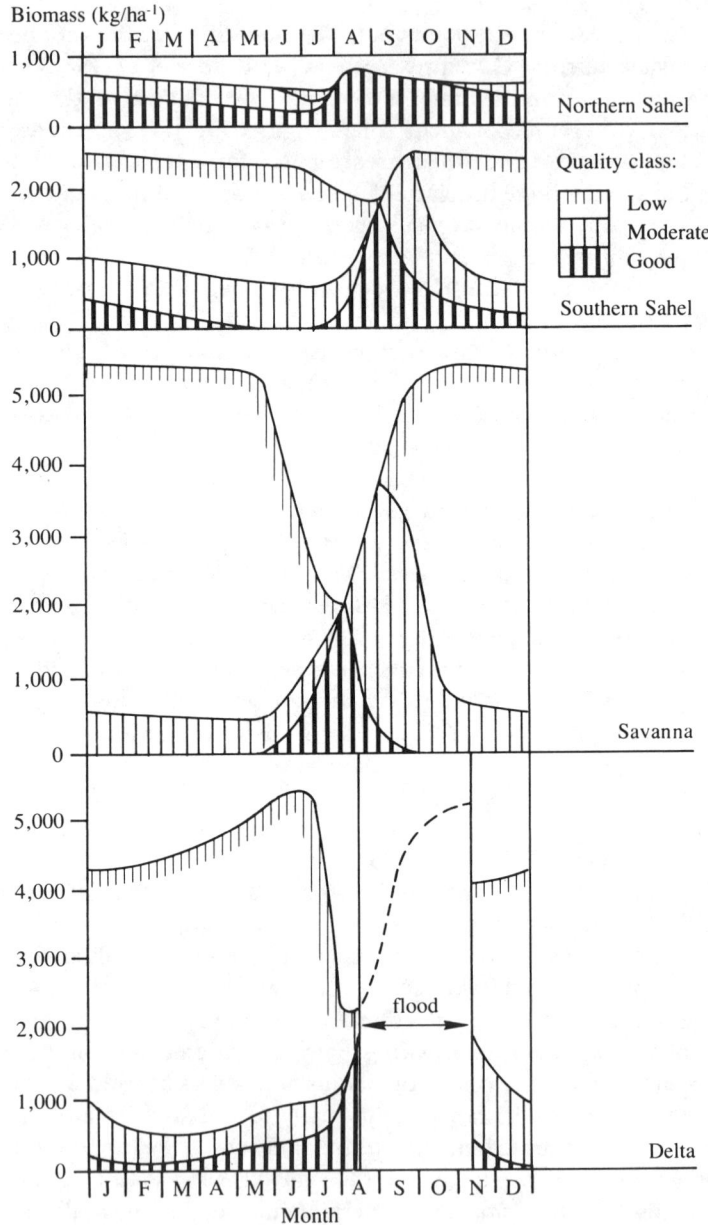

SOURCE: Reprinted with permission from de Ridder et al. 1982.

by converting it into hardy livestock used by humans during the long subsequent dry period.

Fourth, most pastoralists employ *mixed herds* of several species to exploit the different gestation and lactation periods, and variations in timing of livestock births. Empirical data on the demographics of mixed herds are recent (see especially studies contained in Hill 1985). Data presented by Wilson, Diallo, and Wagenaar (1985) show clearly that age of puberty, duration of pregnancy, length of birth intervals, life span, and total fertility vary by species. They note that while it is often said that reproduction rates for animals are low in the arid and semiarid zones, this is strictly true only for cattle. The fact that camels reproduce everywhere and at different seasons from cattle is especially important among milk pastoralists, but sheep and goats also show a number of advantages over cattle. In an important team study of Kenya's Ngisonyoka Turkana, Coughenour et al. (1985) have traced quantitatively the energy flow through the ecosystem. Their measurements find that while the system's energy yield is comparatively low, this production level is roughly equivalent to what others, such as Australian sheep ranchers or Sahelian nomads, achieve in arid environments. In view of the strong seasonality under which such groups operate, the Turkana maintain "relatively high density and biomass" in a marginal landscape, "without inducing discernible degradation of the ecosystem" (Coughenour et al. 1985, p. 624).

Fifth, semipastoralists who keep *sheeps, goats, and camels* gain a considerable degree of flexibility in coping with seasonal food deficits. Goats and camels are browsers and thrive on bushland not well suited to cattle. They can use African bushland species that retain their nutritive value into the dry season, when grassland forage is exhausted. Goats and sheep give milk, of course, but they are also the pastoralists' main source of meat—a dietary preference that accounts for the widespread misconception that African pastoralists avoid slaughtering their stock (true mainly in respect of female cattle). They represent a convenient store of savings, easily converted into either grain or cash (Schneider 1968). Indeed, the author would argue from field observation that they constitute the *main* option for increasing the liquidity of household income in most African settings where banking is not yet effective. Smallstock were often left under the control of women, and the possibility of trading a goat or two could make a huge difference in a household's food situation at the end of a long dry season. These advantages explain the persistence and expansion of smallstock production, even in environments that are favorable to cattle (such as at Ngorongoro in northern Tanzania).

The conversion of livestock into grain remains a necessity for the survival of African pastoralism. Even with mixed herds, the milk supply is extremely variable over the season. A fairly substantial herd must be accumulated before a family can sustain itself on livestock products over the entire year, a point emphasized by Devitt (1979) and clearly evident from production statistics reviewed by Dahl and Hjort (1976). The ability to rely on grain foods during the period of milk deficit is crucial to producers with insufficient milking animals. If, however, the usual seasonal shortfall extends into a major drought, animals begin to die, prices plummet, and grain supplies disappear from local markets. Thus the very success of smallstock-grain conversion in normal seasons leaves livestock producers overextended and more vulnerable in major droughts.

Bush Collection

Other than the growing of smallstock, perhaps the most common hunger-period adaptation shown in traditional African systems was for a household to turn to wild sources of food. Outside of the densely settled heartland areas, most communities lay within walking distance of a bush or woodland "commons." This was the immediate source for firewood and thatching grass, but it also contained an array of wild foods. Often men scattered individually owned beehives through the bush, garnering a prized commodity that could be harvested at will when food and money were exhausted. Then there were numerous edible plants. Lee (1969) records 85 edible species used by the !Kung bushmen; this allowed people to concentrate on only the most palatable and abundant species, such as the prolific mongongo nuts (which provided 33 percent of the group's total caloric intake). Lee also argues that under their traditional regime, !Kung foragers and hunters obtained an estimated 2,140 calories daily per adult, with only a modest expenditure of time. One would expect, therefore, that other people would also turn from cultivation to wild sources during times of food deficit—precisely what Scudder (1962) documents for the Gwembe Tonga of Zambia and Newman (1970) for the Sandawe of Tanzania (formerly hunter-gatherers). Newman (1970, p. 145) warns, however, that while some wild fruits do overlap the critical period, "it is highly unlikely that bush foods are of much value during occasions of severe drought."

Off-Farm Activities

Off-farm activities constitute a prominent feature of many rural African systems. Among the northern Nigerians, for example, it has long been customary for adult males to engage in trading and other off-farm pursuits during the extended dry season when little farmwork occurs. On average, adult men spend 39 percent of their total workdays in this way. Norman, Simmons, and Hays (1982, p. 120) point out that even in the months of peak labor demand, adult men continue their off-farm interests. Such activity may be necessary to provide for a household's immediate needs:

Little income is obtained from farming activities until after the bottleneck period is over. Cash and food resources tend to be low because most crops are harvested between August and December. Therefore, the farming households . . . facing severe depletion of cash and food resources are compelled to work in off-farm employment even though the work needs of their own farms might be high.

These off-farm involvements occur in various economic contexts: livestock trading, "visiting" urban relatives, fishing, long-distance labor migration, and so forth. While the details will vary enormously from one context to another, the point is that such involvement can contribute significantly to farmers' incomes and diet. The timing of such returns is as important as the absolute amounts earned. In northern Nigeria, Wallace (1981) suggests that the government's neglect of farmers' seasonal off-farm pursuits explains why large-scale irrigation schemes that required "free" labor were initially so unpopular.

Social Adjustments to Seasonality

The six options we have discussed constitute the main indigenous *production* strategies for ameliorating seasonal food deficits. However, in any system of livelihood there can be social and economic compensatory mechanisms that cross-cut the production alternatives and also improve a household's ability to cope with the long dry season. Let us review briefly three such adjustments: reciprocal economic exchange, gender-linked allocation of farming tasks, and varying modes of household integration.

The issue of reciprocally based economic exchange is one on which anthropologists have long differed from the prevailing paradigm in conventional microeconomics. Many anthropologists accepted the economic historian Polyani's reasoning that in earlier economic systems, a market mentality was not yet predominant and nonmarket modes of exchange operated in different ways (Dalton 1968). Of particular interest was the apparent importance of reciprocal exchanges ("sharing") within hunting-gathering groups (Service 1966) like the Mbuti pygmies described in Turnbull's classic *The Forest People* (1962). However, a worldwide review by Pryor (1977) of the comparative importance of different modes of exchange showed that while reciprocity was once widely found, it has been replaced generally by market-based exchange nearly everywhere.

Recent studies by anthropologists now actually measure the amount of food shared (among the San peoples and among pastoralists such as the Pakot and Turkana), showing its key importance in systems subject to strong seasonality.[7] It seems that reciprocal food sharing was less central

---

7. See Wiessner 1981 and Lee 1979 on the Bushmen; for the Turkana, see Coughenour et al. 1985, Galvin 1985, McCabe 1984, and Wienpahl 1984.

within agricultural societies, which entered a monetary economy earlier than did either the Bushmen or nomadic pastoralists.

An important advantage of reciprocity over price-governed exchanges is that, provided all members of the group share their food, people will obtain the minimum needed for their survival as long as the group as a whole has not overreached its resource base. In adverse seasons, which arrive unpredictably, all will experience diminished consumption but nobody will starve as long as food supplies hold out. The superior performance of reciprocal exchange under highly uncertain production conditions would explain its prominence to this day among East African milk pastoralists.

Perhaps the food-deficit situation also explains why a gender-linked division of household and farming tasks was prominent in many African traditional cultures. As our example, let us turn to the Kikuyu and Kamba of Kenya, where distinct role orientations were institutionalized for men and women. Put succinctly, the ideal Kikuyu man was somebody quick-witted, adept at convincing others to act, and willing to take risks. The ideal Kikuyu woman, on the other hand, was sober, hardworking, and cautious—attributes suited for conserving household resources and minimizing risks.

These two attitudinal complexes are congruent with different production strategies. Kikuyu and Kamba men managed the household's large stock to exploit trading opportunities in favorable seasons. When crops were in surplus, they enlarged their herds by trading upward from grain into goats and from goats into cattle—exactly as Schneider's (1968) model of East African livestock keeping would predict. In adverse seasons, however, the frugality of Kikuyu and Kamba women became crucial. In Kamba society, it was the compound's senior wife who kept the seed for the coming season in a sealed clay pot buried beneath her bed. She would be the one to decide when the food shortage was so severe that seed must be eaten. (Similarly, among some pastoral groups, women decide how much milk to retain for their children and how much to give to the calves.) A wife cultivated her own garden of maize and beans, from which her children and husband would be fed; she was the one with specialized knowledge about traditional and wild foods. Either orientation by itself would be inadequate under opposite circumstances. The two together gave those in a traditional production unit an incentive to cooperate while retaining the flexibility to activate contradictory strategies in a highly unstable environment.

If the usual end-of-dry season food deficit extended into a major drought, it was not unusual to find large compounds splitting into smaller units. If the family owned livestock, the able-bodied men might leave to try and exchange them for food at a distance. Alternatively, we found that in some eastern Kenyan families, a man might send his wives back to their natal homesteads (also often located at a distance, and thus perhaps less subject to a local drought). Akong'a (1982) mentions the traditional practice of "pawning" a wife or daughters to somebody with food, in expectation that the household head would ransom them back once good times returned. These may seem extreme solutions to food deficit, but at least they were preferable to the complete breakdown of the social system that Turnbull (1972) observed among the starving Ik people of Uganda.

Introduced Seasonalities

In describing the huge influence of natural factors on African agriculture, one tends to overlook introduced seasonalities that have come about because of farmers' participation in the larger political economy. Four are particularly salient: (1) the withdrawal of children from the farm labor force to attend school, punctuated by the holiday periods when they may be at home; (2) the timing of crop payments within the annual calendar in relation to periods of cash need; (3) heightened seasonal deficits accompanying maize farming; and (4) the influence of the government's financial year upon rural service organizations, most evident in remote settings.

In regard to the impact of mass schooling—with universal primary education being an official objective in most East African nations—it is easy to overestimate the direct contribution of children to farming tasks, which was seldom large if measured by the hours worked. Three exceptions should be noted: the strategic assistance older children offered as herders within livestock-keeping systems, a similar role for adolescent boys in fishing communities, and the sometimes significant assistance rendered by teenage girls to their mothers at weeding and harvest. In addition, certain export crops such as tea or cotton are notorious for requiring a heavy seasonal labor input in picking, a task shared by children (Fowler 1982).

However, children often exercise a significant *indirect* influence. The tasks children perform, such as watching crops or scaring birds from the ripening field crops, or the time older girls spend in child care, watching younger children while their mothers do fieldwork, are often vital to agricultural systems. Similarly, livestock-keeping peoples throughout Africa have strongly resisted mass schooling because of the important roles assumed by children.

Seasonality is a feature of mass public schooling in two respects: first, because there are holiday periods when children may return home; and second, because school fees must usually be paid according to some administratively determined schedule. In regard to holidays, there is the obvious possibility that the school calendar might be readjusted to return children at times of peak labor demand. Here the main difficulty appears to be the variability in when peak periods occur, either spatially (from community to community) or temporally (from season to season). In regard to school expenses, subsistence farming is characterized by a marked illiquidity in regard to seasonal cash flow. There will be only certain times in the year when householders have ready cash for paying school fees or other related educational expenses.

In our Kenyan farm surveys, it was found that farmers were anxious about paying educational costs (school fees and uniforms). In extreme instances, a farmer might even sell expensive dairy cattle to keep a son in Harambee school. That a farmer would make this choice at a time when seasons were good and milk in demand indicated a strong preference for diversifying sources of family support, a tendency also found among Yoruba farmers of Nigeria, as documented by Berry (1984).

Another cause of increased seasonality is the switchover from drought-resistant grains to drought-sensitive maize as the staple food, a trend most marked in East and southern Africa (Miracle 1966). Among the Sandawe people, it is significant that, though living in a drought-prone environment, when people took up hoe cultivation maize was their choice of staple (Newman 1970). When Masai pastoralists in Tanzania's Arusha region take up plough farming, it is to plant maize—not sorghum or millet. The Kikuyu, Meru, and Kamba people forced by overpopulation to migrate downslope into the drier lands of eastern Kenya nevertheless plant maize, even where one crop in three fails (Wisner and Mbithi 1974).

It was noted earlier that maize is especially sensitive to the types of moisture shortfall that typically occur within East Africa: delayed onset of the rains, a midseason dry period, and early termination of rainfall. Why then have farmers abandoned the better-adapted indigenous cereals to plant a riskier introduced crop? One clue has already been given: maize is much less vulnerable to bird damage. Farmers in Sukumaland of Tanzania told researchers that this was the principal reason why, when children began attending school, they abandoned millet and sorghum (Hankins et al. 1971). Another may be the ease of storing dried maize on the cob—now the "traditional" technique over much of Africa. A third reason is the preference for maize-based foods in urban areas, either *uji* (thin gruel) or *ugali* (thick maize porridge), probably deriving from the maize rations customary in the large farming and mining sectors from Kenya southward.

Still another influence was the U.S.-supplied famine relief maize accepted by several East African countries in the 1960s, in 1973-74, and again in 1983-84. Then too the national cereal marketing boards established almost everywhere found maize much easier to handle and market.

Maize is now the main staple (alongside rice and wheat in urban areas) over almost all of Africa south of the equatorial forest zone. This means that households have become more prone to food deficits in adverse seasons.

When large numbers of people become dependent on administratively supplied food sources, seasonalities *within* the administrative system become relevant (Chambers, Longhurst, and Pacey 1981). Chambers and others argue that one type of administrative seasonality is the reduced coverage of rural services during the rainy season when travel is difficult and supplies may run short. In East Africa, where the financial year typically ends in June, the end of the rainy season in late May coincides with the period when government offices must close their books for the year. In more remote areas, there can be a six-week period when no expenditures are permitted, followed by an additional wait for the arrival of the next year's funds. This gap exactly overlaps the time when people will be out of food as they await the ripening crops. If a crop failure occurs, a food-deficit crisis may have materialized by the time officials are again mobile.

The worsening economic situation of the poorer African nations accentuates the impact of these seasonalities. Rural services in Africa are particularly vulnerable to disruption of transport and supplies. What we see today in some countries are rusting vehicles lying unrepaired for lack of spares, fuel sales on only certain days of the week, and stringent rationing of travel by civil servants. In the midst of a deteriorating national situation, donor-sponsored projects continue to offer farmers loans and insist they should employ imported inputs in specialized farming. As we shall see below, such "solutions" to seasonal food deficits may in fact *increase* farmers' economic vulnerability because of the high levels of institutional risk associated with the national marketing and input supply systems.

Introduced Solutions to Seasonal Food Deficits

A main thrust of the argument throughout this chapter is that introduced "solutions" to seasonal food insecurity do not in fact outperform indigenous ones except in certain favored situations. A grasp of this fundamental conclusion helps one understand why so many of the official rural development projects sponsored by African governments (and external donors) have failed. There are three main options for farmers that will be examined: (1) to *commercialize* their production through specialization so that they can buy food when needed; (2) to *mechanize* operations in order to produce more food; and (3) to *irrigate* their crops, thereby increasing output, reducing yield fluctuations, and extending the growing season.

Specialized Commercial Production

A general assumption underlying most technical packages recommended to farmers is that specialized commercial production is essential. From colonial times onward, it has been a common practice to divide the African countryside into zones based on a predominant cash crop. Usually such crops were grown for export and constituted a household's main on-farm source of cash income. In the Sahel, for example, one encounters the peanut and the cotton zones; in Kenya, the coffee, tea, and cotton zones; in Tanzania, the cotton versus the tobacco area; and so it goes, country by country. At first, technical packages dealt exclusively with the recommended export crop, and the crop industry also financed the research station. More recently, recommendations for a predominant staple food (usually maize) were incorporated; and increasingly coarse grains and legumes are included.

There is a huge literature on the supply response of African peasant farmers, and on the economics of particular crops, beyond what can be reviewed here. However, we should note that in evaluating individual farm innovations, a major problem arises because the choice of criteria has such a large influence on the outcome. For several decades now, the usual measure of technological superiority in farming has been mean crop yields per unit of land. The employment of this standard biases the comparison, since it pays no attention to the inputs required, especially to increasing levels of purchased inputs and labor. Not only are absolute levels important (with smallholders seeking to minimize both types of input), but also seasonal considerations can be critical.

Again and again in Africa we find that the supposedly superior yields of recommended technical packages either demand a substantial cash outlay at the beginning of the season (when farmers are experiencing food deficit) or else require higher effort at the peak seasons. Since the periods of labor input (in land preparation and weeding) overlap those when cash is needed for improved seed and fertilizers, high requirements of these inputs are interactive. They greatly increase the strain on peasants, particularly among the poorer households, who may already be purchasing foodstuffs.

A doubled or even trebled input requirement is not unusual in "improved" systems of production. Data on returns to labor at the peak periods are often the key to understanding indigenous African farming systems. Alverson (1984) shows that traditional practices in Botswana give a return per unit of labor per hour more than three times that of the "improved" system. Similar results are reported by Gathee (1982) from Kenya. In both of these cases, while officially recommended practices give higher total yields per hectare, the returns per unit of labor required at the peak season are much lower.

Any analysis of gross returns that assumes "free" family labor is therefore likely to miss the key importance to farmers of how they deploy their sharply constrained reserves of cash and labor during the brief growing season. Glaeser's (1984) study of poor Usambara peasants in highland Tanzania found that during average and poor seasons, a "subsistence" strategy based on intercropping and minimal use of purchased inputs gives a higher gross margin per day than does the "improved" farming strategy. This analysis provides a rational basis for explaining why most Usambara smallholders confine their cash expenditures to purchases of seeds and little else.

While in theory commercialized production provides higher yields, it often does not give either higher returns or greater household food security. The following are overlooked negative impacts:

1. It provides relatively low returns per unit of labor input, especially to labor at times of peak demand.
2. It entails higher levels of purchased inputs that are often not affordable or available.
3. It competes for cash at a time when food is most expensive and when household food supplies are likely to be exhausted.
4. It commits farmers to dependence upon outside service agencies, which may suffer from low reliability and have a poor repayment record.
5. It does not guarantee stable prices and food adequacy in the marketplace during the hunger period before a new crop can be harvested.

These adverse impacts are all magnified under production systems characterized by high levels of seasonality.

Of course, *sometimes* African farmers with larger holdings and supportive neighbors do prosper. Anthony et al. (1979) describe their Mazabuka sample in Zambia as being among the best farmers on the continent, much as two decades earlier Elspeth Huxley had observed the same about the Kalenjin farmers of highland Kenya. However, these cases of success for specialized, Western-style mixed farming come from plateau or highland areas. Climatic variability is less than the lowlands, and the holdings cultivated have usually been larger than four hectares, permitting more efficient use of animal power. In such settings, a transition into "yeoman" farming based on commercialized grain production does seem feasible.

Elsewhere, the record in Africa of long-term agricultural change based on specialized grain production or the growing of export crops is less encouraging (Haswell 1981b; A. Shepherd 1981; van Apeldoorn 1981).

Watts (1983) points out that poor households in northern Nigeria are often compelled to sell grain at harvest when prices are lowest to meet their immediate cash needs, but then must buy it back later at high prices when their granaries are empty. O'Leary (1984) argues that a similarly adverse relationship over the season in the terms of trade keeps pastoralists from commercializing their production.

Mechanization

Sub-Saharan Africa is generally regarded as an exception when writers compare worldwide trends in farm mechanization (Binswanger 1984). In black Africa there are still large areas entirely cultivated by hand; and animal traction predominates over tractor farming, even though tractors were introduced well over 30 years ago. When farming has such a pronounced labor bottleneck and so much arable land remains uncultivated in any given season, one might have expected the rapid expansion of mechanized agriculture. This transition would have stimulated the rapid evolution of a commercial economy, and increased yields might have made seasonal food deficits a thing of the past.

Immediately after independence in the mid-1960s, Tanzania, Kenya, Uganda, and Zambia all experimented with tractor "block farming," typically with dismal economic results.[8] Tanzania is currently in its third cycle attempt, urging tractors on reluctant villagers despite an adverse balance of payments and the *total* failure of its two earlier mass campaigns. No doubt a similar story could be documented for many other African countries, such is the allure of tractor farming in a landscape where human energy at the peak periods is sharply constrained.[9]

Mechanization in Africa means either ox- or tractor-powered cultivation, both types being confined mostly to land tillage rather than weeding and harvest operations. The larger the capital investment, the harder it becomes to justify partial mechanization of this nature, since expanding the cultivated area simply increases the weeding and harvest bottlenecks (Newman, Quedraego, and Norman 1980; Norman, Simmons, and Hays 1982). Here it should be noted that in some Francophone countries, official promotion of animal-powered farm equipment has been more systematically pursued. Factories have been established, first in Senegal and later in Cameroon and Mali, to produce locally adapted designs for a multipurpose "tool bar" *(polyculteur)* as well as ploughs, seeders, and carts (Le Moigne 1980). The usual problem with introduced equipment is that it is

8. For accounts of the East African experience with mechanization in the 1960s, see Hall 1968, Metrick 1975, Clayton 1973, de Wilde 1967, and Ahmed and Kinsey 1984.
9. Much the largest of Africa's mechanization schemes have occurred in southern Sudan, where it was hoped that large-scale tractor farming could create an African bread basket (see Ibnouf 1985). The Ghanaian experience is reviewed in A. Shepherd 1981.

too heavy for local animals. Lacking the official support it has enjoyed in Senegal and Mali, animal-drawn equipment has received only sporadic attention in other African countries, where tractors have been emphasized instead.

The crucial difficulty remains the low utilization rates, which have the consequence that technical efficiencies are not translated into economic cost-effectiveness, a point Johnston stresses (1980). Johnston cites several studies from Tanzania that show that with prevailing labor-capital price relationships, animal traction remains more economical than tractor cultivation. The comparison is clouded in countries where imported equipment is cheap because of an overvalued currency, and where tractors are often accorded preferential treatment (waiving of import duties, etc.). The failure of tractors despite these advantages is in part due to the fact that equipment has on average a very short working life. Many factors no doubt contribute: inadequate maintenance, poor supervision, and imports of "orphaned" equipment. The earlier operational lessons are well summarized in de Wilde (1967) and Kline et al. (1969). Two key reports using Tanzania data are Beeny (1975) and Migot-Adholla (1975).

These days the existing problems are compounded by difficulties in obtaining fuel and spare parts, shortages being encountered almost everywhere in sub-Saharan Africa. When farmers cannot be certain of obtaining fuel, the potential contribution of tractors to relieving labor bottlenecks becomes overshadowed by the risk of total failure.

Irrigation

Can adoption of irrigation technologies in their modern form alleviate the seasonal food deficits in African peasant agriculture? Either by shortening the hunger period or by reducing the shortfall in cereal yields, irrigation would seem an ideal technological solution. But what has been the experience in Africa to date?

In answer, the first point to underscore is that Africa's formal irrigation schemes have been very expensive. The figure of $25,000 per hectare (in 1980 prices) given by Toksoz (1981) for Kenya's Bura West irrigation settlement is representative of the actual costs of many African schemes. An FAO (1986) Investment Centre study notes that as a rule African construction costs are higher than those for similar work in Asia. However, another major difference is the presence of already established irrigation bureaucracies in many Asian countries. In African countries, formal irrigation schemes have often required expensive external management. Then there are the unavoidable infrastructural costs that accompany implementation of technologically sophisticated projects in remote environments. Formal schemes usually require that the government build roads, crop storage facilities, staff housing, clinics, and so forth. When irrigation is so expensive and concentrated, it will necessarily be limited to a few favored sites, which then cannot afford to grow lower-value food crops (Moris and Thom 1985).

Formal irrigation in black Africa is employed mainly on only three crops: rice, sugar, and cotton. Of course there are exceptions within individual countries—irrigated sorghum in Sudan, wheat in Nigeria, coffee in Kenya, and irrigated bananas in Somalia—but on the whole these other crops constitute only a minor proportion of the total. To recover their heavy investment costs, formal schemes must grow commercial crops at a high yield level. In particular, irrigated sugar estates in sub-Saharan Africa have an agro-industrial character and a centralized management antithetical to small-scale and peasant production.

Even on rice schemes, farmers are often called *tenants* and enjoy little control over their own operations. In fact rice is not the predominant staple food in black Africa, with the exception of parts of coastal and inland West Africa and Madagascar, with its historic ties to Southeast Asia. In East Africa, rice remains a luxury food enjoyed on special occasions or by urban populations. High production costs of formal schemes mean that locally produced rice often costs much more than imported rice. This necessitates protected local markets and all the distortions these can entail. The same is true for sugar and for wheat, with the added caveat that outside of highland areas, wheat does poorly in tropical Africa and is almost entirely an item of urban consumption (though a rapidly growing preference, particularly in the Sahelian countries).[10]

What contribution do Africa's larger irrigation schemes make to ameliorating seasonal food deficits? As noted above, the specific crops they grow usually are not targeted at reducing food deficits, nor do they save foreign exchange (because of the high import content of the large-scheme mode of production). For participating farmers, the rice they grow can be retained to stabilize the household's food supply, but irrigation schemes will likely involve large payments for participation, eroding a family's ability to ensure food security over the seasons. If schemes incorporate double-cycle cropping, one cycle will overlap farmers' rain-fed crop, a principal reason why several African irrigation schemes have abandoned earlier attempts to institute a double cycle.

The large dams from which the more recent schemes draw their water also cut off the downstream seasonal flood, making the traditional form of *decrue* (or recessional) farming impractical. Here the upset has been greatest in Mali's inland delta, which was severely affected by upstream with-

---

10. The dysfunctional aspects of Africa's increasing reliance upon wheat, accentuated by overvalued currencies and by the capital and foreign exchange–intensive technologies used on the continent for producing wheat, are brought out by Andrae and Beckman (1985).

drawals during the two very dry seasons of 1983 and 1984—an impact Scudder (1980) warned against. In Nigeria, as earlier noted, the loss of traditional *fadama* cultivation has been significant; the hectares taken over to institute a new regime of stabilized flow may actually exceed what was gained by construction of new dams.[11] Furthermore, in a semiarid environment the land flooded *within* the storage basin is often the very best agricultural land in an area, already intensively used. From a sociological viewpoint, implementation of large-scale irrigation has become an excuse to dispossess existing cultivators who were already engaged in small-scale irrigation.[12]

In actuality, FAO (1986) figures indicate that the improvement of small-scale irrigation or minor organizational improvements to the poorly performing large systems would be far more cost-effective than investment in additional large schemes. If we take into account the high opportunity costs of water, capital, and skilled management, and the high recurrent costs of large schemes, most African countries would have benefited if their large schemes had never been built. The outstanding exception is of course Sudan; here the incorporation of sorghum in the official rotation was of critical importance during the extended 1984 drought.

Comparative Effectiveness

To bring to conclusion our extended review of technical options, indigenous and introduced, for ameliorating seasonal food deficits, their comparative performance under contemporary circumstances will now be briefly examined.

First, it should be noted that indigenous solutions to seasonal food deficits have usually been accorded too little attention in technical agricultural research (at least until the farming systems research (FSR) studies in the mid-1970s). Western-trained scientists, whether expatriates or African nationals, were not prepared to take seriously the role of goats, transhumant migration, root crops, and swamp cultivation in the farming system. Now that FSR provides a conceptual paradigm for examining these features of "traditional" cultivation, the key factors are being measured within field studies. Scattered preliminary findings suggest that indigenous technologies often outperform the introduced ones when employed under the constraints that African smallholders face.

---

11. Several useful accounts of the Nigerian experience are found in Adams and Grove 1984. See also Wallace 1981 and Andrae and Beckman 1985.

12. See Tiffen (1985), who reviews land tenure aspects of irrigation; also Adams's horrifying account of land dispossession in the Bakalori scheme, in Adams and Grove 1984. Adams argues that the implementation of the scheme had the same effect for Bakalori farmers as a major drought.

Second, the indigenous strategies are clearly related to farmers' need to outlast the extended energy-deficit period overlapping the end of the long dry season and the beginning of a new cycle. Even under modern conditions, failures in the larger economic system make it highly desirable that farm households pursue their own solutions to this need. It seems likely that the superiority of indigenous solutions is especially evident during the hunger period—though this will not be true for larger, commercialized producers whose monthly income (often from off-farm sources) buffers them from food deficits.

Third, under conditions of increasing population and environmental stress, indigenous solutions are no longer available to all who might desire to employ them. Many solutions require access to common resources, whose productivity falls off rapidly once overuse occurs. In regard to crop diversity and particularly traditional root crops, the disjuncture between genders and between generations may bring about a loss of valuable indigenous varieties and empirical knowledge. In addition, all the indigenous strategies discussed previously are losing their effectiveness under conditions of "labor involution," where a larger and larger rural population depends upon a fixed resource base.

Fourth, under conditions of population intensification, several of the indigenous solutions can leave households extremely vulnerable should a seasonal food deficit be extended into a major drought. The success of such strategies during a typical dry season keeps people in place where the modern economy cannot function well if there is a failure of the entire season. This explains why it is so difficult to estimate in advance the numbers who may require assistance in a bad year, and the catastrophic speed with which a disaster spreads once indigenous solutions collapse.

Fifth, for a variety of reasons not fully reviewed here, the likely response at a governmental level to this situation has been to distribute famine relief obtained on concessionary terms from external donors. Such famine relief can relax the food deficit constraint, becoming a substitute for increasingly ineffective indigenous remedies. However, the activation of famine relief cannot be initiated by individual households. What is known from feedback within African administrations suggests that the process will have delays, random breakdowns, corruption, and inequities. On the other hand, it has already permitted an intensification of rain-fed farming into the drier lands of the Sahel and East Africa. Famine relief during minor droughts is an essential component in the process.

Sixth, under present economic conditions, while recommended technical packages for export crops and maize give higher yields, they may greatly increase farmers' levels of risk. They also often do not match indigenous options in the returns to labor input, particularly labor at the seasons of peak demand. The congruence of farmers' own choices with crops that yield the most per unit of labor input suggests that labor bottlenecks are a significant consideration in their production strategies.

Seventh, while there have been some successful development projects based on export crops—notably tea, tobacco, and cotton—these have been characterized by tight supervision, guaranteed prices, and a well-organized input delivery system. Under such circumstances, smallholders can obtain yields sufficient to repay high input costs and earn adequate profits to ensure access to food in the lean season. Nonetheless, many attempts to copy the parastatal crop-authority model have failed, posing a heightened risk of food insecurity during the dry season. Successful export-oriented cropping schemes appear to depend upon having a highly efficient field support system that can minimize the risks of intensified production.

Eighth, the technical efficiency of mechanized production has not translated into economic cost-effectiveness in circumstances where only a few operations are affected and where individual farmers have very small fields. On their own, both tractor- and ox-ploughing simply shift the labor bottleneck to weeding and harvest operations. Institutional innovations, such as block farms aimed at circumventing the small size of farmers' holdings, require a high level of coordination and management. Tractor farming, because of its higher costs, is severely affected by climatic uncertainty and by low use rates. It is also vulnerable to the shortages of fuel and spare parts that accompany a deteriorating national economic situation. Such blockages affect capitalist as well as socialist approaches. They account for the difficulty of realizing economies of scale in African farming, while also limiting the role of mechanization in reducing the seasonal variability in food production and raising profits to ensure cash resources in the seasons of food stress.

Ninth, the extreme duality in African farming limits the application and diffusion of irrigation technologies as a means of reducing seasonality. To date, irrigated production has been very expensive, necessitating professional management and a concentration on intensive monocropping of rice, sugar, and cotton. Farmers whose only role is as tenants or day laborers cannot be expected to copy the capital-intensive technology they observe in official irrigation schemes. The high costs and many organizational problems associated with these formal irrigation schemes explain why introduced irrigation technologies have not notably increased farmers' food security.

In contrast, African farmers' own small-scale initiatives in making better use of depressional soils have allowed an intensification of farming. With or without formal assistance, this trend will continue. It greatly outperforms the introduced options as a viable approach for ameliorating the impact of seasonal food deficits. However, even this strategy cannot cope with major droughts, which some analysts predict will occur with increasing frequency.

Conclusion

This chapter has been based on a farming systems research perspective. It assumes that African peasant farmers generally respond in rational ways to production opportunities, contingent upon the proximate constraints experienced over the course of each season. The most important of these for subsistence producers is the necessity of securing an adequate food supply to outlast the extended food-deficit period that overlaps the end of each dry season until new crops can be eaten. Because the author's own field research (done some years ago in eastern Kenya) did not specifically address this issue, results of more recent studies in East and southern Africa have been drawn upon also. While the evidence is scattered and somewhat inconclusive, it does support the conclusion that food deficits are a priority concern for African smallholders. Farmers' rejection of introduced technical packages becomes comprehensible when we examine the returns obtained from labor at planting time and cash outlays required during the critical hunger period. Indigenous production choices generally outperform introduced options in these two key areas, which under conditions of increasing insecurity become more rather than less significant.

This finding explains the widespread rejection of recommended technical packages, which in turn accounts for the disappointing performance of so many African rural development projects. By and large, recommended practices have required greatly increased labor inputs from farmers, and often heavy cash outlays as well. They necessitate a dependence upon outside service agencies whose own reliability is decreasing now that many countries are in a continuing state of financial crisis. Thus on top of natural risks, we must also reckon upon high levels of institutional risk within African farming. Under such conditions, entry into "modern" agricultural production makes sense only for those with a cushion of nonfarm income to tide them over periods of food scarcity.

14 The Impact of Technology and Policy on Seasonal Household Food Security in Asia

ROBERT W. HERDT

This chapter focuses on the effects of technology and policy interventions on seasonal food insecurity in Asia. This insecurity arises because the tropical wet-dry seasonal cycle induces food scarcity at the end of a dry season and the beginning of the following wet season, when much of the stored food has been consumed by both humans and animals (Chambers 1983b). Little work is available, wages are low, savings have been depleted, and some people migrate in search of work. When the rains come, cultivation tasks make high energy demands on weak draft animals and on people.

This scenario agrees with many observations of casual social science empiricism. And while it is tempting to construct a simple theoretical model relating the effect of policies or technological change on seasonal income flows and nutrition, it would be misguided. In one recent effort at modeling the direct relationships among food, prices, income, and health, Pitt and Rosenzweig (1984) demonstrated that even without adding the complexity of seasonal variations, the policy implications derived from such models may be misleading if the ultimate goal is to improve the levels or distribution of health in the population. They show that theory offers no prediction even for the direction of the effects of food price changes on health, without complete knowledge of preferences and of the health technology. The challenge of modeling seasonal factors is even greater.

Measuring Seasonality

While identifying the nature and extent of seasonal household food security presents a major challenge, so too does its measurement. The concept of seasonality seems intuitive, especially in a temperate climate where winter and summer form distinctly different seasons, confining crop production to only part of the year. The seasonal temperature swing from the monthly low in winter to the monthly high in summer is reflected in a maximum agricultural production peak in the autumn and a minimum in late winter.

In tropical areas the seasonal temperature swings are hardly noticeable, but in many areas there are distinct seasonal rainfall patterns, with peak rainfall coinciding with monsoon winds during certain months and little or no rainfall in other months. However, while such a distinct seasonal rainfall pattern prevails in much of sub-Saharan Africa and on the Indian subcontinent, it does not hold as distinctly over much of Southeast Asia, equatorial Africa, and equatorial South America. In some areas there are two peaks (bimodal distribution), while in others the distribution is flat.

Seasonality is easier to see than to quantify in a single number. For example, figure 14.1 illustrates a typical seasonal agricultural pattern. The May peak is highest in 1955 and the August–September level is highest in 1977, indicating that the seasonality has declined. Some authors focus on the three-month maximum as a proportion of the three-month minimum as a measure of seasonality. In the data reflected in figure 14.1, the cumulative sum of the harvested amount during the period April–June represented 59.5 percent of the annual total and the amount harvested during the November–January period represented an accumulated 8.4 percent of the annual total in 1955, for a peak-to-trough harvest ratio of 7.0. That ratio declined to 5.6 in 1968 and then increased to 6.9 in 1977. The increase between 1968 and 1977 occurred because the cumulative proportion in the trough period (i.e., November to January) declined (from 10.2 percent to 6.8 percent) even while the cumulative proportion in the peak period (i.e., April to June) declined (from 56.7 percent to 46.8 percent). Thus it is not clear whether seasonality increased or decreased. If one seeks to avoid this ambiguity by choosing peak and trough periods of two months or even one month, there is the problem that over time the peak or trough month may change, resulting in a different length of time between peak and trough.

In areas where there is a bimodal or multimodal pattern of activities, the measurement of seasonality is even more difficult. For example, in the central Visayan region of the Philippines in 1984, the peak harvest month was November, when 27.4 percent of the rice area was harvested. In each of the adjoining months, October and December, 7.6 percent of the area was harvested. In March and April, 20 percent of the area was harvested, and in no other month was more than 5 percent harvested. Thus the harvest was concentrated, with the total harvested area in October–December (42.5 percent) being slightly less than the total harvested area in March–May (45.7 percent) and the harvested area during the other six months of the year being less than 12 percent of the total. The ratio of peak (45.7 percent) to trough (4.4 percent in July–September) was 10.4. In such areas, a measure of month-to-month variability may be the most valuable reflection of seasonality.

Clearly, ratios reflecting the degree of seasonality of food consumption must be much smaller than ratios reflecting the seasonality of harvested area. No one would long survive consuming in one three-month period 10 times the quantity of food consumed in another! And in the case of consumption, one must be concerned with consumption over short periods, say one or two weeks. In fact most consumption data reflect consumption during a single day, and where there are data on consumption at several periods throughout the year, they also rely on several single-day measurements. Seasonal swings in prices, corresponding to the costs of storage and normal profits, are also likely to be smaller than for harvested areas but larger than for consumption.

Thus while there is much clear evidence of seasonal patterns in food consumption, that evidence is not universal, even in low-income developing countries. The phenomenon of seasonality is easy to describe in a graph but more challenging to measure, and one must use care in comparing across different variables, whatever measurements are chosen.

Direct Effects of Policies on Food Security

The factors directly affecting food security have been enumerated as follows (Pinstrup-Andersen 1981): (1) the quantity of available food, (2) the price of available food, (3) the quality of available food, and (4) the income of the concerned consumers. The effect of policy and development projects on each of these factors has been discussed at length elsewhere. Therefore, this chapter will focus on a description of food-related policies and technological change in Asia.

Food availability and prices have distinct seasonal patterns in most developing countries, because food production is typically quite seasonal while food demand is constant and highly inelastic. It is precisely this seasonal fluctuation in production and hence market availability and prices that many policies have sought to soften.

Most Asian countries have instituted policies to provide consumers with a reasonable degree of food security, although few would claim that they have been completely successful in achieving that goal. There are many reasons. For one, the policies have largely been confined to actions affecting the principal staples, wheat and rice. For another, their successes have been largely confined to urban areas. Third, in many cases the countries have not been able to maintain their policy objectives in the face of global food shortages such as occurred in 1964-65 and 1973-74.

Government interventions have had both short-run and long-run objectives (Barker and Herdt 1985). The short-run objectives generally have been stated as "ensuring stability" in the market or preventing "undue price variability," objectives that would, if successful, be reflected in decreased seasonality of prices in the market. In most cases prices are the principal direct indicator of the effect of the policies, and governments have been most concerned about retail prices paid by consumers. Long-run objectives, when stated, relate to farmers' income levels and future production, although governments do evaluate levels of availability as indicators secondary to prices.

Countries attempt to enforce their policies through controlling sales or through other interventions in the market. Rationing is a seemingly attractive approach because it appears to be the only way one can maintain low prices when faced with a quantity inadequate to meet market demands. Black markets, smuggling, and other evasions have overcome most attempts at rationing after even brief periods.

The most widely used approach to enforcing food policies is sales by government or by government-controlled shops at fixed prices, backed by a government wholesaling agency that also controls imports. The Philippines, India, Indonesia, Bangladesh, Sri Lanka, Nepal, and other countries have all used some modification of this approach. Precise details change from time to time, and sometimes only some parts of the population, such as those living in a particular region or those with purchasing privileges, have access to the subsidized food (Ahmed 1981; Ray, Cummings, and Herdt 1979; Barker and Herdt 1985; Arjyal and Poudyal 1982; Bouis 1983; Pinstrup-Andersen 1988). The key to making such systems work is adequate supplies to provide to distribution outlets.

Policies that mandate uniform farm procurement prices throughout the year are also often employed. They tend over time to undercut private marketing channels and leave greater and greater proportions of marketed surplus in the hands of the government purchasing agency, as occurred in the Philippines and Indonesia (Mears 1981). The same effect is achieved even more rapidly when the government prohibits movement of food stocks from one region to another, as it did in India. The public agency made most of its purchases in the surplus region and transported them to the consuming regions. "The faster rate of increase in the food grain output of these surplus pockets and a policy of uniform prices throughout the procurement season has resulted in a shift toward a concentration of the pattern of market arrivals which in turn has increased the management problems of the public sector agency" (Tyagi 1982). Over time the burden on the government to store and handle the food clearly becomes unmanageable, while at the same time it retards the development of competitive markets operated by private traders.

Indirect Effects of Technology on Seasonality of Food Security

While most policies have attempted to promote seasonal food security through market interventions that affect sale prices or quantities offered, technology can affect the seasonality of production as well as some aspects of marketing in a way that has direct effects on the seasonality of food availability and on employment and income, which affect people's ability to purchase food.

Varieties

The new semidwarf wheat and rice varieties that spread throughout Asia in the 1960s and 1970s have several characteristics that might have affected the seasonality of production. First, while most traditional varieties are sensitive to day length, the new varieties are insensitive to the number of hours of daylight, and for that reason flower and ripen any time during the year. This characteristic becomes more important the farther from the equator one goes because of the changing length of day through the seasons outside the tropics. Second, the new varieties could, in many locations, produce a crop in fewer days than the old varieties, which required 180 days or more. IR8, one of the first new varieties of the 1960s, took 150 days to mature. IR36, a variety widely grown in the early 1980s, took 110 days, and newer rices mature even slightly faster. Their insensitivity to day length makes it possible to plant any time during the year, while the shorter ripening time can mean more crops harvested from a field in a year. Third, the new varieties, being semidwarf, could profitably take advantage of higher rates of fertilizer application, especially when grown with adequate water. Irrigation investments increased rapidly as a number of countries attempted to capitalize on the increased productivity. This led to greater dry-season production. In some cases irrigation authorities attempted to enforce increased uniformity in the seasonal pattern of cropping in order to use water most efficiently.

Agricultural researchers have also experimented with chemicals to help control weeds and insects. These have the potential effect of changing the seasonal pattern of demand for labor.

Technological changes may also affect seasonality in other ways. Prior to semidwarf varieties, most rice grown in Java was tightly bound to the stalk, and the traditional method of harvesting such rice was to bundle the entire stalk and carry it to the farmyard, where it was dried and threshed at a convenient time. The new rices are less tightly attached to the straw, making it less practical to transport them unthreshed, so they must be threshed in the field and transported in bags, possibly increasing the demand for labor during the peak harvesting season.

These kinds of technological changes might affect food security by changing the seasonal patterns of market deliveries, thereby reducing seasonal fluctuations in availability and prices. Because they affect the pattern of production activities, they might also have an impact on the seasonality of labor use and thereby on the income of laborers and farmers. Data on production timing, intensity, and employment will now be examined to see whether such effects are apparent.

Timing

An illustration of changes in the seasonal pattern of harvested area between 1955 and 1977 is shown in figure 14.1 for Indonesia and in table 14.1 for Indonesia and the Philippines. In Indonesia, May was the peak

FIGURE 14.1 Pattern of total rice area harvested in Java and Madura, Indonesia, 1955, 1968, and 1977

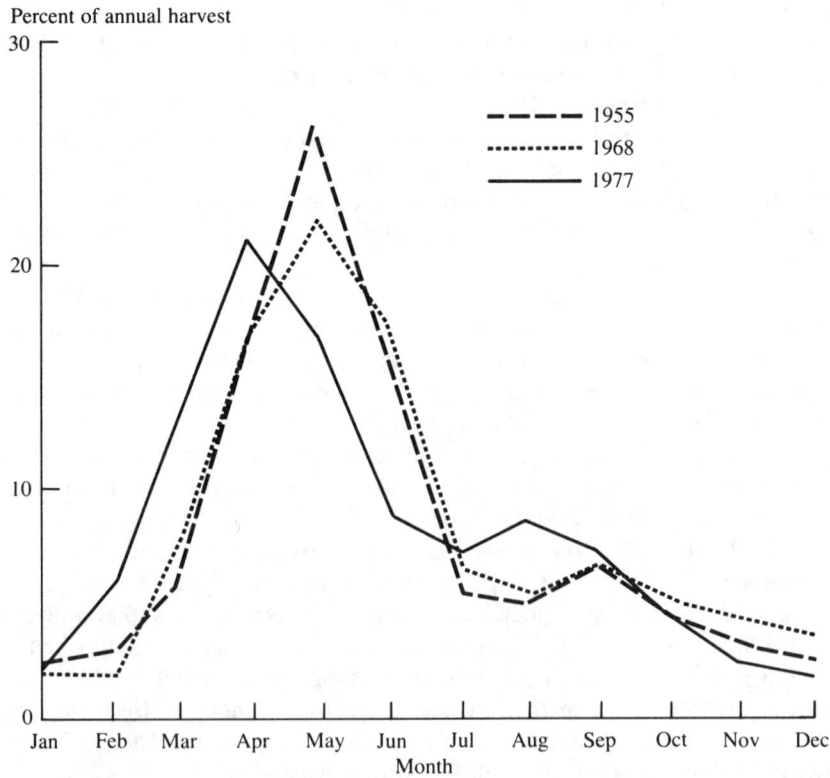

SOURCE: Adapted from Mears 1981, p. 40.

TABLE 14.1 Percentage of rice area harvested by month, Indonesia and the Philippines

	Java, Indonesia			Philippines		
	1955	1968	1977	1968	1975	1983
January	2.4	1.7	2.2	3.8	2.6	3.0
February	3.0	1.7	5.7	4.3	3.8	4.5
March	7.8	8.0	13.4	4.6	8.1	10.3
April	16.9	17.3	21.3	12.3	10.5	10.4
May	26.4	22.2	16.6	5.9	5.6	5.8
June	16.2	17.2	8.9	3.6	3.2	2.1
July	5.3	6.2	7.2	1.1	1.4	1.1
August	4.9	5.4	8.6	3.3	3.4	1.8
September	6.7	6.7	7.2	4.3	7.6	4.7
October	4.3	5.3	4.2	15.5	18.3	14.5
November	3.4	4.4	2.7	21.6	22.0	23.8
December	2.6	3.9	1.9	23.1	12.6	14.2
Percent MVs	0[a]	3[a]	53[a]	20	64	82

SOURCES: Monthly harvest data for Indonesia from Mears 1981; monthly harvest data for the Philippines from BAEcon 1985; and modern variety data from Herdt and Capule 1983.

[a] Refers to percentage of entire national rice area in modern varieties, while monthly harvest data refer only to Java.

harvest month in 1955, when 26 percent of the area was harvested. That peak was reduced to 22 percent in 1968, and in 1977 the peak of 21 percent occurred in April. It appears that the harvest has become earlier and slightly more uniformly spread as the area in modern varieties has increased—to 53 percent of Indonesia's total in 1977. One analyst attributes this to the expansion of irrigation as well as to new varieties, indicating how policy actions and technology have reinforced each other (Mears 1981).

Data for the Philippines reveal changes similar to those observed in Indonesia. There is a difference in that the Philippine data show two peaks—a small one in April and a main one in November (table 14.1). The proportion harvested during the principal peak (October–December) was 60.2 percent in 1968 when only 20 percent of the area was planted to the new rices; this decreased to about 53 percent in 1975 and 1983. The trough months of July–August went from 4.4 percent in 1968 to 2.9 percent in 1983. One also observes an increase around the secondary peak—the March harvest increased from 4.6 percent to 10.3 percent of the total. The three-month peak-to-trough ratio in the Philippines went from 7.5 in 1968 to 6.6 in 1975 to 6.9 in 1983, while in Indonesia it went from 7.4 in 1955 to 7.7 in 1968 to 7.5 in 1977 (recognizing the earlier peak in 1977).

TABLE 14.2 Patterns of market arrivals in three-month peak season (percent of total market arrivals)

	1961-62	1970-71	1979-80
Wheat			
Punjab	53	70	92
Uttar Pradesh	54	43	58
Rice			
Punjab	72	87	96
West Bengal	22	18	23

In the Punjab, India, the rapid increase in wheat production from new varieties—and to a lesser extent the increased production of rice—resulted in rapid shifts in the patterns of market arrivals over a period of two decades, as shown in table 14.2. Between 1961-62 and 1979-80, the increasing concentration of market deliveries following the harvest period of wheat (April-June) and rice (October-December) in the Punjab is startling, but in Uttar Pradesh and West Bengal there was almost no change. Over the period, the Punjab became a major surplus-producing area while Uttar Pradesh and West Bengal remained marginal and deficit states, respectively (Tyagi 1982). It is evident from these data and that of other states that seasonality of market deliveries increased more sharply in the surplus state.

Cropping Intensity

The typical cropping pattern in unimodal rainfall areas is highly seasonal, with cropping concentrated in the rainy seasons. New varieties that permit an increase in cropping intensity would appear to be one way to reduce seasonality of agricultural production, prices, and labor use. The potential for such an increase depends on the specific characteristics of the crops that can be grown in an area. One alternative is to grow a second crop of rice or wheat; another alternative is to produce other crops in the off-season. Several sets of data reflect changes in this dimension of production in Asia.

An examination of harvesting and planting dates of the main crops grown in the Kosi, Bihar (India), area found that the new varieties were transplanted earlier and had significantly shorter cultivation dates, leading to much earlier harvest dates (Hoskins 1984). However, the varieties available at the time of the study were not short enough in growth period to permit intensification beyond two crops in a year.

A series of surveys of rice farmers in two progressive rice producing areas of the Philippines illustrate the changes in cropping intensity that have resulted from technological change and irrigation investments. In

1966 a survey was conducted of a sample of farmers along the main highway in the central Luzon area of the Philippines. Modern varieties had not yet been released to farmers, and only about 20 percent of the land was irrigated and cropped twice a year. By 1970, modern varieties were being grown in two-thirds of the crop area during the wet season, but double cropping was confined to that 20 percent of the area that was irrigated. In 1973, major new irrigation facilities had become available and the proportion of the area with two crops of rice had increased to about 50 percent.

A similar pattern of increasing cropping intensity occurred in Laguna province, south of central Luzon. It had more irrigation in 1965, when about 65 percent of the sample farmers' land produced two crops per year. By 1978, when modern rice varieties were planted on virtually 100 percent of the area, 93 percent of the area was double cropped to rice (Herdt forthcoming).

In another location, in the south-central province of Iloilo, cropping intensity increased through the introduction of new cropping systems involving nonrice crops in the dry season following rice. In 1974 about 80 percent of the land of a sample of farmers in the study area was used to produce one crop of rice per year and was fallow thereafter. After four years of on-farm research on alternative systems, about 85 percent of the land was producing two or more crops per year, with the second crop about equally split between a second crop of rice and a nonrice upland crop (E. Price 1982).

Employment

The literature of development is littered with references to "peak labor" periods, usually cited as limitations to production intensification. There is little comparative evidence on the direct effect of technological change or policy on seasonality of employment in agriculture, however, because labor use data are usually obtained from one-time interviews with farmers.

The periodic survey of Philippine rice farmers referred to above showed that labor used for dry-season rice production increased as the area under dry-season crops increased from 20 percent of the total area in 1966 to 50 percent in 1973. Labor use during the main (wet) season crop increased from about 65 to about 85 days per hectare between 1966 and 1974 and remained at about that level in subsequent surveys through 1982. In another area further south, labor use increased from 90 days per hectare in 1965 to about 100 days per hectare in 1978.

Delgado and Ranade (1987) present data from Ryan et al. (1980) that suggest that 12–16 percent of farm labor is used during the peak labor month in three villages of India's semiarid tropics; the minimum monthly use was 2 percent. It is easy to recognize that new technologies should not make labor peaks worse, but usually those peaks are associated with good agricultural conditions. Despite the desire to flatten the peaks, research scientists often design technologies that would use higher and more peaked labor rates than existing technologies.

A comprehensive examination of the seasonality of employment, income, and consumption expenditure for a set of rural families is provided by Hayami (1978). His analysis of a rice village in the Philippines illustrates the complex interrelationships and the wealth of information needed to trace these factors over the course of an entire one-year crop cycle. A somewhat less comprehensive view of the same phenomena for selected households in a different part of the Philippines is provided by Ledesma (1982). Both studies illustrate seasonal income peaks coinciding with sales of the main and secondary rice crops. Consumption expenditures, on the other hand, show much smaller seasonal differences, with somewhat smaller and more variable levels among landless workers. This is of course consistent with the observation made above about relative seasonality of income and consumption.

Summary

This brief review has not turned up overwhelming evidence for a widespread seasonality effect on food consumption in Asia, although the evidence indicates that seasonal effects are present in Bangladesh, one of the lowest-income countries of the region. In some other countries it appears that marketing systems or policies have minimized the effects of highly seasonal food production.

There is considerable documentation of the seasonality of employment and earnings in agriculture. Daily record-keeping projects in the Philippines have documented large fluctuations in total consumption expenditures (not expenditures on food alone), but even there expenditures did not fall below minimum rice requirements. In six villages of the semi-arid tropics of India, where agricultural production is highly seasonal, there was also no evidence of significant variations in seasonal food consumption.

There is widespread government intervention in the market for rice, the principal staple in Asia. Despite this intervention, or perhaps because of it, there are large seasonal swings in market deliveries and prices received by farmers. Where governments have vigorously pursued the objectives of seasonal and temporal price stability, they have purchased increasing proportions of the main staple over time. This can be the cause of a budgetary crisis and may in turn lead to an eventual inability to meet the ambitious targets of price stability, as in the Philippines where surplus production in 1977–82 led to falling real prices and eventually to a slowing in the rate of output increase, leading in turn to the need to import rice in 1984–85.

Even in Indonesia, where the government has a major program of intervention in the rice market, significant seasonal price fluctuations continue. And it is clear that even an extensive and well-financed program cannot keep prices from rising when internal demand moves ahead of internal supply, or in an import-dependent situation when world prices increase sharply. In such conditions, seasonal price rises are exacerbated.

However, the case of Indonesia, as well as other countries in Asia, indicates that well-financed, extensive (and hence expensive) systems for distributing and selling the major food staple seem to have contributed to stabilizing consumption to a considerable extent. This is certainly one explanation for the apparent lack of large seasonal consumption swings in most Asian countries. The degree to which one should attribute this to direct government intervention, rather than to market forces that operate in spite of intervention, is still open to examination on a country-by-country basis. In a market environment with adequate roads and communications, private marketing systems have contributed much to stabilizing consumption, and policies should be designed to complement those institutions rather than replace them.

15 The Effect of Irrigation on Seasonal Rice Prices, Farm Income, and Labor Demand in Thailand

DOW MONGKOLSMAI AND MARK W. ROSEGRANT

Monsoon Asia is generally characterized by distinct wet and dry seasons, which are associated with wide seasonal fluctuations in agricultural production, farm income, labor use, wages, and prices. Irrigation is perhaps the most obvious instrument for attempting to reduce the fluctuations between peak and slack seasons. This chapter examines the impact of irrigation in Thailand on seasonal food availability and prices, farm income, and labor demand.

At an aggregate level, expansion of irrigated area may reduce seasonal rice price variation. Two related effects may contribute to reducing the magnitude and increasing the predictability of seasonal price spreads. First, an increase in the absolute level of rice production in the dry season may reduce the storage requirements and cost of holding much of the wet-season production through to the following wet season. Second, seasonal price changes are influenced by the deviations in rice production from expectations. If it can reduce variability in production, and hence the deviations of actual from expected production, irrigation will contribute to mitigating the interyear instability in seasonal price changes.

At the farm level, irrigation may have large effects on seasonality in cropping practices, yield and production, farm income, and labor demand. When irrigation permits dry-season cropping, it will tend to smooth the seasonal pattern of production, income, and labor use.

The first section briefly describes the development of irrigation in Thailand, while the second explains the effect of irrigation on aggregate rice availability and prices. The next section uses a farm-level analysis to examine the impact of irrigation on production, farm income, and labor use, with implications for seasonal labor migration. General conclusions are presented in the final section.

Irrigation Development in Thailand

Early investments in irrigation schemes in Thailand were concentrated on large projects and major water distribution systems in the Central Plain. The benefits of the investments were largely to increase yield of the wet-season rice crop in the central region by improving the reliability of the water supply. In the early 1970s the emphasis was shifted to smaller projects to provide water distribution networks at the farm level in the already existing large irrigation schemes in the Central Plain, and to other regions of the country. Although there has been a rapid expansion of dry-season irrigated area, particularly in the Central Plain, this remained only 19 percent of the irrigated area in 1982.

The regional disparity in irrigated areas as shown in table 15.1 suggests that most irrigation benefits go to rice in the central region. Upland crops, especially those grown in the north and northeast, receive relatively fewer benefits from irrigation.

The benefits of irrigation include not only double cropping of rice in the Central Plain and a modest expansion of upland crop cultivation in certain parts of the north and northeast, but also an increase in rice yield per hectare in both the wet and dry seasons. The increase in rice yield has come about because irrigation has induced a shift in planting technique from broadcasting to transplanting, a shift from traditional to modern rice varieties, and increased use of fertilizer and other chemical inputs. The wet-season rice yield per hectare in irrigated areas during the latter half of the 1970s averaged 2.7 tons, almost double the rain-fed yield of 1.5 tons.

TABLE 15.1 Irrigated and rice-planted areas, Thailand, 1982

	North	North-east	Central	South	Whole Kingdom
Irrigated area (1,000 ha)	746	438	1,845	291	3,320
	(22.5)	(13.2)	(55.6)	(8.8)	(100.0)
Rice-planted area (1,000 ha)					
Wet season 1982–83	2,112	4,257	1,964	654	8,987
	(23.5)	(47.4)	(21.8)	(7.3)	(100.0)
Dry season 1983	66	35	507	26	634
	(10.4)	(5.5)	(80.0)	(4.1)	(100.0)
Percentage irrigated area/wet-season planted area 1982–83	35.3	10.3	93.9	44.5	36.9
Percentage dry-season planted area 1983/irrigated area	8.8	8.0	27.5	8.9	19.1
Percentage double cropping rice area	3.1	0.8	25.8	4.0	7.0

SOURCE: Adapted from Ministry of Agriculture and Cooperatives 1984.
NOTE: Numbers in parentheses indicate percentages of area in whole kingdom of Thailand.

Moreover, rice yield in the dry season, which averaged 3.8 tons per hectare, was significantly higher than the wet-season yield, because of the greater use of modern inputs and high-yielding varieties, combined with higher solar radiation and better water control in the dry season.

Total rice production increased from an average of 13.9 million tons in 1971-75 to 15.9 million tons in 1976-80 and 18.1 million tons in 1981-83. The contribution of the dry season increased from 6.3 percent of the total crop to 10.5 percent and 12.4 percent during the same periods. The difference in yield per hectare between the wet and dry seasons as well as an expansion in the area double-cropped to rice are the factors that account for the increase in dry-season production.

Rice Availability and Prices

The increased rice output resulting from irrigation has increased the supply of rice available for domestic and foreign consumers. The growth rate in rice production has permitted modest growth in per capita domestic rice consumption, together with large increases in export volume. The proportion of rice output that was exported increased from 9 percent in 1971-75 to 15 percent in 1976-80 and 19 percent in 1981-83. The increase in rice exports has been mainly contributed by the dry-season crop, as can be seen by its output growth rate of 16.1 percent per year during 1976-83, compared with 2.8 percent per year for the wet-season crop.

While the modern variety dry-season harvest is mostly exported, the preferred local variety wet-season crop is mainly supplied to the domestic market. The effect of the dry-season crop on seasonal rice price variations will depend partly on the size of the dry-season crop relative to that of the wet-season crop. If the dry-season crop is small, the large supply of the wet-season output that is placed on the market after the harvest will push prices down. In each month that follows, the price will increase by at least the cost of storage, until the wet-season harvest of the following year.

When a larger dry-season crop is marketed, this may dampen the seasonal price increase of the wet-season output, or may even cause prices to fall slightly. Hence the difference between the highest price relative to the annual average and the lowest price relative (that is, the monthly deviation from average) in any crop year can be expected to be smaller with than without the dry-season crop.

In the Thai case, since most of the dry-season harvest is exported, it is expected to have a small effect on the seasonal price pattern. Variations in dry-season production, which are mostly due to the availability of water, on the other hand, may cause some variations in domestic prices, because an unexpectedly small dry-season harvest for export in any year would mean that a larger portion of the wet-season crop will be drawn from the domestic market if export agreements have already been made. If export demand competes for the wet-season crop, this will affect domestic prices.

To examine seasonal price variations, the monthly retail price data for 5 percent broken white rice between 1966 and 1983 are used to compute the seasonal price rise (defined as the difference between the highest and the lowest real monthly prices relative to the 12-month centered moving average, and expressed as a percentage of the lowest price relative). The seasonal price rise declined from 29 percent of the lowest price relative during 1966-70, when dry-season cropping was insignificant, to 20 percent in 1971-75, 11 percent in 1976-80, and 13 percent in 1981-83, as the dry-season rice output became increasingly important.

However, it cannot be readily concluded that double cropping of rice has caused the reduction in seasonal variations in rice prices during a crop year. The price relatives exhibit no clear pattern of the high and the low months. For some years the low month precedes the high month, and for other years the reverse takes place. The number of months between the high and the low price relatives in different crop years also varies greatly. Moreover, the difference in price relatives between high and low months is rather small, especially since 1975. This is likely to be a result of the government rice export tax policy, which tends to stabilize the domestic rice price at a low level, rather than a result of a larger dry-season crop entering the market.

When the world price of rice increases, the government raises the rice export tax (rice premium) rate in order to curtail exports and maintain adequate rice supply for domestic consumption. The domestic price is thus kept down below the world price for domestic consumers. Conversely, the rice premium rate is reduced to promote exports when there is a rice surplus and the world price of rice decreases. The rice premium stabilizes domestic prices by preventing them from varying fully with the world price. The frequent change in the rice premium rates in some years, especially in early 1970s in response to greater fluctuations in world prices, has dampened the seasonal variations in domestic rice prices and has resulted in the fluctuating pattern of seasonal price changes.

In his study of seasonal rice prices in Indonesia between 1953 and 1969, Goldman (1974) used an anticipatory price model to examine the influence of the dry-season crop and other variables on seasonal prices. According to this model, the seasonal price rise in any market year is hypothesized to be influenced by the deviations from expectations of both wet- and dry-season rice production, as well as by the yield of maize, which is a major substitute for rice in Indonesia, and by the amount of rice injected by the government for price stabilization. Production expectations play an important role because the changes in expectations concerning supply and demand conditions may lead to a change in commodity prices over a period of time, as market participants decide whether to build up inventories for future sale or release a portion of stocks for current consumption.

The early-season price reflects the market's estimate of the size of the wet-season crop as well as the level of stock carried over from the previous market year. Since part of this large early-season supply is carried into the dry-season part of the market year, expectations about the dry-season crop also affect the early-season price level.

A similar approach is suggested here to explain seasonal variation in rice prices in Thailand. Since irrigation is expected to reduce year-to-year variations in rice output caused by weather conditions, as well as to reduce variations from the expected production levels for each crop, a look at the effects of the latter variations on price rises would indirectly reflect the effect of irrigation on seasonal price variations.

The Thai rice production data suggest that up to 1975, the deviations from expectations of wet-season rice output were larger than the dry-season deviations. After 1975, however, the dry-season deviations tended to increase somewhat and become larger than the wet-season deviations in some years. But with the dry-season output being rather small (9–13 percent of the total annual output), it is important to test whether the effect of dry-season variability on seasonal price spreads is significant.

Since there is no close substitute for rice in Thai domestic consumption, yields of other crops are not included in the anticipatory price model being tested for the Thai case. However, along with deviations from expectations of production in each cropping season, deviations of rice exports and of rice export prices are also included in the model. These replace the market operation variable used by Goldman for Indonesia.

To amplify, the definition and detailed computation of the variables in the anticipatory price model are presented in appendix A, and regression results are given in table 15.2. The dependent variable in the regression estimate is the percentage seasonal price rise (DEPBK), which is a function of the independent variables, including deviations of wet-season rice production (QRCEWI) and of dry-season rice production (QRCEDI) as a percentage of crop production; deviations of rice export from expectations, as a percentage of expected export (QRXEXI); and rice export price deviations, as a percentage of expected export prices (CRXTR). The latter two variables are used together and in separate equations, since they measure nearly the same effect. The variance in exports may capture the effect of any variation in the export price on domestic prices. The results indicate that the export variable is more significant than the export price variable in explaining seasonal price rises, but the export price variable also has an independent effect. The deviations from expectations of wet-season production and of rice export were significantly related to seasonal prices rises

TABLE 15.2 Regression results

Dependent Variable	Constant[a]	QRCEWI	QRCEDI	QRXEXI	CRXTR	R^2
DEPBK (1966-79)	0.1067[b] (3.3159)	0.0102[b] (2.7719)	−0.0031 (−0.2471)	0.0035[b] (2.7787)	0.0003 (1.6185)	0.3416
DEPBK (1966-79)	0.1036[b] (3.9384)	0.0103[b] (2.9292)	—	0.0036[b] (3.1133)	0.0003 (1.6818)	0.4034
DEPBK (1966-79)	0.1055[b] (3.7152)	0.0079[b] (2.2738)	—	0.0024[b] (2.4314)	—	0.3042

SOURCE: Calculated.
[a]t-statistics are in parentheses.
[b]Significantly different from zero at 99 percent confidence level.

(table 15.2). Therefore, the effect of rainfall, which largely influences variability of the wet-season crop, and the deviations in export, which are directly related to domestic availability, tend to be more important factors explaining the seasonal price rise for consumers than the dry-season production deviations or export price deviations.

Thus it seems that the dry-season cropping and the stability in output made possible by irrigation in limited areas in the Central Plain have not exerted much influence on the seasonal movements in the domestic price for rice consumers. This is attributed to the small area devoted to dry-season irrigated cropping. It is rather at the farm level that the impacts of irrigation on labor use and farm income are likely to show their greatest effect.

Rice Production, Labor Use, and Farm Income: Farm-Level Analysis

Irrigation can be expected to increase cropping intensity and alter production methods, thereby affecting the level of use and earnings of inputs. Among these inputs, changes in labor requirements will be treated in the most detail, because of their implications for the problem of seasonal labor migration, and because of the impact of labor demand on the groups of rural people who are most vulnerable to food insecurity problems—that is, landless laborers and farmers dependent on rainfall.

Irrigation facilities are not well distributed among the regions of Thailand. Unirrigated areas, which depend entirely on rainfall for cultivation, are mainly located in the northeast and, to a smaller extent, in the north and south. It is therefore in the central region that changes in cropping intensity and planting methods having the greatest impact on labor utilization have taken place. Thus the benefits of irrigation discussed in reference to the central area should not be overstated or generalized for the entire country.

The impacts of irrigation on output and labor use in rice cultivation can best be seen by comparing rain-fed and irrigated areas, in the wet and dry seasons. Since about 80 percent of the total dry-season rice is grown in the Central Plain, data from a field survey undertaken in this area will be used as the basis for analysis. The survey was undertaken by the Chanasutr Project, under the Greater Chao Phya Project, in the northern part of the Central Plain. The Chanasutr Project covers an area of 84,320 hectares (ha) in four provinces, and an intensive water distribution network had been constructed in about 28.5 percent of the total project area by 1982.

The total sample size from irrigated areas in Chanasutr was 60 farm households, half of them with and half without on-farm development or land consolidation.[1] These samples were chosen randomly from the specified irrigation zones and canals. The 65 samples from rain-fed farms were also randomly selected from the Takli district in Nakornsawan, where soil and weather conditions are similar to those in the irrigated areas.

The comparison between the two sample groups in the 1982 dry season and 1982–83 wet season yields the following results:

Cropping Intensity

Cropping intensity, defined as cultivated area as a percentage of total landholding area, has significantly increased with irrigation. The rain-fed farms grew rice only in the wet season on 77 percent of the landholding area and maize and sorghum on 8 percent of the area. Since no crops were grown in the dry season, the total cropping intensity was 85 percent in 1982.

Irrigated farms generally grow rice as the main and second crops. In areas without land consolidation, wet-season rice took up 92 percent and dry-season rice 73 percent of the total land holding. The other crop grown in a small area was sugarcane. The total cropping intensity was 168 percent.

In land-consolidated areas, cropping intensity was even higher, at 192 percent, with rice accounting for 95 percent in the wet season and 89 percent in the dry season, plus some sugarcane and bean (see table 15.3). As the level of water control increases, the cropping pattern changes from one rice crop toward greater double cropping of rice or rice and upland crops.

---

1. Land consolidation involves construction of minor irrigation systems to supply water to each farm, minor drainage systems to remove excess water from each field, rearrangement of farm holdings between newly constructed irrigation ditches and drains, and land leveling to improve on-field water control.

TABLE 15.3 Cropping intensity (cultivated area as a percentage of total land holding)

	Rain-fed	Irrigated	Irrigated with Land Consolidation
Dry season, 1982			
Rice	—	73	89
Sugarcane	—	1	4
Total	—	74	93
Wet season, 1982–83			
Rice	77	92	95
Maize	6	—	—
Sorghum	2	—	—
Sugarcane	—	2	4
Total	85	94	99
Total cropping intensity	85	168	192

SOURCE: Survey data.

Planting Methods

In all the rain-fed areas, only dry soil broadcast planting is practiced, but in irrigated areas, there has been a shift in rice planting method to transplanting and, to a small extent, to puddled soil broadcasting system. With land consolidation, however, there is a further shift from transplanting to puddled soil broadcasting. This new broadcasting method, coupled with increases in application of fertilizer and chemicals, provides a high yield per hectare. This technique requires a well-leveled field and a good water control system to drain out water before broadcasting in order to increase the water level with the height of the growing plants. Hence it has been adopted in over 70 percent of the land-consolidated areas in Chanasutr. In irrigated areas without land consolidation, puddled soil broadcasting can be, and was, practiced in a more limited area (28 percent in the wet season and 30 percent in the dry season) depending on the degree of water controllability.

Production and Yield

The wet-season irrigated yield in survey areas of 2.9 tons/ha of planted area was more than twice the rain-fed yield of 1.2 tons/ha. With land consolidation, the wet-season yield (3.8 tons/ha) was even greater than in irrigated areas without land consolidation. The dry-season yield (3.9 tons/ha) in irrigated areas was generally higher than the wet-season yield and higher in areas with land consolidation than without (table 15.4).

TABLE 15.4 Production and yield of rice

	Rain-fed	Irrigated	Irrigated with Land Consolidation
Wet season, 1981–82			
Production (tons)	335.85	423.41	323.56
Planted area (ha)	294.72	155.84	92.16
Yield (ton/ha planted area)	1.14	2.72	3.51
Harvested area (ha)	196.88	145.51	85.20
Yield (ton/ha harvested area)	1.71	2.91	3.80
Dry season, 1982			
Production (tons)	—	349.75	336.93
Planted area (ha)	—	89.44	79.68
Yield (ton/ha planted area)	—	3.91	4.23
Harvested area (ha)	—	82.08	74.72
Yield (ton/ha harvested area)	—	4.26	4.51
Wet season, 1982–83			
Production (tons)	286.72	322.71	320.48
Planted area (ha)	244.80	111.68	84.80
Yield (ton/ha planted area)	1.17	2.89	3.78
Harvested area (ha)	181.76	100.16	77.28
Yield (ton/ha harvested area)	1.58	3.22	4.15

SOURCE: Survey data.

Distribution of Paddy Output

Farm households in irrigated areas tend to sell most of their dry-season rice output and retain less than 10 percent for seeds, consumption, or rent payment. The wet-season crop was retained in a higher proportion than the dry-season crop, and about 85 percent was sold. Rain-fed farms, having only the wet-season crop, sold half of the output, retaining 30 percent for consumption and 11 percent for seeds, with the remainder being stored for future sale and rent payment.

Farm Household Gross Cash Income

Table 15.5 shows that the gross annual cash income per household in 1982–83 was about 68,000 baht in irrigated and 11,000 baht in rain-fed areas. While the wet season and dry season contribute an almost equal share to gross annual cash income in irrigated areas, the dry-season income provided only 7 percent of the annual cash income of the rain-fed farms.

The largest source of household income was the sale of paddy, contributing over 90 percent of the total income in each cropping season in irrigated areas without land consolidation, 80–85 percent in land-consolidated areas, and only 65 percent in the wet-season rain-fed areas. Other crops and livestock accounted for 24 percent of total cash income in the rain-fed areas in the wet season, and livestock alone contributed about 23 percent in the dry season. Livestock and nonfarm employment were important sources of income for the rain-fed households, not only in the dry season when rice cannot be grown, but also in the wet season, because paddy sales brought in only 65 percent of the total income.

For the land-consolidated farms, the other important sources of income were employment on other rice farms, especially in the wet season, and tractor rentals, mainly in the dry season. Since the broadcasting techniques used in land-consolidated areas require less labor than the transplanting method, household laborers were hired to work on other farms in irrigated areas without land consolidation. This hiring out occurred before or after broadcasting and harvesting on these laborers' own farms. Income from tractor rentals is mainly derived from renting out small tractors for land preparation, not only on rice farms but also in areas growing upland crops in the dry season. Consequently, cash income from sources other than paddy was much larger for land-consolidated farms than for households in other areas.

Net Cash Income of Households

After deducting production costs from the gross annual cash income received from the sale of paddy, the net cash income from paddy is found to be about 28,000–30,000 baht per household in irrigated areas but slightly negative for the rain-fed farms. Cash income from other sources, although approximately the same in rain-fed and irrigated areas, was only 4,300 baht per household, resulting in a very low net cash income for rain-fed farms (4,200 baht), compared with 35,000 baht in irrigated and 39,000 baht in land-consolidated farms (table 15.5).

Labor Use in Rice Production

Where the broadcasting technique is applied, the man-days of labor per hectare planted will be smaller than if transplanting is practiced. Thus fewer man-days are employed for planting in rain-fed and land-consolidated areas than in irrigated areas without land consolidation, where transplanted rice is predominant.

In the wet season, total man-days per hectare in rain-fed areas was 34, compared with 62 in irrigated areas and 48 in land-consolidated areas. Labor used for planting was 12.5 man-days per hectare in irrigated areas, compared with only 1.0 in rain-fed and 4.3 in land-consolidated areas. The man-days used for caring for crops, however, tend to be larger in land-consolidated areas (23.0), where large amounts of fertilizer, herbicides, and other inputs are used relative to other areas (table 15.6).

Man-days per hectare are higher in the dry season than in the wet

TABLE 15.5 Farm households' gross cash income by source

	Rain-fed		Irrigated		Irrigated with Land Consolidation	
	Baht	Percent	Baht	Percent	Baht	Percent
Dry season, 1982						
Farm income	10,500	23	874,232	98	861,953	94
Sale of paddy	—		817,762	91	731,223	80
Other crops	—		49,970	6	125,730	14
Livestock	10,500	23	6,500	1	5,000	
Off-farm income	—		1,600		32,880	4
Employment in rice farms	—		400		6,480	1
Tractor rentals	—		1,200		26,400	3
Nonfarm income	36,000	77	19,500	2	17,765	2
Employment	33,000	71	11,300	1	15,665	2
Trade	—		3,000		—	
Other	3,000	6	4,300		2,100	
Total cash income	46,500	100	895,432	100	912,598	100
Cash income per household	788		34,400		32,593	
Wet season, 1982–83						
Farm income	548,730	89	838,695	97	914,020	93
Sale of paddy	398,640	65	831,695	96	849,870	86
Other crops	107,090	17	500		22,150	2
Livestock	43,000	7	6,500	1	42,000	4
Off-farm income	14,315	2	4,500		46,800	5
Employment in rice farms	10,315	2	4,500		30,800	3
Tractor rentals	4,000	1	—		16,000	2

Nonfarm income	53,445	9	22,000	3	23,425	2
Employment	49,945	8	14,200	2	19,225	2
Trade	—		3,000		—	
Other	3,500	1	4,800	1	4,200	
Total cash income	616,490	100	865,195	100	984,245	100
Cash income per household	10,449		33,277		35,152	
Year						
Total annual cash income	662,990		1,760,627		1,896,843	
Annual cash income per household	11,237		67,716		67,744	
Household's annual cash income from sale of paddy	6,873		63,441		56,468	
Annual cash costs of paddy production per household	7,024		33,033		28,674	
Net cash income from paddy	−151		30,408		27,794	
Cash income from other sources	4,364		4,275		11,276	
Total	4,213		34,683		39,070	

SOURCE: Survey data.

TABLE 15.6 Use of inputs for rice cultivation

	Dry Season, 1982		Wet Season, 1982–83		
	Irrigated	Irrigated with Land Consolidation	Rain-fed	Irrigated	Irrigated with Land Consolidation
Land					
Area planted (ha)	89.44	79.68	244.80	111.68	84.80
Owned (percent)	95.89	80.92	90.13	91.83	82.08
Rented (percent)	4.11	19.08	9.87	8.17	17.92
Area harvested (ha)	82.08	74.72	181.76	100.16	77.28
Damaged planted area (percent)	8.23	6.22	25.75	10.32	8.87
Labor					
Total man-days/area planted (ha)	71.36	57.57	34.03	61.95	47.77
Family labor (percent)	45.31	60.16	60.43	40.94	58.37
Hired labor (percent)	51.58	37.32	28.56	54.48	39.92
Exchanged labor (percent)	3.11	2.52	11.01	4.48	1.71
Man-days/hectare of land used in each operation					
Land preparation	2.52	1.60	1.19	2.11	2.08
Water pumping	2.88	0.80	—	5.67	0.37
Planting	14.00	5.33	0.95	12.46	4.29
Crop caring	22.60	28.89	12.62	14.65	22.97
Harvesting	28.59	20.32	26.10	30.07	18.47
Threshing	4.35	2.89	0.88	4.04	2.26

Machine					
Land preparation					
Machine-hrs/area planted (ha)	12.70	9.64	5.29	11.17	10.66
Machine-hrs					
Own (percent)	83.80	63.54	19.14	80.77	63.72
Hired (percent)	16.20	36.46	80.86	19.23	36.28
Threshing					
Machine-hrs/ton of rice output	0.72	0.59	1.00	1.81	0.65
Machine-hrs					
Own (percent)	9.09	28.30	17.14	18.18	39.13
Hired (percent)	90.91	71.70	80.00	79.92	60.87
Exchange (percent)	—	—	2.86	1.89	—
Chemical Inputs					
Fertilizer (kg/ha)	256.60	369.20	72.77	216.06	351.41
Insecticides					
Kg/ha	9.38	9.43	0.94	—	9.90
Litre/ha	1.04	0.81	0.28	0.96	0.70
Herbicides					
Kg/ha	4.67	11.99	0.62	1.35	9.94
Litre/ha	0.69	1.60	—	0.84	1.60
Pesticides					
Kg/ha	0.68	0.30	0.78	0.78	0.42
Litre/ha	0.31	0.36	—	1.54	0.29
Other					
Kg/ha	—	0.27	14.13	—	0.27
Litre/ha	2.23	0.12	0.43	2.05	—

SOURCE: Survey data.

season (71.4 in irrigated and 57.6 in land-consolidated areas). Crop care takes more man-days per hectare in the dry season than in the wet season. This is because the area planted is less and the problems of insects and pests are usually more severe in the dry season, so farmers tend to spend more time in the fields between planting and harvesting than during the wet season.

Hired labor is used mostly for transplanting and harvesting, while family labor is mainly used for crop care. Therefore, in the rain-fed and land-consolidated areas where the broadcasting technique is mostly practiced and crop care does not require hired labor, the proportion of family labor used relative to hired and exchanged labor was highest. In the irrigated areas without land consolidation, where the transplanting method is predominant, on the other hand, there was a greater requirement for hired labor (see table 15.6). In the dry season, this large demand for hired labor is satisfied mostly by labor from nearby villages in the same province, and also from other nearby provinces in the areas where dry-season crops are not grown. In the wet season, however, hired labor comes from the surrounding rain-fed and land-consolidated areas either before or after planting and harvesting on these laborers' own farms.

These results clearly indicate that irrigation has an effect on seasonal labor requirements in two respects. Double cropping of rice in certain intensively irrigated areas in the Central Plain has increased labor employment in crop production during the dry season. Moreover, the increased use of fertilizer and chemical inputs required for high-yielding varieties and the shift to the transplanting method in irrigated areas also result in a larger labor requirement for planting and crop care in both seasons. The variation in employment per farm between seasons has been reduced, but the extent to which this is so in aggregate terms depends on the area brought into cultivation in the dry season.

Rain-fed farms exhibit a strong seasonality in employment, since labor is engaged in agriculture only in the wet season and seeks nonagricultural employment in the dry season. During the wet season, moreover, there is a peak labor requirement for transplanting and harvesting rice during which hired labor is employed to supplement family labor. For other operations, such as land preparation and crop care, however, family labor is mainly used.

Interestingly, labor tends to migrate seasonally from the northeast to the irrigated areas of the central region for farm or other work in the dry season and return for the planting season for both rice and upland crops in May–June.

Because of higher labor demand in irrigated areas in the central region, rural wages tend to be higher than in the northeast, but even in the central irrigated areas, rural wages have been rather stable and real income of labor has increased only slightly.

Conclusion

This discussion has pointed out that as a direct result of irrigation, there is double cropping of rice and an increase in rice yields in both the wet and dry seasons. Thus there has been a significant increase in rice production in the past decade. This increase in rice output, especially that of the dry season, has been absorbed largely by an expansion in rice exports, while the larger wet-season crop mainly serves domestic consumption. Hence the impact of irrigation on the seasonal availability of food for domestic consumers, especially those in urban areas, tends to be small.

Despite the rapid expansion in rice double cropping in the Central Plain of Thailand during the 1970s, the double-cropping area is still very small compared with wet-season irrigated areas and the total areas planted in rice during the wet season. The dry-season output contributed only about 12 percent to total annual rice output in 1981–83. Although there was a larger variation in dry-season output in the latter half of the 1970s than in the previous years, such deviations do not seem to exert any significant influence on the seasonal price spread in Thailand.

The impacts of irrigation can be more clearly seen at the farm level. The data from the Chanasutr Project field survey in the Central Plain for the crop year 1982–83 reveal that there has been a change toward greater cropping intensity in irrigated areas and an increase in double cropping of rice. In the area studied, cropping intensity nearly doubled with the introduction of irrigation. In rain-fed areas, residual moisture is inadequate even for secondary nonrice crops in the dry season.

In addition to a second crop, irrigation provides wet-season yields that are two to three times greater than those in rain-fed areas. As a result, despite higher variable costs, net farm income is much higher in irrigated areas and is distributed more evenly across seasons. Rain-fed farms are proportionately more reliant on nonfarm and off-farm income, receiving nearly half of their gross cash income from these sources, compared with an average of less than 10 percent in irrigated areas. However, while rain-fed farms rely proportionately more on off-farm and nonfarm income, in absolute terms income from these sources is no higher than for irrigated farms. Rain-fed farmers are unable to compensate for low farm incomes with income from other sources.

The relatively low yields and reliance on a single crop cause rain-fed farmers to retain nearly half of production for home consumption, while irrigated farms retain only 10–15 percent of each crop.

In addition to the direct impact of irrigation on yield and production, irrigation shifts the seasonal pattern of labor use in the wet season. Planting methods and labor requirements for planting are directly related to irrigation quality. In rain-fed areas, dry soil broadcasting is used exclusively, while in irrigated areas without land consolidation, a mix of methods is used, with transplanting predominant. Labor use for land preparation and planting is therefore substantially larger in irrigated areas without land consolidation. In land-consolidated areas, the better quality of water control permits a shift to puddled soil broadcasting, which reduces peak-season labor use relative to other irrigated areas. Labor use remains, however, larger than for rain-fed areas.

The labor requirement for crop care also increases with irrigation, owing to increased use of fertilizer and chemical inputs. Overall labor use in irrigated areas is thus much larger. This greater demand for labor is largely filled by hired labor, so wet-season irrigation increases the wet-season total wages received by hired laborers, in addition to raising dry-season labor use and, thus, incomes. The rain-fed farms exhibit a greater seasonality in employment since they have to resort to nonfarm work, which often leads to unemployment in the dry season. In contrast, irrigation represents an opportunity for agricultural employment throughout the year.

The stronger demand for labor increases average wages as well. Thus labor from the rain-fed areas in the northeast tends to migrate seasonally to the Central Region, and especially to Bangkok, for employment in agriculture, industry, or services. In the Central Region, on the other hand, high labor requirements for dry-season rice cropping are met mostly by labor from the nearby provinces in the same region. This suggests that irrigation stabilizes seasonal demand for labor, boosts wages for hired labor, and reduces seasonal out-migration of agricultural workers.

Appendix A: Definition and Computation of Variables Used in Regression to Explain Seasonal Price Rises

Variables in Regression

DEPBK = percentage seasonal price rise, Bangkok, in crop year (April–March)

QRCEWI = deviations of wet-season rice production from expectations, as a percentage of expected crop-year production

QRCEDI = deviation of dry-season rice production from expectations, as a percentage of expected crop-year production

QRXEXI = deviations of rice export from expectations, as a percentage of expected export

CRXTR = rice export price deviations, as a percentage of expected export prices

Computation of Seasonal Price Rises

DEPBK = (PRELBK$_H$ − PRELBK$_L$)/NM
PRELBK = REALBK/MAR
REALBK = (MRPBK/FPIXR) × 100

where:

PRELBK$_H$ = highest price relative within the crop year
PRELBK$_L$ = lowest monthly price relative within the crop year
NM = number of months between the low and high price relatives (between PRELBK$_L$ and PRELBK$_H$)
REALBK = real monthly price of rice, Bangkok
MAR = 12-month centered moving average of real prices
MRPBK = current monthly price of rice, Bangkok
FPIXR = monthly food price index, excluding rice

Computation of Deviations of Rice Production from Expectations

QRCWP1 = exp($a + b$ YEAR) from LN(RPROTW) = f(YEAR)
QRCDP1 = exp($a + b$ YEAR) from LN(RPROTD) = f(YEAR)
QRCTP1 = QRCWP1 + QRCDP1
QRCWP2$_t$ = (QRCWP1$_t$ + RPROTW$_{t-1}$)/2
QRCDP2$_t$ = (QRCDP1$_t$ + RPROTD$_{t-1}$)/2
QRCTP2$_t$ = QRCWP2$_t$ + QRCDP2$_t$

where:

RPROTW = rice production, Thailand, wet season
RPROTD = rice production, Thailand, dry season
QRCWP1 = expected rice production, wet season, alternative 1
QRCDP1 = expected rice production, dry season, alternative 1
QRCTP1 = expected rice production, crop year, alternative 1
QRCWP2 = expected rice production, wet season, alternative 2
QRCDP2 = expected rice production, dry season, alternative 2
QRCTP2 = expected rice production, crop year, alternative 2
LN = natural logarithm

Deviations of Rice Export and Export Prices from Expectations

Same methodology as for computation of deviations of rice production from expectations.

16 The Impact of Drought and Technological Change in Rice Production on Intrayear Fluctuations in Food Consumption: The Case of North Arcot, India

PER PINSTRUP-ANDERSEN AND
MAURICIO JARAMILLO

A major cause of malnutrition among the rural poor in many regions of developing countries can be found in the seasonal and other intrayear fluctuations in energy intakes (Chambers, Longhurst, and Pacey 1981). Energy intake is influenced by the ability of low-income households to acquire food. The ability to acquire food at a particular time, in turn, is influenced by household food stocks and purchasing power, as well as food prices and availability in local markets. Changes in the food system that alter the levels of any of these four factors will therefore affect food security.

Technological change, such as the replacement of traditional varieties with modern ones and the introduction of irrigation, has been very effective in increasing the yields and production of various crops (notably rice and wheat) as well as increasing incomes of farmers in developing countries (Pinstrup-Andersen, 1982; Pinstrup-Andersen and Hazell 1985). However, the impact of technological change on food consumption and nutrition, including seasonal fluctuations, is poorly documented.

This chapter presents the results of a study of rice farmers and landless labor households in North Arcot, India, over a 10-year period. During this period, a significant technological change took place in rice production in the region. An attempt is made here to estimate the effect of this change on intrayear fluctuations in calorie consumption by using standard deviations and coefficients of variation around the trend lines. In addition to direct comparisons of these magnitudes over a period of years, attempts are made to estimate the relative contributions of various factors to intrayear fluctuations in calorie consumption in order to identify ways of reducing such fluctuations through government policies and further technological change.

The chapter is organized as follows. The survey area and households are described briefly, followed by a presentation of the results of the direct comparisons of calorie consumption levels and intrayear fluctuations for the survey years. Although technological improvement in rice production was undoubtedly the most powerful change that took place, the region was affected by other factors during the 10-year period. Attempts are made to isolate the effects of other factors, including a major drought, in order to identify factors that influence the effect of technological change. Three such factors (the level of and fluctuations in rice production, rice prices, and household incomes from various sources) are analyzed in considerable detail. This is followed by an attempt to determine the effects of each of these factors by means of regression analysis. A brief summary and conclusions complete the chapter.

Description of Survey Area and Sample Households

The North Arcot district, located in the southern Indian state of Tamil Nadu, is the area of concern in this study. The average farm size is 1.6 hectares (ha), and more than 60 percent of the farmers cultivate less than 1 ha. It is an important rice-producing area, with paddy being grown on more than 40 percent of the total arable land.

A light southwest monsoon from July to September and the northeast monsoon from October through December provide more than 90 percent of the average 1,000 millimeters (mm) of annual rainfall. The long dry season (January-June) provides strong sunshine, which is favorable, given good irrigation, to rice production. The weather regime allows three crops to be planted each year:

1. season 1 (Samba)—sown in July-August and harvested in December-January;
2. season 2 (Navari)—a smaller crop that stretches through the dry season from December-January through May; and
3. season 3 (Sornavari)—which goes from May-June through September.

Monthly data were collected from a sample of households during three years. The first survey was undertaken during 1973-74 by Madras and Cambridge universities (Farmer 1977). The subsequent two surveys were done during 1982-83 and 1983-84 by the Tamil Nadu Agricultural University and the International Food Policy Research Institute (IFPRI) (Hazell and Ramasamy forthcoming).

Eleven villages were included in the first two surveys. The third survey included a subsample of five of these villages. In order to facilitate comparative analysis among the three survey years, the households from the subsample villages were analyzed separately for all three years, while the total sample from the 11 villages could only be analyzed for the two years for which they were surveyed. The five resurveyed villages have poorer infrastructure and less intensely irrigated agriculture than the six villages that were not resurveyed in 1983-84; indeed, they were the five poorest villages in the sample.

Each year's sample was divided into five groups, three of which were separately analyzed: (1) small rice farmers whose principal crop was rice and who operated 1 ha or less; (2) large rice farmers whose principal crop was rice and who operated more than 1 ha of land; and (3) landless laborers. The term *large* should be interpreted in a relative rather than an absolute sense, since the average farm size in that group was less than 3 ha. Nonrice farmers and nonagriculturalists, the other two groups, were included in the total sample but were not analyzed separately.

The impact of a drought during 1982-83 was augmented by the fact that the area had not completely recovered from a similar drought two years earlier. The rainfall in 1980-81 was 27 percent below the normal 1,000 mm per year; 1981-82 was a normal year; but in 1982-83 the rainfall fell again below normal (about 24 percent below). As a result, 1982-83 yields were estimated to be 40 percent lower than normal, and only 25 percent of the mean paddy area was planted with rice (Hazell and Ramasamy forthcoming). Season 1, the monsoon season, was negatively affected by the lack of rain; but it was seasons 2 and 3 that were affected the most, since they depend heavily on tank irrigation. There were only 30 days of tank water at the end of the rain, as opposed to 120-140 days in a normal year. The 1983-84 rainfall was about 50 percent above normal—so high that some of the season 2 rice yields were reduced by excess rain.

The rice yields are discussed elsewhere (Hazell and Ramasamy forthcoming) and will not be presented here.

Household Energy Consumption

Dramatic increases occurred in household energy and protein consumption during the study period. Average calorie consumption increased by 65 percent and protein consumption doubled between 1973-74 and 1983-84 (table 16.1). The negative effect of the drought in 1982-83 was offset by positive effects brought about by other factors during the period from 1973-74 to 1982-83, leaving the 1982-83 consumption level slightly above the 1973-74 level. However, data limitations preclude isolating the negative effect of the drought from other effects, such as increased availability of wheat and rice from public distribution outlets.

As expected, both calorie and protein consumption were higher among large farmers than among small farmers and landless laborers. However, during the 10-year period from 1973-74 to 1983-84, small farmers in the five villages increased consumption more than large ones, thus narrowing the gap between those two groups. The explanation is that

TABLE 16.1 Daily energy and protein consumption

	5 Villages			11 Villages	
	73–74	82–83	83–84	73–74	82–83
Calories/person					
Small paddy farmers	1,420	1,483	2,606	1,634	1,597
Large paddy farmers	1,707	1,861	2,884	1,725	1,929
Landless laborers	1,419	1,495	2,154	1,612	1,642
All households	1,483	1,639	2,448	1,670	1,727
Grams of protein/person					
Small paddy farmers	29	32	64	35	34
Large paddy farmers	37	40	69	37	42
Landless laborers	29	34	57	34	36
All households	31	36	62	35	38
Calories/adult equivalent					
Small paddy farmers	1,666	1,714	2,953	1,949	1,875
Large paddy farmers	2,038	2,232	3,456	2,046	2,275
Landless laborers	1,662	1,783	2,572	1,913	1,962
All households	1,746	1,950	2,909	1,980	2,048
Grams of protein/adult equivalent					
Small paddy farmers	35	37	73	41	40
Large paddy farmers	44	48	82	43	49
Landless laborers	33	40	67	40	43
All households	36	43	74	42	45

large farmers have now reached calorie consumption levels that suffice for most households, and further increases in incomes are spent largely on more expensive calories and nonfoods rather than on more calories. The increase in calorie and protein consumption among the landless in the five villages was also substantial but was considerably less than the increase among farmers. Daily consumption among the landless, which was similar to that among the small farmers in 1973–74, fell in relative terms to 83 percent of small-farmer consumption, or 450 fewer calories, in 1983–84.

In order to estimate seasonal and other intrayear variations across years and household groups, attempts were made to remove two other sources of variation from the data. First, month-to-month variations expected from lumpiness of purchases were reduced through the development of two-month moving averages. Second, the effects of yearly trends were isolated by linear regression.

As shown in figures 16.1–16.3, strong downward trends were observed for 1982–83, and even stronger upward trends were found during 1983–84.

FIGURE 16.1 Two-month moving averages of daily calorie consumption per adult equivalent, small paddy farmers, five villages

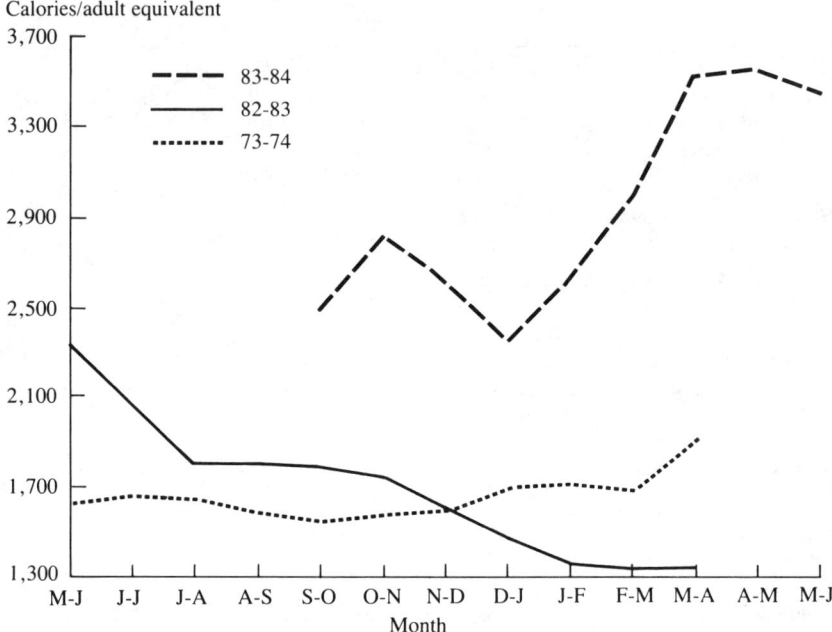

The principal reason for these trends was the drought, which gained momentum during the latter half of 1983–84, and the subsequent recuperation during 1983–84.

The intrayear fluctuations defined as the fluctuations around the trend line of two-month moving averages show clear seasonality, although other sources of fluctuations appear to be significant. Consumption is generally low right before the harvests in December–January, April–May, and August–September. In order to quantify the magnitude of the intrayear fluctuations, the standard deviation and the coefficient of variation of detrended two-month moving averages were estimated (table 16.2).

The standard deviation and coefficient of variation for calorie consumption were generally lowest among the producers during the first survey year and highest during the last. Differences between years and between household groups were tested for statistical significance.[1] As shown

---

1. Since the test is valid only for normally distributed variables, a normality test was run. The results indicate that the distribution was sufficiently close to normal. Since standard deviation and coefficient of variation are not expected to be normally distributed, however, the results should be interpreted with caution.

FIGURE 16.2 Two-month moving averages of daily calorie consumption per adult equivalent, landless laborers, five villages

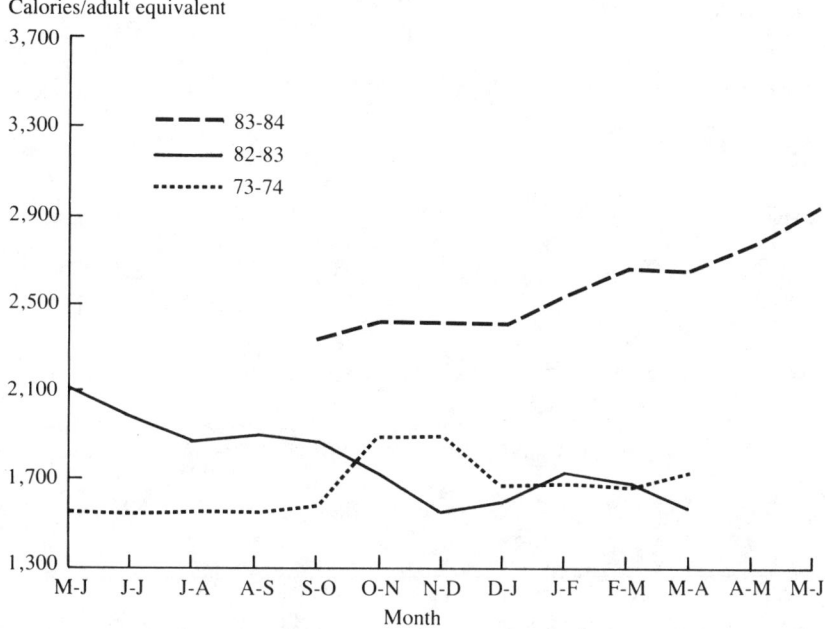

in table 16.2, increases in the standard deviation from 1973–74 to 1982–83 were statistically significant for both producer groups. Further increases from 1982–83 to 1983–84 were significant only for small farmers.

As discussed earlier, a comparison between 1973–74 and 1983–84 is likely to provide a better indicator of the effects of technological change than comparisons with 1982–83. Both small and large farmers show large and statistically significant increases in the standard deviation between those years. Except for the drought year, no significant difference in the standard deviation in calorie consumption was detected between small and large farmers. During the drought year, both the 5 resurveyed villages and the total sample of 11 villages show a significantly higher standard deviation among large farmers. No such difference was found in the coefficient of variation. Thus the difference in standard deviation was associated with higher levels of consumption by large farmers.

The standard deviation for calorie consumption by landless households did not differ significantly among the three years. The lack of increase with increasing consumption level is probably explained by higher dependence on off-farm income sources during the latter years, which are

FIGURE 16.3 Two-month moving averages of daily calorie consumption per adult equivalent, large paddy farmers, five villages

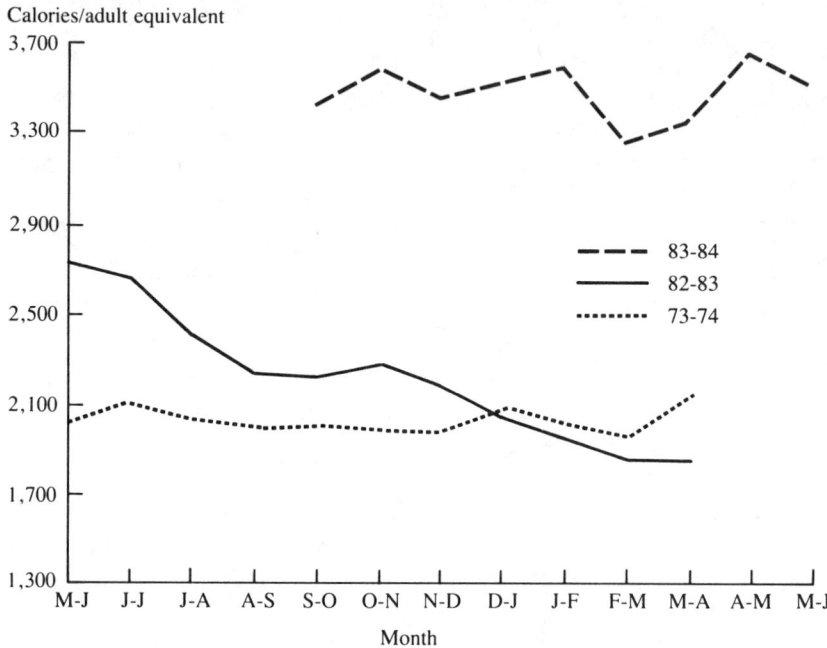

less seasonally determined. This explains the fall in the standard deviation during the drought year, when one-third of the incomes of landless labor originated outside farming, as opposed to 15 percent during 1973-74 (Pinstrup-Andersen and Jaramillo forthcoming).

The increase in intrayear variations among farmers was accompanied by increasing annual average consumption. This implies that the magnitude of variation relative to the mean consumption (CV) need not be increased. As shown in table 16.2, the seasonal fluctuations relative to the mean were significantly higher in 1983-84 than in 1973-74 only for small farmers. Thus it may be concluded that technological change increased absolute intrayear fluctuations in calorie consumption among paddy farmers, particularly among small ones, whose relative fluctuations also increased.

In order to identify ways in which the intrayear fluctuations may be reduced, we analyzed the effects of factors hypothesized to be important determinants of intrayear fluctuations in calorie consumption—that is, the level of and fluctuations in rice production, rice prices, and household incomes and total expenditures.

TABLE 16.2 Mean household-level variance (standard deviation and coefficient of variation) of detrended two-month moving averages of calorie consumption per person per day

	Resurveyed Villages (5)			Whole Sample (11)	
	73-74	82-83	83-84	73-74	82-83
Standard deviation					
Small paddy farmers	145.7 $bc\gamma$	287.0 $bd\alpha$	491.4 cd	179.3 $a\gamma$	298.7 $a\alpha$
Large paddy farmers	165.8 bc	391.4 $b\alpha\beta$	476.5 c	271.3 a	370.5 $a\alpha$
Landless laborers	342.4γ	312.5β	380.2	288.2 $a\gamma$	343.3
All households	246.3 bc	342.0 bd	423.7 cd	251.7 a	348.9 a
Coefficient of variation (percent)					
Small paddy farmers	10.7 $c\gamma$	16.4	24.5 c	11.7 $a\gamma$	16.2 $a\gamma$
Large paddy farmers	10.5 b	19.3 b	17.5	15.4	17.3β
Landless laborers	22.0γ	19.9	19.2	17.8γ	19.7$\beta\gamma$
All households	16.0	19.2	19.4	15.3 a	18.2 a

NOTE: The following pairs of estimates were tested for statistically significant differences using Tukey's test (Tukey 1982): (1) within each group of households in each sample, every year was tested against every other year; and (2) within each year, every subgroup of households was tested against every other one. Significant differences ($\alpha \leq 0.1$) are shown in the table with matching letters as follows:

Letter	Sample	Year		Sample	Year
a	whole	73-74	versus	whole	82-83
b	resurveyed	73-74	versus	resurveyed	82-83
c	resurveyed	73-74	versus	resurveyed	83-84
d	resurveyed	82-83	versus	resurveyed	83-84

αSmall versus large farmers.
βLarge farmers versus landless laborers.
γSmall farmers versus landless laborers.

Levels and Intrayear Fluctuations in Rice Production

Mean rice production per paddy farmer was slightly higher during the drought year (1982-83) than during 1973-74 among small farmers but considerably higher on larger farms (table 16.3). The increase among the latter was due primarily to larger rice acreage. Spectacular increases in production between the drought and the following year illustrate the large negative effect of the drought. Furthermore, a comparison of the first and last survey years provides an indication of the large effect of technological change. The amount of own-production rice consumed by small farmers stayed roughly the same over the three periods studied, but increased significantly for large farmers after the drought. As a percentage of total production, rice consumption decreased among rice-producing households from about two-thirds in 1973-74 to about one-third in 1983-84, indicating a large increase in the marketed surplus.

TABLE 16.3 Rice production, consumption, share of production consumed (means across paddy-producing households), and number of net paddy-buying households

	5 Villages			11 Villages	
	73-74	82-83	83-84	73-74	82-83
Small paddy farmers					
Production (kg/hh/yr)[a]	751	844	1,510	909	946
Consumption[b] (kg/hh/yr)	442	533	503	419	457
Consumed/production	0.59	0.63	0.33	0.46	0.48
Net buyers					
Number	5	3	0	6	11
Percent of all producers	38.5	25.0	0.0	20.7	22.5
Large paddy farmers					
Production (kg/hh/yr)[a]	991	1,567	3,271	2,366	2,630
Consumption[b] (kg/hh/yr)	627	712	877	668	713
Consumed/production	0.63	0.45	0.27	0.28	0.27
Net buyers					
Number	2	8	1	2	20
Percent of all producers	28.6	22.9	4.8	8.0	24.7
All paddy farmers					
Production (kg/hh/yr)[a]	835	1,383	2,785	1,583	1,995
Consumption[b] (kg/hh/yr)	507	666	774	535	617
Consumed/production	0.61	0.48	0.28	0.34	0.31
Net buyers					
Number	7	11	1	8	31
Percent of all producers	35.0	23.4	3.5	14.8	23.9

[a]hh = household.
[b]From own production only—i.e., excluding purchases.

Thus it appears that most of the increasing rice production was sold, while the amount of production kept for home consumption changed only slightly.

The intrayear fluctuations in the quantity of rice produced during each month were lowest during the drought year and highest during 1983-84. Although large, the increase in intrayear fluctuations of rice production was less than the increase in production levels as reflected in reduced coefficients of variation. Thus it may be concluded that between 1973-74 and 1983-84, intrayear production fluctuations increased in absolute terms and decreased when expressed as percentage of average production.

Levels and Intrayear Fluctuations in Incomes and Total Expenditures

As shown in table 16.4, total household expenditures and incomes increased dramatically during the 10-year period from 1973-74 to 1983-84. Accounting for inflation, average total expenditures for all sample households tripled. Small farmers experienced the largest absolute and percentage increase. Among the three groups of households, landless laborers had the lowest income at the end of the period. Their real incomes and expenditures increased in absolute terms during the 10-year period, and increased relative to those of large farmers, but decreased relative to those of small farmers. It appears that landless laborers were more severely affected by the drought than rice farmers. The least affected appear to be large farmers not in the resurvey villages. This would be expected because of the better water control they possessed.

The deterioration in incomes for the landless during the drought year appears to be due to a large decrease in farm employment, a decrease that was only partly compensated for by increasing nonfarm incomes. Off-farm incomes also decreased dramatically among rice farmers, while employment outside farming increased (Pinstrup-Andersen and Jaramillo forthcoming).

After the drought, the relative importance of off-farm incomes among the landless regained its 1973-74 level. This was not the case for the rice farmers, who derived a much larger share of their incomes from their own farms in 1983-84 than 10 years earlier, the reason being the increased production of rice, groundnuts, and other crops (Hazell and Ramasamy forthcoming).

Intrayear fluctuations in incomes and total expenditures for each of the three survey years were also examined. During 1973-74 and 1983-84, survey farmers obtained a large share of their incomes during the month of January. This corresponds to the peak of the major rice harvest. Considerable shares of their annual incomes were also gained after each of the other two rice harvests—that is, in April-May and August-September. The rather large share of total incomes obtained immediately after the rice harvests reflects the heavy dependence on rice as a cash crop and very limited storage of rice for subsequent sale.

Total expenditures, on the other hand, showed much smaller intrayear fluctuations. Thus it appears that survey farmers preferred to convert the rice harvest into cash and to hold cash rather than converting rice stocks as needed to meet cash needs throughout the year. It also appears from the data that the harvests in April-May and August-September have become much more important income sources. This is undoubtedly in large part due to the spread of modern varieties and improved water control, which made rice growing during the Navari and Sornavari seasons more advantageous than before.

During the drought year, very little rice was grown during the Samba season, and the January income peak did not occur. Instead, farmers planted upland crops, notably groundnuts, which were sold during October-November.

TABLE 16.4 Total annual expenditures and incomes (rupees per capita)

	5 Villages			11 Villages	
	73–74	82–83	83–84	73–74	82–83
Expenditures					
Current rupees					
Small paddy farmers	309.9	596.4	1,843.9	345.5	682.1
Large paddy farmers	505.9	908.7	1,954.9	522.7	967.3
Landless laborers	311.6	574.4	1,438.9	379.8	625.5
All households	339.2	712.8	1,720.2	407.2	777.7
1974 rupees					
Small paddy farmers	309.9	356.5	1,106.1	345.5	407.7
Large paddy farmers	505.9	543.2	1,172.7	522.7	578.3
Landless laborers	311.6	343.4	863.2	379.8	374.0
All households	339.2	426.1	1,031.9	407.2	464.9

Incomes				
Current rupees				
Small paddy farmers	246.5	561.4	1,285.9	319.0	728.6
Large paddy farmers	430.7	1,405.1	1,535.2	568.7	1,741.8
Landless laborers	301.4	423.7	1,014.6	384.6	416.0
All households	312.1	766.2	1,236.4	408.3	888.3
1974 rupees				
Small paddy farmers	246.5	335.6	771.4	319.0	435.5
Large paddy farmers	430.7	840.0	920.9	568.7	1,041.3
Landless laborers	301.4	253.7	608.6	384.6	248.7
All households	312.1	458.0	741.7	408.3	531.0

NOTE: The expenditure and income figures above are not directly comparable. Incomes were calculated with *farmgate prices*, while the value of the own-production share of expenditures was calculated with *consumer prices*. This explains why expenditures are greater than incomes in some cases. Deflation to 1974 values was based on the price index reported in Tamil Nadu, *An Economic Appraisal*, various issues.

Seasonal fluctuations in incomes of the landless households were smaller than those for farmers. During 1973-74, the monthly incomes of the landless were rather constant except for a significant increase in November that corresponded to the harvest of groundnuts. A more marked seasonality in the incomes of landless laborers was found in 1983-84, with peaks in January (corresponding to the Samba rice harvest), June (corresponding to the main planting time for rice), and October (corresponding to the harvest of upland crops, primarily groundnuts).

Intrayear fluctuations in total expenditures among farmers reflect these income fluctuations. Total expenditures generally peaked during the months of January, April-May, and August-September. It is interesting to note that although large incomes did not materialize during January of the drought year, expenditures were high. This is explained by customary harvest celebrations during that month, which took place in spite of the poor rice harvest.

The magnitudes of the intrayear fluctuations in incomes and total expenditures are shown in table 16.5. Intrayear fluctuations of net incomes for farmers increased between 1973-74 and 1983-84, both in absolute terms and relative to income levels. Landless laborers experienced increases in absolute fluctuations that were less than the increase in income levels, resulting in decreased fluctuations relative to income levels (CV). This reflects the earlier-mentioned shift toward income sources that are less sensitive to seasonal variations. Small farmers experienced a large increase in intrayear fluctuations of total expenditures, although it was less than the increase in the level of total expenditures. Fluctuations in total expenditures are virtually the same for the two years being compared, for both large farmers and landless laborers. In view of the large increases in total expenditures during the 10-year period, this implies a large reduction in fluctuations relative to total expenditure levels.

Factors Causing Intrayear Fluctuations in Calorie Consumption

Regression analysis was applied to estimate the impact of selected variables hypothesized to influence intrayear fluctuations in calorie consumption. These variables were the levels and intrayear fluctuations of household income and total expenditures, intrayear fluctuations in the price of rice, and household size. Intercept dummies for the year were also included to test for differences among years that were not accounted for by these variables. Finally, an intercept dummy was included to test for differences between farmers and landless laborers. The regressions were run with the standard deviation and the coefficient of variation for calorie consumption as dependent variables. The results are shown in tables 16.6 and 16.7.

Intrayear fluctuations in calorie consumption are estimated to in-

TABLE 16.5 Mean household-level variance (standard deviation and coefficient of variation) of two-month moving averages of net income and detrended total expenditure (rupees per person per day)

	5 Villages			11 Villages	
	73–74	82–83	83–84	73–74	82–83
Total Expenditure					
Standard deviation					
Small paddy farmers	0.21	0.31	0.60	0.21	0.32
Large paddy farmers	0.66	0.56	0.58	0.67	0.52
Landless laborers	0.30	0.26	0.34	0.31	0.25
All households	0.32	0.37	0.49	0.35	0.38
Coefficient of variation (percent)					
Small paddy farmers	23.5	22.6	19.9	22.6	23.0
Large paddy farmers	45.1	30.6	19.1	40.6	28.0
Landless laborers	46.9	22.2	16.4	35.5	21.9
All households	38.1	25.1	18.2	32.8	25.1
Net Income					
Standard deviation					
Small paddy farmers	0.82	0.89	2.30	0.97	1.44
Large paddy farmers	2.16	3.50	5.60	2.85	4.00
Landless laborers	0.32	0.27	0.58	0.43	0.24
All households	0.80	1.44	2.31	1.00	1.58
Coefficient of variation (percent)					
Small paddy farmers	116.6	99.1	151.5	109.9	122.9
Large paddy farmers	163.1	229.3	223.6	162.9	192.3
Landless laborers	37.8	35.0	32.2	39.8	33.5
All households	84.8	113.0	108.0	79.2	102.3

crease with higher net income and total expenditures. The rate of increase was about the same or slightly smaller than the increase in the level of calorie consumption. Thus the effect of changing incomes on the coefficient of variation was negative but very small. Intrayear fluctuations in net incomes appear to have a significant impact on fluctuation in calorie consumption among farmers but not among the landless. Thus efforts to reduce intrayear fluctuations in calorie consumption through policy interventions aimed at lowering the intrayear income fluctuations are likely to be effective only for paddy producers and not the landless. The principal reason for this difference between farmers and the landless is that intrayear fluctuations in incomes are much smaller for the landless (table 16.5). The policy implication is that measures that would bring intrayear fluctuations in incomes of farmers closer to those currently experienced by land-

TABLE 16.6 Results from regression analysis aimed at explaining variation in the standard deviation (SD) of calorie consumption

	Farmers		Landless		Total Sample	
	(1)	(2)	(3)	(4)	(5)	(6)
Household net incomes[a]	23.7 (0.01)[b]	—	80.6 (0.01)	—	43.2 (0.01)	—
Household expenditures[a]	—	128.1 (0.01)	—	124.3 (0.01)	—	140.8 (0.01)
SD of net income or household expenditure	2.8 (0.04)	44.5 (0.09)	−34.0 (0.28)	144.7 (0.01)	0.08 (0.98)	74.0 (0.01)
SD of rice price	−129.6 (0.30)	−7.1 (0.95)	154.2 (0.37)	58.3 (0.71)	9.1 (0.93)	28.1 (0.77)
Household size	−16.6 (0.01)	−9.6 (0.01)	−33.3 (0.01)	−22.8 (0.01)	−23.0 (0.01)	−13.2 (0.01)

	(1)	(2)	(3)	(4)	(5)	(6)
Intercept dummies						
1973-74 = 1	−229.0	−101.2	−99.0	−12.6	−165.6	−41.1
	(0.01)	(0.02)	(0.04)	(0.82)	(0.01)	(0.22)
1982-83 = 1	−85.8	12.7	16.2	72.4	−45.3	53.6
	(0.01)	(0.72)	(0.73)	(0.16)	(0.09)	(0.07)
Farmers = 1, landless = 0	—	—	—	—	12.5	−15.6
					(0.51)	(0.36)
F-ratio	13.5	23.9	10.1	19.3	16.8	33.8
R^2	0.22	0.33	0.19	0.31	0.18	0.30
No. of observations	296	296	259	259	555	555

NOTE: For each of the three groups (farmers, landless, and total), two functions were run: one (reported in columns 1, 3, and 5) with household income and all other variables except household expenditures, and one (reported in columns 2, 4, and 6) with household expenditures and all other variables except household income.

[a] Expressed in logarithms.
[b] Level of significance (0.01 means 0.01 or less) based on a t-test. Since SD is not expected to be normally distributed, the results of the t-test should be interpreted with caution.

TABLE 16.7 Results from regression analysis aimed at explaining variation in the coefficient of variation (CV) of calorie consumption

	Farmers		Landless		Total Sample	
	(1)	(2)	(3)	(4)	(5)	(6)
Household net incomes[a]	−0.61 (0.27)[b]	—	−2.00 (0.11)	—	−0.82 (0.12)	—
Household expenditures[a]	—	−2.47 (0.03)	—	−2.28 (0.12)	—	−2.48 (0.01)
CV of net income or household expenditure	0.00 (0.67)	0.14 (0.01)	0.02 (0.44)	0.31 (0.01)	0.00 (0.98)	0.23 (0.01)
CV of rice price	0.15 (0.09)	0.10 (0.25)	0.30 (0.01)	0.13 (0.18)	0.20 (−0.41)	0.11 (0.09)
Household size	−0.32 (0.07)	−0.37 (0.03)	−0.72 (0.03)	−0.98 (0.01)	−0.41 (0.01)	−0.55 (0.01)

Intercept dummies						
1973-74 = 1	−7.44	−11.00	−4.11	−10.68	−5.34	−10.63
	(0.01)	(0.01)	(0.09)	(0.01)	(0.01)	(0.01)
1982-83 = 1	−3.11	−5.74	−1.91	−4.03	−2.02	−5.05
	(0.04)	(0.01)	(0.42)	(0.07)	(0.12)	(0.01)
Farmers = 1, landless = 0	—	—	—	—	−1.19	−1.08
					(0.23)	(0.18)
F-ratio	4.4	8.2	2.0	22.1	5.0	24.0
R^2	0.08	0.15	0.04	0.34	0.06	0.24
No. of observations	296	296	259	259	555	555

NOTE: For each of the three groups (farmers, landless, and total), two functions were run: one (reported in columns 1, 3, and 5) with household income and all other variables except household expenditures, and one (reported in columns 2, 4, and 6) with household expenditures and all other variables except household income.

[a]Expressed in logarithms.
[b]Level of significance (0.01 means 0.01 or less) based on a t-test. Since CV is not expected to be normally distributed, the results of the t-test should be interpreted with caution.

less laborers would reduce intrayear fluctuations in calorie consumption among farmers.

Intrayear fluctuations in calorie consumption and total expenditures are closely correlated. This is expected because food occupies a relatively large share of total expenditures.

Intrayear fluctuations in rice prices are shown not to have a statistically significant influence on intrayear fluctuations in calorie consumption when both are measured in terms of standard deviations (see table 16.6). However, when fluctuations were measured in terms of the percentage of the annual average level—that is, when the coefficient of variation of the rice price is used to predict the coefficient of variation of calorie consumption—the relationship was significant (see table 16.7). The impact is considerably larger among the landless than among farmers. This implies that policy measures aimed at reducing fluctuations in the price of rice over the year may be effective in reducing intrayear fluctuations in calorie consumption, particularly among the landless. Neither the standard deviation nor the coefficient of variation of the rice price appears to have a significant effect on intrayear fluctuations of calorie consumption when total expenditures are used instead of net incomes in the regressions. Fluctuations in rice prices are a source of fluctuation in total expenditures, and it appears that the effect of the former on calorie consumption is probably accounted for in part by the total expenditure variable.

Larger households were found to have lower intrayear fluctuations in calorie consumption. Since calorie consumption and its fluctuations are expressed on a per capita basis, this finding is caused primarily by lower per capita consumption in larger households.

Intercept dummies were included to detect differences between survey years that are not explained by the above variables. The results indicate that there were significantly higher intrayear fluctuations in calorie consumption in 1983–84 than in 1973–74. These fluctuations could not be explained by this study.

Conclusion

Based on this study, it may be concluded that incomes, total expenditures, and food consumption of survey households increased dramatically during the 10-year period from 1973–74 to 1983–84. These increases were due in part to technological change in rice production.

Intrayear fluctuations in calorie consumption by paddy farmers also increased during the period, both in absolute terms and relative to average consumption levels. Among landless households, the intrayear fluctuations increased, but at a rate below the increase in average annual consumption.

The standard deviation in rice production among producers was lowest during the drought year but increased considerably between 1973–74 and 1983–84. However, the increase was less than the increase in production levels, thus resulting in lower coefficients of variation. Similarly, increases in the absolute level of the intrayear fluctuations in rice prices were less than the increases in price levels.

Intrayear fluctuations in net incomes of paddy farmers increased considerably during the 10-year period in absolute as well as in relative terms, while fluctuations in total expenditures relative to expenditure levels decreased. The increase in total fluctuations among landless laborers was less than the increase in their income level.

Intrayear fluctuations in both net incomes and rice prices appear to contribute significantly to intrayear fluctuations in calorie consumption. Income fluctuations appear to be important only for paddy farmers, while price fluctuations are important for both farmers and the landless. A considerable portion of the interyear difference in intrayear fluctuations in calorie consumption cannot be explained by levels and fluctuations in production, incomes, and prices. However, policy measures aimed at reducing fluctuations in the incomes of rice farmers and in rice prices are likely to result in reduced intrayear fluctuations in calorie consumption. Nonfarm employment opportunities have clearly been successful in reducing intrayear fluctuations in food consumption among landless laborers. Policies and programs that enhance employment during periods when incomes from farming are very limited would reduce intrayear fluctuations in food consumption among small farmers. Such measures should be based on solid information about the timing of farm incomes as well as labor demand in order to counter intrayear fluctuations in both. This type of information is also needed for households with different characteristics.

The nature and timing of policy measures and programs to address the problem of seasonal food insecurity may differ depending on the characteristics of targeted households. For example, small farmers and landless laborers are unlikely to face income shortfalls during the same period of the year. Furthermore, price fluctuations will affect households that are net sellers of rice differently from those that are net producers. Nevertheless, concerted efforts to reduce intrayear fluctuations in food consumption through policies and programs that promote crop diversification should also be explored. The analyses reported in this chapter further indicate the technological change has all but eliminated energy deficiencies among small farmers, even during the lean season, and has greatly reduced such deficiencies among landless laborers. The principal policy implication of this chapter, therefore, is the need to pursue technological change that further increases rice yields. This can be achieved through better water control and improved modern varieties and production practices.

This will serve the interests of small farmers directly if extension and credit services ensure their access to this technology. Small farmers and the landless also stand to benefit directly through expanded demand for labor and higher wages.

17 The Role of Agricultural Research and Secondary Food Crops in Reducing Seasonal Food Insecurity

RICHARD LONGHURST AND MICHAEL LIPTON

This chapter explores the role of agricultural research and secondary food crops (SFCs) in reducing seasonal food insecurity. The importance of this subject derives from three facts. First, the concept of secondary food crops (which are defined in greater detail in the next section) is primarily relevant to poor countries; only there do many people function with diets, and agricultures, so concentrated that the concept of a main staple food crop, and thus of its complement, a secondary food crop, makes sense. Second, poor people in poor countries generally have much more seasonal variation in household food security than rich people in rich countries. Third, SFCs are often either less seasonal in production and consumption than a main-season major food staple, or counterseasonal to staple crops, either in production and or in consumption. Given these facts, SFCs seem a promising source of food security for poor people in poor countries, in lean seasons.

We begin by seeking a usable definition for SFCs. We then examine the role of SFCs in the life of poor people, in a rather a priori way, supported at this stage only by some highly aggregated (and for Africa, somewhat doubtful) tables. Next, we examine how SFCs, thus defined, affect seasonal fluctuations in access to food or income. This is followed by a review of current research into SFCs. Finally, some implications for research and policy are tentatively suggested.

The discussion develops three propositions about SFCs:

1. Specific secondary crops are much more important for poor households and vulnerable groups within those households (e.g., women and children) who are either consumers or producers of these foods than is suggested by their role as measured in terms of average output figures or budget shares.

The authors are grateful to Ann Watson for typing this chapter with great speed and efficiency, and to Alison Pyle for research assistance.

2. Some SFCs may be highly significant in diets in ecologically marginal areas, because they are well adapted to survive extended rainfall deficiency in semiarid conditions.
3. Some SFCs become very important at critical times: intra-annually, before the main food crop is harvested, or interannually, when the main staple crop has largely failed.

These three propositions, if valid, make a strong case for increasing food crop research on SFCs. The remainder of the chapter presents evidence to support these propositions and discusses their implications for agricultural research.

Defining and Describing Secondary Food Crops

Generally, "secondariness" in terms of food intake overlaps with "secondariness" in production and employment generation. Most secondary food crops are grown as a subsidiary food source for home consumption, though even some gathered foods are traded quite vigorously in the market (Fleuret 1979b). Secondary crops, however, are rarely cultivated in stands for commercial purposes but are usually intercropped with the main staple.

A subjective, although operationally useful, definition is that for the average household over the entire year, an SFC (1) provides less than 20-25 percent of the household's dietary energy;[1] (2) provides less than 20-25 percent of the value of the household's income, including the value of produce consumed by the household plus income either from its sale or from wage employment involved in its production; and (3) uses up less than 20-25 percent of a scarce primary factor for cultivation (e.g., water, absorbable nitrogen, seasonal labor, or arable land).

This chapter's emphasis is on seasonal issues, and on a crop's direct role in enabling a household to consume adequate food energy. However, before we turn to this, there are a number of further ways in which agricultural research and policy might underemphasize crops that are secondary in general but vital in certain key respects.

First, secondary crops provide large proportions of micronutrients, especially vitamins: green leafy vegetables have iron, calcium, and vitamin A; legumes, vitamin B; fruits, vitamin C. Some also provide protein, al-

---

1. According to Food and Agriculture Organization food balance-sheet data (FAO 1980), with 1977 as the latest year, excluding cereals, roots and tubers, sugar, and vegetable oils, the proportion of energy availability in developing countries from the remaining vegetable sources (pulses, roots and oilseeds, vegetables, and fruit) is 10 percent. This varies little between continents, but is highest in Africa (12.9 percent).

though pulses usually provide less protein per acre than high-yielding cereals (Lipton 1985).

Second, a food crop may supply barely 10 percent of dietary energy yet provide more than 25 percent of a household's income to buy food. Such food crops may also be valued largely for economizing on land or for "noncaloric" reasons (taste, straw, crop rotation).

Third, SFCs have important features and by-products, apart from adding diversity and seasonal stability to the diet. Some assist soil conservation and reduce desertification, while others contribute to soil fertility (by nitrogen fixation and water retention) or to the reduction of pests and weeds (by influencing the microclimate within crops).

And finally, these positive features of SFCs are offset by some negative aspects. Many SFCs contain antinutritional factors that harm taste or absorbability into the gut.[2] Most of these antinutritive factors can be removed by prolonged boiling or soaking, but food preparation of this nature is costly. Consequently, in seasons when major staples are readily available, farmers may not harvest their secondary crops, saving them for a period of dearth or allowing them to be eaten by livestock or turned into green manure.

With these points in mind, it is possible to identify a wide variety of crops that fall within the definition. At one extreme is the low-status contingency crop that usually absorbs little labor time, being planted around the compound with other crops to fill in seasonal food shortages (e.g., taro and the cereal fonios—minor millets—in some West African countries). The other extreme is the high-value crop, usually vegetables, intensively cultivated throughout the year, usually with supplementary water in the the dry season, for sale to high-income urban consumers in nearby towns or even overseas (e.g., mushrooms in Taiwan). The extremes occasionally meet; for example, vegetables in Sri Lanka are grown for both subsistence and commercial purposes. Nonetheless, it is quite clear that SFCs can be defined to include a broad range of commodities that have diverse characteristics in different agroeconomic settings.

Secondary Crops and Seasonal Food Availability at the Local Level

Secondary food crops play a role in farmers' strategies to reduce fluctuations in the food supply that is much larger than their share in food

---

2. For example, cyanogenic glucosides are found in some roots and tubers, in some leaves of these crops, and in some legumes. Legumes also often contain trypsin inhibitors (these inhibit the digestibility of legumes by enzymes in the stomach). Oxalates are contained in many green leafy vegetables, and these reduce calcium availability. Phenols are present in some legumes, and these make the plant protein partially unavailable. Red quinua seeds have bitter-tasting saponins.

energy production. In this regard, they are essential components in many mixed-farming systems, being intercropped with staples, either on farms or in small cultivated gardens close to the compound.

The seasonal role of SFCs is obviously most important where seasonality is most marked, and will vary between the unimodal areas of West Africa, southern Africa, and India, the bimodal areas of East Africa, and the trimodal seasonality of Bangladesh. These various roles of SFCs are exemplified in Gambia by the fonios (millets), such as findi *(Digitaria exilis)*, which are used both as a mixed crop and in small stands immediately adjacent to the compound to provide food before the maize and early millet harvests. Findi has low labor requirements and matures in only 60 days. This low workload is important in view of the labor peaks for women, who are responsible for transplanting rice in the midrains of Gambia. In one village in Gambia, findi covered 11 percent of cultivated area but "cost" only 1.3 percent of total human energy input in subsistence farming. Late millet, on the other hand, covered a similar area, but total human energy input was 7.2 percent (Haswell 1981b). Of course, the total food yield per hectare of findi is only about one-third that of millet, but the availability of small amounts of food energy at low labor cost early in the cropping season is an essential part of the farmers' strategies to ensure food security.

Not only cereals but also some legumes fill seasonal gaps. Bambara groundnuts *(Voandzeia subterranea)* are short-maturing legumes that can be intercropped with a cereal and harvested during a second weeding. The beans of the African locust *(Parkia biglobosa)*, a tree legume, mature in the dry season of February–March, and if there is a drought the deep rooting system of the tree still allows it to provide a crop (Campbell-Platt 1980). These beans form the basis of the traditional food *dawa dawa*, and small quantities supplement soups with protein and fat.

Roots and tubers are counterseasonal simply because their harvests are not fixed. A crop such as cassava can be harvested at any period from four months to over two years (although cassava is not secondary but a staple in many parts of Africa). The introduction of *Xanthosoma taro* in Papua New Guinea has smoothed out many of the seasonal variations in consumption (Spencer and Heywood 1983). Also, the sweet potato *(Ipomoea batatas)* can be harvested flexibly. Most of these tubers normally rank higher than cereal crops in terms of calories produced per acre (Bouwkamp 1984), but not in terms of value-added per acre because of high processing costs and low value/weight ratio. However, as an SFC, cassava can stabilize farmers' and laborers' income even if used mostly outside the subsistence sector. In Indonesia, some 70 percent of produced cassava is sold, a good deal of it for urban processed foods (Falcon et al. 1984).

Green leafy vegetables (GLVs) do not contribute greatly to energy intake, but are a major source of vitamins A and C and are often used as relishes to enliven monotonous diets. Some are planted by farmers in vegetable gardens; others are collected in the bush and eaten as a relish. Most can survive periods of drought; yields are much reduced, but proportionately less than for most "major" cereals and even tubers.[3] In Mexico, GLVs are cultivated in depressions in areas with a high water table, making it possible to have greens throughout the year (Messer 1972).

Fruits too have a marked seasonality that often offsets that of main food staples. Mango, papaya, and citrus fruits become available for periods of about six weeks and often supply calories when most needed. Tree crops are less affected by moisture stress than main annual crops, and many tree fruits become available at the end of a dry season, well in advance of the main staple.

Most of the evidence for assessing the importance of SFCs in ensuring seasonal food security comes from West Africa. In Senegal among pastoralists, energy intake just met requirements throughout the cool season (January to March), but was inadequate in the wet season (Benefice, Chevassus-Agnes, and Barral 1984). Our calculation shows that the proportion of calories provided by legumes, vegetables, and fruit was highest in the cool season (7.5 percent), and the absolute amounts and proportions at other seasons were small (about 1–3 percent).

In Burkina Faso (see chapter 8), the consumption of SFCs (pulses, leaves, and fonio) is always highest in the hungry (harvest) period of October–November. There was no clear distinction in consumption trends between rich and poor groups.

In northern Nigeria, SFCs (seeds, nuts and legumes, vegetables and fruits) contributed 5.7 percent to energy intake year-round in nutrition surveys in Zaria (Simmons 1976a) and 7.4 percent in Malumfashi (Longhurst 1984). These crops provided 11.2 percent and 12.1 percent of protein intakes, respectively. A recalculation of Simmons's data for three northern Nigerian villages allows a breakdown by bimonthly groups (Longhurst 1985). This shows that the contribution of SFCs to energy intake is at its highest in the cereal preharvest period of August–September (8.6 percent) compared with the postharvest periods of October–November (4.7 percent) and December–January (6.6 percent) (see table 17.1). However, the sample sizes are too small to disaggregate rich and poor to see whether the more vulnerable poor households gained from this modest seasonal use of SFCs. In three villages, energy intake rose during the pe-

---

3. Common GLVs in Africa are species of *Bidens, Corchorus, Oxalis, Solanum, Portulaca, Colocasia,* and *Sonchus* (e.g., in Swaziland, Antonsson-Ogle 1984; for Tanzania, see Fleuret 1979b).

TABLE 17.1 Sources of calories by season in three Zaria villages in northern Nigeria, 1970-71 (percent)

	Apr.-May	Jun.-Jul.	Aug.-Sep.	Oct.-Nov.	Dec.-Jan.	Feb.-Mar.
Cereals	77.4	67.8	67.4	75.5	65.2	63.4
Cereal products	6.0	12.7	8.5	5.5	10.4	16.2
Starchy roots	0.6	2.2	5.1	2.2	1.6	1.8
Milk	1.0	1.3	1.2	1.0	0.8	0.9
Meat	0.4	1.0	0.4	1.1	1.0	1.5
Poultry, fish, eggs	0.0	0.0	0.0	0.5	0.0	0.1
Seeds, nuts, legumes	2.9	4.7	6.0	2.7	5.0	3.5
Fats and oils	9.1	7.5	7.6	8.1	12.6	9.9
Vegetables, fresh	0.2	0.5	1.6	0.8	0.4	0.3
Vegetables, dry	1.2	1.1	1.0	1.1	1.2	1.1
Fruits	0.4	0.0	0.0	0.1	0.0	0.0
Sugar, sweets	0.2	0.4	0.2	0.4	0.7	0.5
Salt, spices	0.0	0.0	0.1	0.1	0.0	0.0
Snacks, misc.	0.6	0.9	1.3	1.0	1.1	0.7
Total intake (kcal)	2,457.0	2,311.0	2,456.00	2,274.0	1,951.0	2,137.0

SOURCE: Calculated from Simmons 1976a, pp. 11-129.

riod of weeding and high energy expenditure (2,456 calories per capita) in the middle of the rainy seasons, and was lowest in the postharvest season (1,951 calories per capita). It is difficult to say for which people any given period was a hungry season; but plainly SFCs played a part in allowing people to raise food intakes when work needs were greatest, although this must have been when main staple stores were running low and prices high.

In Anambra state in Nigeria, where the wet season extends from April to November, no difference in energy intake was observed in the two seasons (Nnamyelugo et al. 1985). How did people maintain energy intakes long after the harvest? SFCs provided the answer. The main food staples were maize, yam, cassava, and cocoyam, but these were supplemented by various legumes. Reduced consumption of tubers and cereals in the wet season was compensated for by increased intake of legumes, which contributed 16 percent of energy intake in the wet season compared with 10 percent in the dry season.

In the Ivory Coast, SFCs appear more important in diets in the postharvest season of July-September, with pulses, seeds, vegetables, leaves, and fruits contributing 18.2 percent of energy supply, compared with 11.8 percent in the harvest period (Berio, Francois, and Odovafa 1985).

In the eastern province in Zambia, there is a unimodal rainfall pattern from November to March, with the main harvest in May-June. The major crops, maize and groundnuts, are the main farm activities, main income sources, and main dietary ingredients. SFCs are largely intercrops, such as beans, cowpeas, and pumpkins, and provide greens, seeds, and fruit for relish ingredients (Kumar 1985). Sweet potatoes are also consumed and are an important seasonal calorie source, mainly as a snack in the harvest period. Other SFCs do not have a significant hungry-season role. Poor and not-so-poor alike depend largely on maize for food in the hungry season; other major crops are not replaced by SFCs, with the exception of sweet potatoes.

A dietary survey was conducted in central India (Pingle 1975) on two tribal groups, the Koyas and Maria Gonds, to assess the seasonal contribution of cultivated and wild SFCs to dietary intake. Once again, main staples such as millets contributed a major proportion of energy (82-89 percent) and protein (68-79 percent) to the dietary intake of both tribes in the post- as well as preharvest seasons. Other cultivated foods, such as legumes, pulses, and vegetables, form an important source of minerals and vitamins in the postharvest rather than preharvest season.

Wild foods, however, contributed a significant proportion of nutrients in the preharvest season for both tribes, with the seasonal trend more marked among the Maria Gonds, the poorer of the two groups. Gathered foods contributed 12 percent of energy intake and 24 percent of protein intake before the cereal harvest (September), compared with 2 percent and 4 percent, respectively, postharvest (January-March). The less poor Koya community relied on wild foods for 10 percent of energy and 11 percent of protein intake preharvest and for zero in both cases postharvest. In this case, wild SFCs proved more important for the poorer community. This finding on the "diet" side is confirmed on the "income" side in Rajasthan villages (Jodha 1983), where common property resources, including access to wild foods, formed a large (although declining) share of income sources for poorer villages and households.

The evidence then allows only a partial response to the propositions presented at the beginning of this chapter. It also allows us to identify research gaps.

First, some types of SFCs (e.g., gathered foods, inferior staples, and root crops) are more important for the poor than the rich. This probably applies especially to SFCs that do not have a time-bound harvesting period, and that can easily fit into cultivation of a staple (with respect to timing of canopy cover, planting dates, and so forth). Many tree fruits come to harvest at the end of the dry season, providing food at this time. A longer rainy season gives more scope for a wider variety of SFCs, and hence they can provide a contribution to consumption and incomes.

Little evidence was found to support the proposition concerning the role of SFCs within the family. This is a great surprise considering the large amounts of money spent to encourage children to eat some types of foods, especially vegetables.

Third, the proposition concerning the ecological significance of SFCs has been confirmed. Many (such as legumes, fonios, tree foods, and roots and tubers) survive in semiarid areas and contribute to the diet.

Finally, as suggested earlier, SFCs do have important roles inter- and intra-annually. Their by-products, principally fodder, can also act as important sources of income.

The Current State of Agricultural Research on Secondary Crops

A comparison of research funds allocated to different crops at international and national levels reveals a number of surprises. Generally, SFCs such as those discussed here (minor roots and tubers, pulses, fruits and vegetables) receive more funds than their role in the diet appears to warrant (see table 17.2). For example, pulses contribute 3.7 percent to energy intake but receive 15 percent of Consultative Group on International Agricultural Research (CGIAR) funding. Although vegetables receive negligible funds in the CGIAR system, they are funded in the Asian Vegetable Research and Development Centre (AVRDC) in Taiwan. Rice and wheat appear to be underfunded; cereals contribute 60 percent to energy intake in less developed countries, but receive only 49 percent of international research funding.

TABLE 17.2 Food energy sources in developing countries (1980) and approximate annual allocation of CGIAR centers expenditures (1983) by commodity

	Energy		CGIAR Funding Level
	Kcal/day	Percent	Percent
Cereals	1,412	60.1	49
Roots and tubers	213	9.1	12
Pulses	87	3.7	15
Livestock and products	146	6.2	23
Vegetables	35	1.5	<1
Oilseeds	68	2.9	<1
Sugars and honey	150	6.4	<1
Fish and seafood	15	0.6	<1
Alcoholic beverages	30	1.3	<1
Other foods[a]	193	8.2	<1
Total	2,349	100.0	100

SOURCE: FAO 1984; TAC 1984; and CGIAR 1985.

[a]Fruit, nuts, animal oils and fats, stimulants, and spices.

However, at the national level in sub-Saharan Africa, the situation is completely different. In this case, the congruence ratio (the ratio of a commodity's proportion of research expenditure to its share of economic value in the crop mix; Judd, Boyce, and Evenson 1983; Lipton 1985) is 0.19 for sweet potatoes. The ratio is 0.09 for cassava, but in Africa cassava is hardly an SFC. Similar ratios are found for these two crops when data are aggregated for all developing countries. At the national level in Africa, rice is adequately researched and wheat is overresearched, with congruence ratios of 1.05 and 1.30, respectively; the ratio for soya beans is 23.6. Considering their importance in the diet, cassava, and sweet potatoes and almost certainly millet and sorghum, therefore, receive much less national funding than their importance deserves, even given the prospects for progress in the various crops.

All these are major staples in many African countries. That is one reason why their neglect at national levels is being partly redressed by international centers like the International Institute for Tropical Agriculture and the International Center for Agricultural Research in Dry Areas. In these cases, however, the research environment is usually far removed from the often semiarid, mixed-crop realities of farm life.

The real problems appear to lie with the low level of research funds allocated to subsistence, seasonal, and calorie-important SFCs: yams, melons, pumpkins, some bean sprout types, and bananas and plantains are key examples. Such SFCs receive research outlays far below "congruence" levels. Moreover, it is probable that even congruence would represent insufficient levels of resource allocation, for three reasons.

First, there will likely be high marginal returns to research on SFCs because of past neglect.

Second, these SFCs are often essentially "helper" crops, which in mixes or rotations carry out valuable functions such as fixing or restoring nitrogen or binding soil, and therefore help to raise yields for major crops and staples. Indeed, research into SFCs need to avoid overstressing yield enhancement *for them* at the expense of such functions, to avoid the danger that the farmer loses more by sacrificing the special advantages of SFCs than he gains from their higher yields.

Third, even if such SFCs "on average" contribute little to energy consumption, they help reduce the variability of energy intake by providing valuable calories in years and seasons when the staples fail.

For all three reasons, "overcongruent" investment in several SFCs may well be justified. Also, such an effort by CGIAR centers would compensate for the "below-congruent" national investment levels that are "justified" by (1) perceived low returns (for one country) on investing in a crop grown in marginal areas and perhaps adapted to produce low and stable yields in less fertile soils, and (2) low political returns on crops needed only seasonally, in remote places by poor people.

Research Requirements

An important consideration in the design of research projects is the preferred direction and ultimate goal of research. Should the focus be on increasing yields of robust traditional varieties grown in safe environments, or increasing the hardiness of already high-yielding crops grown in marginal areas? Should research focus specifically on SFCs or on major staple crops? These decisions have major implications for SFCs. If high-yielding but high-risk major crops were made safer (or if the yields of hardy but low-yielding major crops were increased), the importance of and gains from research into SFCs would diminish.

Indeed it might be asked if agronomic research, rather than research into storage or processing techniques, or improved credit and marketing facilities, is the most cost-effective means of reducing seasonal food insecurity. This question is especially pertinent to many SFCs that are inferior goods: as income rises, their consumption decreases. Therefore, significant technological advances that result in increased incomes reduce the marginal returns to further research on SFCs.

The answer to the question of whether a greater share of research resources should be devoted to SFCs, especially for reasons such as reducing seasonal malnutrition, depends on the product (in a Markov sequence) of three expected values or probabilities: (1) that of *success;* (2) that of *adopting area,* given success; and (3) that of *value-added* per unit of adopting area. Each of these factors should include delays and risk when calculating the research costs incurred. The factors may, of course, be interpreted discretely or continuously; and interactions among them, and among research efforts specific to different crops, should be allowed for. Our nutritional and seasonal emphasis indicates, moreover, that "adopting area" or "value-added" should be so weighted to favor crops providing sale or employment income, or those directly grown for own-consumption in hungry seasons and by poor people.

For some research decisions, information is available to estimate the probabilities and delays for each crop. However, in all cases, implicit weightings are made by research station directors in determining what crops, areas, and so on to emphasize, and the sums are thus done implicitly. We can, then, reword our earlier question: Are there any general characteristics of secondary crops, relative to major crops, that affect the probabilities and risks involved with each of the three factors? In particular, given a priority on improving nutrition, do secondary crops deserve higher or lower "hungry-season" or "poverty-group" weightings than major crops, for their expected values of adoption area or value-added?

What of the probability of research "success" for secondary crops? In the initial phases of research, the existing pool of available knowledge about SFCs is usually smaller than that for major crops. Fixed costs to acquire this knowledge will be high for each variety of SFCs.

Furthermore, secondary crops may be grown because they are perceived as less risky, or as insurance in case of an insufficient primary crop harvest. That risk-averting quality inherent in secondary crops may prevent the rapid adoption of innovations, since any change can be seen as increasing risk. Also, since by definition secondary crops are grown on a smaller portion of total cultivated area, the expansion induced by technological advances) and hence the returns to research investment) will be relatively low, even with adoption rates of 100 percent.

Overall, the range of environmental conditions under which a given secondary crop is cultivated is commonly greater (more heterogeneous) than for most major crops (except for cassava), making research breakthrough more difficult. Also, environments for secondary crops are commonly more hostile (semiarid, swampy, or otherwise marginal) than those for major crops. Therefore, the probabilities of success and benefits may be small.

These factors are offset by seasonal, and to some extent by poverty, considerations. Secondary crops grown for consumption must smooth out the seasonal distribution of food supplies if only a single-season main crop is harvested. This logical necessity disappears, of course, if the main crop is continuously harvested (e.g., cassava) or multiple-cropped; but even in such cases, the secondary crop is often selected for its seasonally corrective capabilities (for own-consumption or for income from sales). Secondary crops are thus seasonally corrective at best, neutral at worst.

High storage costs for major cereals and imperfections in capital markets discourage storage, which leads to marked seasonalities in food security and reinforces the need for more research into secondary crops than is suggested by the unfavorable values of the critical factors discussed above.

To assess their impact on poverty, secondary crops can be divided into three groups: those for which the income elasticity of demand in a given location at a given time is one or more ("luxury" crops), between zero and one, and negative ("inferior" crops). The first group includes exotic vegetables and fruits, such as asparagus and strawberries: no upgrading of research outlays, other than has already been indicated, is required in such a case on poverty grounds. Such "extra" research would not help the poor, except to the extent that poor farmers grew the crop for income.

The third group, "inferior" foods, comprises exactly the sort of crop for which research outlays enjoy little political backing, even considering the previous discussion. A case can be made for upgrading the attention to these crops on poverty grounds. However, many inferior foods are more or less gathered foods, used as major food sources only when cultivated crops fail. The normal approaches of agricultural research imply cultivation, not gathering, and in view of the large variety of gathered crops and their ecosystems, we doubt whether "research into gathering" is in most cases a sensible goal for agricultural science. However, many "inferior" foods that are secondary in a given locality are cultivated, not gathered. Examples are cocoyams and taro in much of West Africa; intercrops such as *hulga* and *matki* bean sprouts in the millet areas of western India; cassava in Kerala, Sri Lanka, and parts of Thailand; almost all nonrice, nonwheat cereals in Bangladesh; millets and sorghums, where farmers choose between these and low-yield but high-value local rice varieties. In such cases, the concentration of likely research benefits to poor and underfed growers and consumers of secondary crops justifies higher proportions of research outlay than the above arguments alone would indicate. Where inferior foods are gathered, however, despite their relatively greater importance for the poor, especially in the lean season, it may be more cost-effective nutritionally to develop marketing systems for major crops.

Many secondary food crops are neither of these two "ideal types" (luxuries like strawberries, inferior goods like cocoyams). The income elasticity of demand for most crops lies between zero and one. This is often true for "seasonal staples" such as fonio millets in West Africa. It is often, but not automatically, sensible to increase the share of research into these secondary crops on poverty grounds.

This chapter has suggested (1) that the existence of seasonal food insecurity normally favors increased research expenditures on SFCs; and (2) that poverty favors research on some secondary crops (like the cocoyam) but disfavors others (like the strawberry). But do (1) and (2) interact, to strengthen or to weaken the case for "more research" into some secondary crops? In other words, are SFCs more counterseasonal for poor than for nonpoor farmers?

A prior question to ask, however, involves the probability of research success: to ask whether research into secondary crops in likely to lead to unfavorable outcomes. Perhaps these crops are secondary not because they are underresearched, but because they are unresearchable. Are the crops too varied in growing conditions, or restricted to bad conditions? Or are they low yielding mainly because they are grown by poor people, who command little research (or inputs) because they have little economic and political power? If the former, is something else required, instead of improved varieties or practices for secondary crops, such as better extension, marketing infrastructure, or processing techniques?

It may be possible to justify the development of secondary crops for efficiency as well as equity reasons, even if the rate-of-return calculation looks unfavorable. Where they are largely poor people's crops or improve seasonal food security, their development can contribute through nutrition to the volume and quality of labor, and even to life expectancy.

Finally, it should be asked if staples could be adapted to do what SFCs do, rather than make the latter higher yielding or more versatile. The fixed costs of starting up research on some secondary crops may be very high, given the lack of knowledge on their ecology and physiology. The unit cost of research progress in secondary crops may also be very high in view of their heterogeneity, or because such crops' low yield is adapted to low-quality, or heterogeneous, environments. An agriculture that becomes more input-intensive, market-linked, research-based, and externally oriented may lead to a weeding out of many inferior secondary crops and encouragement of superior ones.

Would it be more cost-effective to spend research resources on "giving" maize the desirable seasonal characteristics of the fonio millets (i.e., greater stability of yield in response to rainfall variations) than on making the fonios more like hybrid maize.(i.e., tripling their yields)? Or is the second option a nonstarter in view of consumer preference for maize? Would improved processing techniques for these minor cereals and legumes benefit their poor consumers more, especially in slack seasons, than similar outlays on yield-increasing research?

These questions have to be evaluated in the context of an analytical frame such as that suggested at the beginning of this section. This means asking how many people are and will remain dependent solely on legumes and fonio millets in the hungry season, so that increased research on maize would dangerously divert research benefits from them, and how close we are to increasing fonio yields compared with stabilizing those of maize.

Certainly, there has to be a greater understanding of the way in which farmers grow secondary crops, and their place within the overall crop mix, together with a careful assessment of what is needed. It may be more important to ensure that researchers and policymakers, even if their main interest is in major crops, adequately recognize and allow for the role of SFCs and crop mixes within farmers' seasonal strategies, rather than that they initiate a crash program of SFC yield improvement. The main "poor people's crops" (cassava, yam, sorghum, millet, maize, upland rice, rainfed wheat) are still the main priorities. But at present, international research may not be doing enough to redress the biases of national systems against the remote, the minor, the "inferior," and the seasonal.

PART VI

Policy Implications

18 Policy Recommendations for Improving Food Security

DAVID E. SAHN

Agricultural growth and market development represent the logical long-term means of reducing seasonal food insecurity. One need only witness the almost total lack of seasonal cycles of malnutrition and food energy intake in developed countries. Likewise, the magnitude of seasonal fluctuations in consumption is generally smaller in middle-income countries, where agricultural and market development are most advanced. This reduction in transitory seasonal food insecurity occurs despite the fact that agriculture itself remains a highly seasonal endeavor.

Conversely, seasonal undulations in food security are most pronounced in the lowest-income countries, where agricultural progress has faltered; infrastructure is most limited; and markets remain poorly integrated, inefficient, and selective as to whom they effectively reach and serve. It is therefore among the lowest-income African countries that a combination of environmental factors (e.g., unimodal rainfall, soils that do not hold moisture) and economic stagnation contributes to making seasonal fluctuations in food security most acute. Likewise, the evidence from Asia indicates that in the poorest countries like Bangladesh, seasonal food insecurity is predominantly a characteristic of poor rural communities, especially in areas where the agricultural calendar is highly seasonal in terms of the timing of food production and labor demand.

Prior to the introduction of modern agricultural techniques or the influence of a monetary economy, agricultural societies experienced considerable seasonal variability in production, work, and income. Traditional social, agricultural, and biological adaptations, however, were often quite effective in buffering households from the hardships of the lean season.

Consequent to the process of agricultural development and commercialization, labor, commodity, and credit markets represent another set of mechanisms through which households can increase and stabilize food intakes, in spite of the seasonality inherent in agricultural activities and food prices. Nevertheless, there is some evidence that the dissolution of traditional practices of coping with seasonal stress, in favor of increasing reliance on the market for food sales and purchases, may represent a transitional risk factor of seasonal food insecurity. The dislocation that often accompanies the transition from subsistence production to commercialization increases significantly the risk of seasonal shortfalls in food consumption.

This vulnerability may arise in a number of forms, under a variety of circumstances. There may be a loss of traditional coping mechanisms such as tribal and community networks of sharing and transferring food resources to households in greater need during a period of food stress. This may be attributable to population pressures, greater land scarcity, change in patterns of access to land, environmental degradation, or simply an inevitable process of social evolution that occurs with commercialization and modernization. Likewise, there may be a shift away from traditional patterns of cultivation and foraging, food processing, income diversification, and other means of adapting to the local ecosystem that protected households during seasons of dearth. If this occurs prior to a sufficient level of market development and integration, the farmer, pastoralist, and wage earner may face the risk of more catastrophic losses and acute seasonal shortfalls in food availability, incomes, and consumption.

Any assertion that modernization and monetization contribute to increased seasonal shortfalls in food consumption, even in the short term, must be evaluated with caution. Such changes are inevitable and represent the greatest scope for long-term economic prosperity, as well as reductions in transitory food insecurity. If there has not been a successful change to market-oriented strategies to avoid seasonal food deficits, any loss of traditional strategies and cultural knowledge for coping with stress may have deleterious consequences. Certain households may therefore face a transitional problem if (1) the shift from cultivating for home production to cash earnings involves increasing reliance on poorly functioning markets for credit, inputs, wages, and commodity sales, and (2) the transformation to the monetized economy is accompanied by an erosion of the broad range of social distribution networks, cultural practices, and traditional technologies for ensuring access to food.

In much of the developing world, the major contributors to seasonal variations in food security are corresponding variations in production, work, income, and prices. Of primary importance, then, is to ascertain what policies can be pursued in order to improve seasonal food security and enhance the ability of households to cope with seasonal variability in their environment. This is especially important since agriculture often accounts for 70–80 percent of the gross domestic product of poor countries and is inherently seasonal, in terms of both the output produced and the employment generated.

Policies to improve food security in periods of heightened risk cover a wide terrain. They reflect the complex etiology of levels of food consumption and living standards. In considering these policies, it is useful to distinguish between (1) untargeted projects and policies that address the problems of transitory food insecurity through, for example, programs to improve the efficiency of markets and productivity of resources, and (2) targeted food security interventions that are generally designed to mitigate directly the consequences of the seasonal variations observed in the household's pattern of income, expenditures, consumption, and nutrition.

Untargeted Interventions

Price Stabilization

Price policy and marketing interventions are widely applied in order to improve food security during the preharvest season of shortage. Under the auspices of state-owned or state-operated public enterprises, governments often intervene directly to achieve the policy objective of seasonal (and interyear) price stabilization.

Well-financed and professionally managed stabilization schemes in Indonesia and the Philippines, for example, have contributed to a reduction in the magnitude of seasonal price spreads during the past few years. The question remains, however, as to the applicability of relatively successful stabilization efforts in Asia to other countries, especially in sub-Saharan Africa. Experience clearly illustrates the high budgetary costs involved in directly reducing seasonal price spreads. Further impediments are the prerequisites of highly trained manpower and considerable investment in well-developed storage, transport, and communication infrastructure. Likewise, if governments face serious credit constraints, logistical difficulties, and a shortage of information (e.g., on crop yields by region), these factors will also hinder the ability to defend announced price spreads and may even contribute to further instability as private traders speculate on government actions.

In addition, direct government intervention in commodity markets will inevitably squeeze out private traders; and so too will attempting to control food prices through limiting movement of food stocks across regions, as was done in India. Such initiatives may discourage development of a competitive private sector grain trade and end up aggravating the extent and instability of seasonal price increases, since markets will be served by fewer suppliers operating in less diverse agroclimatic zones. These considerations admonish policymakers to proceed with caution before embarking on a procurement and storage scheme to reduce seasonal price spreads.

Government efforts to reduce seasonal price spreads and instability should concentrate on improving private sector market performance. Such an approach should be accorded a high priority, even if it is deemed necessary for governments to retain a complementary role as a procurer and storer.

There are three key ways for government to reduce seasonal price increases indirectly. The first is to improve the quality of information, which will (1) promote more accurate anticipations of supply and demand for foodgrains in the future, (2) reduce speculative pricing, (3) foster competition, and (4) generally improve the performance of the private sector marketing agents.

The second is to conduct trade policy as it relates to the quantity and timing of imports. Speculative behavior will be reduced by improvements in the extent to which both the amount of imports entering commodity markets and the timing of such market operations coincide with anticipations. This will likely result in seasonal price movements being more predictable.

The third, which is expanded upon below, concerns infrastructure development. This is necessary in order to increase productivity, thus raising incomes, and to improve the functioning of commodity markets, thus moderating and stabilizing off-season price increases.

Infrastructure

Infrastructure development is of critical importance in any coordinated strategy to overcome seasonal food insecurity. In this regard, enhancing market infrastructure for food crops, agricultural inputs, and credit is paramount.

Improvements in the economic infrastructure will contribute to ensuring access to food through a number of means. From the input side, much of the threat of seasonal food insecurity stems from the inability of farmers to exploit improved agricultural technology because of limited availability or high prices of inputs. For example, the increased seasonal demand for credit that arises from the introduction of improved seed varieties needs to be accompanied by greater access to banking facilities that can provide credit at reasonable costs. Likewise, the demand for other inputs, such as fertilizer, is highly seasonal and must rely on the availability of credit as well as extensive physical infrastructure, such as rural roads.

Tangible reductions in seasonal food insecurity are also likely to occur as a result of investments in infrastructure that improve performance of output markets. For example, improved market integration and price transmission will act to increase production incentives to farmers and their levels of marketing. In addition, investments in market infrastructure will enhance food security by encouraging risk sharing over a broader population. Integrating commodity markets will reduce the link between the supply and demand situation in a given locality. Seasonal fluctuations in prices and supplies that follow from local production patterns will be reduced.

The policy relevance of improving market infrastructure is also clear when one considers that intertemporal price spreads largely reflect the costs of storage. To the extent that improved market facilities lower storage losses and costs, seasonal price increases can be reduced. It should be noted, however, that the costs of capital constitute the majority of storage costs, and that the cost of borrowing must be moderated to reduce seasonal price increases.

Investments in physical infrastructure, especially in sub-Saharan Africa, remain controversial because of their high cost.[1] While the debate continues as to whether such funds are better invested in projects to reduce seasonal food insecurity directly or in research on improving agricultural technology, there is little doubt that both avenues are facilitated by improved infrastructure. The returns to further technological breakthroughs to increase farm income and raise the marketed surplus, thereby reducing shortages and precipitous seasonal price increases, are predicated on infrastructure development. The availability of infrastructure that increases the productivity of resources and improves the incomes of farmers and the purchasing power of consumers is therefore a logical and necessary component of a larger strategy to reduce transitory food insecurity.

Technological Change and Agricultural Research

Technological change promoted by advances in agricultural research and facilitated by economic incentives and extension services is one of the major components of any set of public policies designed to improve food security. Given that transitory episodes of seasonal food insecurity appear to be another manifestation of poor performance of the food and agricultural sector, especially of factor, product, and credit markets, there is all the more reason to focus efforts on technological and institutional changes that promote rural and agricultural development. In placing a priority on accelerating agricultural growth, policymakers must recognize that adoption of new technologies alters not only the average annual level of production but also its seasonal pattern, and the seasonal pattern for labor demand and consequently for wages. Thus the choice of technology, in combination with farm management decisions, affects seasonal variability in work, incomes, food production, and consumption. This has important

---

1. For a thorough and enlightening discussion of the role of infrastructure and other investments to accelerate agricultural growth in Africa, see Mellor, Delgado, and Blackie 1987.

policy implications for both the direction of agricultural research and its eventual consequences for the poor.

Scientific breakthroughs on new varieties that increase cropping intensities are needed to reduce the seasonality of production, prices, and labor use. New crops are required that are less susceptible to infestation or other sources of losses during shortage. More and improved irrigation schemes also hold promise for increasing output during what was traditionally the dry season, particularly small-scale irrigation projects that do not rely on expensive investments beyond the means of most countries, especially in sub-Saharan Africa. Large-scale formal irrigation schemes have proven generally unsuccessful in improving household food security in Africa, and usually have shown an unsatisfactory rate of return. Other technologies, such as those that alleviate bottlenecks during peak periods of labor use or spread out the demand for labor and create new opportunities for work and wage income during the slack season also deserve further exploration.

In addition, there is a need to explore and improve traditional systems of averting the risk of seasonal food insecurity, especially among rural populations where markets are not well developed. More research is required on methods of (1) diversification, such as staggered planting, intercropping, relay cropping, and so forth; (2) exploitation of vertisol areas; and (3) production of starchy staples, including root and tree crops that serve as famine-breaking foods available for consumption during the height of the dry season and the preharvest period. Concerning this last point, there is considerable justification for expanding the research agenda on the role and scope for nonprincipal staple crops, especially coarse grains and non-cereal crops, which represent secondary staples in Asia, in Latin America, and increasingly in Africa, where they are still the primary source of calories in many regions.

The rationale for increasing research expenditures on secondary food crops derives from a number of points raised in this book: (1) The ratio between these commodities' proportion of national research budgets and their share of total production (congruence ratio) is low. (2) Related to the first point, the past neglect of these crops in research presents greater scope for significant advances. (3) They play important roles as helper crops (e.g., reducing soil erosion, fixing nitrogen), resulting in high cross-elasticities of supply with major cereals. (4) They have a high marginal utility for producers and consumers because they provide income or a source of food at critical periods of seasonal scarcity, primarily for poor households. (5) They often provide high quantities of micronutrients that are otherwise deficient in the diet.

As with most broad policy prescriptions, there are qualifications attached to the need for further analytic research. Specifically, it may prove more cost-effective to concentrate on breeding into a major cereal the robust characteristics of millet or the famine-preventing characteristics of cassava, for example, than to try to increase its yield. The calculus required to resolve such a question must consider an array of issues. These range from who produces the various crops to the expected cost in terms of investments in plant breeding that are required to achieve the desired crop characteristics.

One further compelling question is whether and to what extent research efforts and the accompanying technological progress involve trade-offs between (1) increased yields or output and (2) reduced seasonal variability in production or consumption. Pinstrup-Andersen and Jaramillo (chapter 16) present evidence that the higher yields that drive higher incomes are accompanied by greater intrayear fluctuations in consumption. Although their findings confirm earlier work (notably Simmons 1976a), they cannot be generalized. Even where higher yields are acquired at the expense of greater seasonal variation, technological progress is arguably still beneficial. If overall levels of production and consumption are raised sufficiently so that the minimum available and consumed during the lean season is greater than it was prior to the technological change, households are better off regardless of the variability in any of the parameters that affect food security. On the other hand, it is harmful if technological change contributes to higher yields, incomes, and consumption during certain seasons, but also to fluctuations in which lean-season consumption is lower than prior to technical change. This increased variation can be the result of poorly functioning markets that cannot adjust to the changing demands caused by technical change; or to the discontinuation of practices (e.g., secondary crops, diversification) intended to alleviate shortfalls in food availability. A review both of the technical package and of the appropriateness of introducing it in a given environment is warranted. However, it will likely prove to be more propitious to concentrate efforts on measures such as increasing access to savings institutions and markets, improving extension services, or adjusting the technology than to roll back the clock to a time before technological change. Decisions on how to proceed need to arise out of careful analysis of the circumstances particular to each context.

In sum, important seasonal caveats apply to promoting technological change. Prominent among them are accounting for levels of risk, including the risk of becoming seasonally hungry; and considering not just technical efficiency, in terms of yields per unit of land, but the timing of the demand for and supply of purchased and nonpurchased inputs required for agricultural development. This latter point is especially important, and suggests that farming systems research gives more attention to the seasonal dimensions, by examining factors such as returns per unit of labor of alternative technological packages during the peak season. This will enhance the effectiveness and acceptability of new technology and make it more likely to address the problem of seasonal food insecurity.

Finally, while the promotion of household food security during the lean season is paramount, new technology must be designed with respect to the seasonal scarcity of productive resources. Adoption of new technology will be impeded if it is not sensitive to the risks farmers confront. These include the seasonally determined periods of food deficits that, depending on the choice of technology, crops, and so forth, may be mitigated or exacerbated by new technology. For example, a new technical package for hybrid maize production may raise yields, but it may not provide the same level of food security during seasonal shortages as, for example, a mixed cropping strategy that ensures some food output throughout the year. Thus the focus on profit maximization and risk minimization should be broadened to include the farmer's concern with transforming lumpy input requirements, production, and profits into access to food in all seasons.

In a similar vein, resource constraints facing the farmer may slow the adoption of modern techniques of production. For example, technologies that alleviate labor bottlenecks in one season may simply create more severe ones during other periods of the year. Animal traction may relieve labor-short areas of Africa during planting, only to exacerbate the scarcity of workers for weeding and harvest activities. Furthermore, the extent to which tractors are an appropriate technology for meeting a seasonal bottleneck depends on factors such as whether their "appropriateness" is attributable to distortions, such as overvalued exchange rates, and whether their use is impeded by the shortage of fuel and spare parts, and by the risks inherent in such large capital expenditures. Likewise, farmers will be reluctant to use seed varieties that do not store well even if yields are higher; and seasonal shortages of other inputs such as credit may also impede technological innovation. These types of problems are especially relevant in parts of Africa, where labor, credit facilities, and storage infrastructure are most limited. Agricultural research and extension will therefore be most successful if they respond to seasonal factors by (1) designing technologies that extend the length of the harvesting period, and so take advantage of seasons of low labor demand; (2) selecting cropping strategies that minimize input requirements during periods of scarcity; (3) identifying methods to increase labor efficiency at peak seasons; and (4) ensuring available credit through well-organized and accessible financial institutions.

Targeted Interventions

Even if a government succeeds in stabilizing aggregate demand or normalizing seasonal price increases, certain households in certain regions may still face the scourge of seasonal food insecurity. Therefore, seasonally targeted interventions for vulnerable regions or households are recommended for coping with marketing inefficiencies, dislocations, and other factors that may imperil households during the lean season. These have the potential not only for reducing suffering, but for promoting gains in worker productivity through raising consumption during the season of peak labor requirements. In addition to the real economic benefits that accrue from supplementing workers' food intake when the work load is greatest, supplementing maternal diets during the lean season also results in better pregnancy outcomes.

There are a variety of options for seasonally targeted interventions to ensure access to food. These may be divided into three categories: those that generate income through productive work; those that transfer income directly to the household; and those that affect the prices faced by consumers.[2] All of these projects can in theory be targeted temporally, thereby augmenting household resources when they are most meager. The most cost-effective form of intervention and the accompanying means of targeting, however, must be determined on a case-by-case basis. What works in Indonesia with its highly developed rural health care infrastructure or in Sri Lanka with its literate population will clearly not be applicable in much of Africa, where both physical and social infrastructure are extremely limited, and where government services often do not reach into rural areas where the seasonality of food security is most acute. While the diversity of the environments in which projects are implemented makes generalizations difficult, certain types of projects lend themselves to seasonal targeting while others do not, for either logistical or administrative reasons.

One reason for targeting interventions to specific vulnerable households during periods of seasonal stress is that it is more cost-effective than projects that operate throughout the year and attempt to incorporate all households in the community. However, there are financial costs of designing and implementing programs that are seasonally targeted; it may prove more expensive to start and stop programs than to allow them to operate throughout the year. Likewise, there are often administrative burdens and financial costs to excluding from a program those households that are not vulnerable to seasonal food insecurity—this in addition to the risk of inadvertently excluding those in need of assistance. While there is little docu-

---

2. For a more thorough review of targeted intervention schemes to ensure access to food, see Kennedy and Alderman 1987 and Sahn 1986.

mentation of the costs of targeting interventions (Kennedy and Alderman 1987), there is some generalizable information to assist policymakers.

For example, among the most promising interventions to generate employment and income during the slack season are labor-intensive public works projects and home gardens. Concerning the former, the role of employment schemes in reducing food insecurity is well documented (Thomas 1985; Clay 1986; Sahn, Rogers, and Nelson 1981). Not only can employment schemes provide jobs during the slack season, but they can serve as means of instituting and supporting a wage floor. This is to say nothing of their long-term potential to promote self-sustaining development through the creation of physical assets, social and economic infrastructure, training, and improved agricultural practices.

Most labor-intensive works projects involve at least partial payment in kind, usually in the form of food; although other forms of payment, such as fertilizer or other inputs, also warrant consideration. The primary advantage of working for food, rather than cash, is that the infusion of cash wages into a region during the slack season may create pressures on food markets, aggravating or resulting in further increases in off-season price levels. Expanding demand through cash wages in the face of inelastic supply would create a seasonal bottleneck, raising prices and further imperiling the poor. In such a case, providing partial payment in the form of food (preferably along with cash) will offset the potential for undesirable food price increases. Likewise, partial payment in fertilizer may alleviate seasonal shortages of the input itself, or credit to purchase it, thereby breaking the cycle of low fertilizer application leading to low yields, low incomes, and low fertilizer use.

Household gardens may also be an effective intervention for augmenting incomes and food availability beyond what is earned and available from normal field agriculture.[3] To the extent that wage income and food availability are limited during certain seasons by the climate and by agricultural practices, home gardening may provide secondary food crops, especially fruits, vegetables, pulses, and roots and tubers, during seasons of food and income shortages.

Of course, the distinction between household gardens and field agriculture, their complementarity, or the extent to which home gardening declines during the process of agricultural development must all be examined more closely. It seems quite clear that tending home gardens, exploiting resources such as waste water from the household and leisure or low-pro-

---

3. For a discussion of experiences with home gardens, see, for example, Niñez 1985 and Soemarwoto et al. 1986; and for a more general review of home gardening as a means of improving food security, see UNICEF 1982 and Bittenbender n.d.

ductivity time during the slack season, can represent a valuable source of income and food. There is considerable latitude for research, extension, and other services to encourage and improve such endeavors.

Food price subsidies can also be targeted temporally to improve food security of poor households during the preharvest period.[4] Targeted interventions to reduce transitory food insecurity among the poor must be distinguished from general explicit and implicit subsidies that serve the entire population. Food subsidies targeted to vulnerable groups and that employ a flexible price wedge (relatively larger in the lean season and smaller in the postharvest period) represent a particularly good strategy for dealing with seasonal food deficits. Another viable approach may be to subsidize a specific commodity, such as cassava or other roots, tubers, and coarse grains, which might be self-targeting toward households that consume and produce the majority of such goods during seasons of high food prices and low earnings. Such subsidies should not, however, be paid for by taxing the farmer; otherwise, supplies of these commodities will be reduced and their effectiveness compromised.

There is also scope for seasonally targeting various forms of income transfers, including food rations and food stamps. As with temporal price interventions, the administrative and fiscal requirements of such efforts will largely condition their appropriateness and viability. Also of great interest are targeted efforts that provide assistance in repaying loans and procuring credit. These may prove quite effective in assisting the poor who have little collateral and avoiding the scenario whereby a household mortgages its future in order to survive the lean season. This is especially important given that the need to borrow capital for the purchase of inputs for the next crop generally coincides with the period of highest prices and lowest savings and food stores. This peak in demand for credit is not likely to be met unless governments undertake special initiatives. Doing so will decrease the possibility of moneylenders exploiting the poor through harsh terms for those who borrow, thereby threatening further declines in living standards.

Mother and child supplementary feeding projects that include a primary health care component are also a likely candidate for seasonal application. While the study presented by Lawrence and his colleagues in this volume (chapter 2) presents tangible evidence of the benefits of supplementary feeding on pregnancy outcomes and nutrition, the extensive literature indicates considerable variability of the effectiveness of these pro-

---

4. For a complete review of targeted food subsidy and income programs, their strengths and limitations, in terms of both feasibility and effectiveness, see Pinstrup-Andersen 1988.

grams in raising the calorie intake of household members, and thus the nutritional status of mothers and pre-school-age children.[5] Nevertheless, some important elements of success have been observed. Most prominent include (1) providing a level of supplementation that accounts for the expected sharing of the food ration among all family members, or for the substitution of the food ration for foods normally allocated to the mother and child; (2) choosing commodities for distribution that have the highest ratio of the incremental impact on the incomes of the household receiving the food ration to the given cost of transferring the food, including the cost of the food itself and shipping and distribution;[6] and (3) ensuring that the delivery of food be accompanied by the provision of health care and nutrition education.[7] If the latter is adhered to, the food serves as an incentive for the utilization of existing primary health care facilities, while the nutrition education is designed to alter the nature of the household preference function in favor of child nutrition. This clearly implies that in areas not served by primary health care services, such as rural villages in many areas of sub-Saharan Africa, there will not be the externalities of improved health and knowledge that would result from an integrated mother and child feeding program, thereby dramatically reducing the cost-effectiveness of such interventions.

Therefore, of equal if not greater importance is that policymakers place a premium on improving the provision of primary health care, with a focus on responding to the seasonality of disease and hence malnutrition. Whether it be programs of malaria control or oral rehydration therapy, integrating a seasonal component into health care delivery systems is a vital adjunct to any food security scheme. This is especially urgent given that there is a tendency to neglect and underuse health care delivery services, which require augmentation during the season when disease is most prevalent. For example, during the rainy season, when the prevalence of diarrhea is highest, roads are most difficult to negotiate and voluntary health workers most busy working on their farms. This impedes the demand for and delivery of health care because of competing time requirements, and the higher opportunity costs of both the sick who need health care and the voluntary health workers who supply the services. Thus explicit government initiatives may be necessary to counterbalance the inherent seasonality of health services and institutions, and their budgets. These initiatives may involve (1) increasing the remuneration for health workers, (2) im-

---

5. For a comprehensive review of the costs and effectiveness of various supplementary feeding programs, see Beaton and Ghassemi 1982 and Anderson et al. 1981.
6. For a complete discussion of commodity selection, see NRC 1982.
7. For a thorough analysis of the value of nutrition education and primary health care as complements to a feeding program, see Berg 1984.

proving health infrastructure so that it is more accessible, responding to the changing spatial pattern of demand that may derive from the seasonal migration of labor, and (3) placing greater emphasis on preventive public health measures to reduce the incidence of seasonally predictable diseases like malaria.

A few other areas of intervention involve broadening the purview of potential responses to seasonality. For example, consider factors like the seasonality of school tuition fees and taxation. Government efforts to generate revenue are strongly seasonal; they may represent an added stress that compromises food security. Spreading out such collections or preventing their coincidence with other acute demands for cash outlays, such as for purchasing farm inputs, will reduce the introduced seasonality inherent in institutional structures.

Considerations should also be given to making other minor adjustments in administrative procedures to reduce seasonal stress, especially where manpower constraints exist. Such adjustments might include avoiding a school year that conflicts with periods of peak agricultural labor demand, or providing leaves from military service or other government jobs in seasonal periods of acute labor shortages. Another important policy measure would be ensuring that budgetary resources are not exhausted, which would reduce services and programs coincident with seasonal cycles of depleted food stocks and personal savings. Likewise, efforts such as facilitating seasonal migration through improved child care services and housing, and providing schools for migrants and their families, are the types of initiatives that will not only improve seasonal food security in the short term but protect the integrity of the family unit in the long term.

Conclusions

A few conclusions deserve to be highlighted from the previous discussion. First, the literature concerning the human ability to adapt biologically to seasonal reductions in food energy intake without physical damage and impaired functional performance remains ambiguous. However, there is considerable evidence that many population groups confront sufficiently acute periods of seasonal stress that the decline in food consumption levels contributes to malnutrition and threatens further impoverishment as a consequence of disaccumulation of assets. This suggests that policymakers give greater emphasis to preventing and mitigating the consequences of adverse seasonal cycles of food insecurity.

Second, in order to ensure that households have the ability and desire to acquire adequate food throughout the year, there is a need to build upon existing and indigenous strategies to cope with seasonality, while promoting other compensatory mechanisms that revolve around improved storage and marketing infrastructure, savings and credit schemes, more efficient labor markets, and greater labor mobility to ease seasonal migration.

Third, the role of government should primarily be viewed as creating an economic and social climate that will encourage private sector initiatives to improve seasonal food security. This role involves gathering and disseminating information to improve competition and the efficiency of intertemporal arbitrage among traders. There is also a need to assist farmers in increasing production through promoting systems that (1) ensure the availability of inputs (e.g., fertilizer) at reasonable prices, (2) encourage the development of small-scale irrigation and water management schemes to reduce the seasonality of production, (3) support national agricultural and farming systems research institutions that promote technological change to alleviate seasonalities in agriculture, and (4) disseminate knowledge and experience through extension services. In addition, government plays a key role in expanding the infrastructure that will allow the private sector to carry out its storage and interperiod marketing and processing functions more effectively. The development of institutions that facilitate, rather than impede, the functioning of credit markets is fundamentally important, and can most effectively be achieved by adherence to policies that reward savers and encourage borrowers through interest rate reforms.

Fourth, to the extent that traditional arrangements for coping with seasonal stress are inadequate, or deteriorate as a consequence of greater integration with inefficient or distorted markets, temporally targeted interventions to ensure access to sufficient food warrant consideration. These are also needed to smooth out the transition from indigenous coping strategies to increased reliance on market and new technologies, when households are at greatest risk. Governments and donors must mobilize resources for such purposes in order to protect and sustain any country's most valuable resource—its human capital.

In doing so, there is a legitimate cause for concern that government not assume responsibilities for doing what social networks traditionally may have done, and what markets could do if they functioned efficiently. This is not, however, to be confused with abdicating responsibility for enabling all households to have access to food. Rather, a problem arises when government policy impedes traditional coping mechanisms or technological and market-oriented solutions, through, for example, the inappropriate use of food aid.

Fifth, food aid represents an important resource for direct intervention programs that address seasonal troughs in consumption. The use of food aid in seasonally targeted programs not only represents a resource with a low opportunity cost that can be transferred to households during a period of seasonal dearth, but it also reduces potential price increases that would accompany nonfood transfers. While the wage-good argument is not always applicable, it is especially meaningful in regions where markets are functioning poorly and the price flexibility is high.

Of concern, however, is that the history of food aid is replete with examples of poor targeting, both temporal and spatial. For example, there is much evidence that food aid flows are countercyclical to the needs of developing countries. This has resulted in situations where food aid was reduced precisely during periods of greatest need (Sahn and Alderman 1987). Given that there are acute periods of seasonal stress, which are made worse in years of low production and food availability, the timing of food aid arrivals or distribution must avoid the postharvest season, especially in years of good harvest. In a similar vein, the myriad examples, including the case of Burkina Faso presented in this volume (chapter 8), of food aid being distributed to the wrong regions, communities, and households, represents a serious shortcoming in such interventions. This brings to the fore the policy issues of improving targeting and instituting nutritional surveillance and early warning systems. Identification of appropriate indicators and systems for collecting and analyzing them is paramount. Of even greater importance is developing a framework for action whereby information leads swiftly to the provision of relief and development efforts to the appropriate households, and during the correct seasons.

Sixth, as with many problems related to economic development, it appears that rural populations are most susceptible to the deleterious effects of seasonality. Rural agricultural and wage employment, and therefore rural incomes, are linked to variability in the agricultural calendar. In addition, rural consumers are more likely to confront greater fluctuations in market prices because imperfections in marketing systems are likely to be greatest in remote areas. This is partially attributable to the dearth of available physical and social infrastructure as one moves further from major urban settlements. Furthermore, action is required to address the pro-urban bias in targeted measures to ensure access to adequate food, such as subsidies and transfers.[8]

Seventh, seasonal problems do not necessarily demand seasonal solutions. This theme, implicit throughout this book, has important policy implications for reducing seasonal food insecurity. A great deal of the difficulty in buffering the inevitable seasonal pattern in agriculture is attributable to the element of uncertainty in the seasonal undulations in production and prices. Seasonal uncertainties for producers and consumers are largely due to interyear fluctuations in output that arise from

---

8. For a discussion of the fact that food subsidies and food-related income transfer schemes are often planned or implemented to serve the urban population more than rural dwellers, see Pinstrup-Andersen 1988.

unpredictable weather patterns. That is, interyear fluctuations in the volume and timing of production and imports result in seasonal unpredictability in food availability, earnings, and prices. Likewise, it is in the years of low output and earnings, and relatively higher prices, that the period of seasonal dearth is most perilous.

All of this suggests that policymakers should recognize that reducing interyear instability in agriculture is a means of reducing seasonal food insecurity. This should be coupled with addressing the problem of poorly functioning capital and commodity markets that reduce the ability of producers and consumers to respond to uncertainty; and promoting the infrastructure and the flow of information that enhance the ability to find seasonal employment and sustain earnings throughout the year.

Eighth, the transitory food security that corresponds to the seasonal undulations is most severe in sub-Saharan Africa—this because Africa is most affected by (1) indigenous seasonalities, such as unimodal and unpredictable rainfall, porous soils, and seasonal labor shortages; and (2) introduced seasonalities caused by inefficient government institutions, lack of infrastructure, use of primitive and inappropriate production technology, and ill-advised investments in efforts such as large-scale rather than small-scale irrigation projects.

Although the problem is worst in Africa, it is important to recognize that seasonalities vary from year to year, country to country, region to region, village to village, and household to household. This has important implications in terms of the need for decentralizing strategies and promoting versatile responses to seasonal stress. Programs and policies designed in distant capitals to improve seasonal food security must therefore place emphasis on creating a social, economic, and political environment at the local level that will facilitate initiatives among villages and households to cope with the changes and hardships brought by the seasons. Whether it be through indirect efforts, such as development of technologies that allow the farmer greater flexibility in planting and harvesting dates, or direct interventions that provide consumption credit to households when they themselves determine the need for such assistance, accommodating the wide range of seasonalities should be the central theme of initiatives to improve seasonal food security.

References

Publishing organizations are abbreviated in the references as follows:

AMREF = African Medical and Research Foundation
FAO = Food and Agriculture Organization
ICRISAT = International Crops Research Institute for the Semi-Arid Tropics
IFPRI = International Food Policy Research Institute
IRRI = International Rice Research Institute

Abdullah, M. 1983. "Dimension of intra-household food and nutrient allocation—A study of a Bangladeshi village." Ph.D. diss., University of London.

Abdullah, M., and E. F. Wheeler. 1985. "Seasonal variations and the intra-household distribution of food in a Bangladeshi village." *American Journal of Clinical Nutrition* 41:1305–13.

Adams, W. M., and A. T. Grove, eds. 1984. *Irrigation in tropical Africa.* Cambridge African Monographs no. 3. Cambridge: African Studies Centre, Cambridge University.

Ahmed, I., and B. Kinsey, eds. 1984. *Farm equipment in innovations in eastern and central southern Africa.* London: Gower.

Ahmed, R. 1981. "Foodgrain distribution policies within a dual pricing mechanism: The case of Bangladesh." In *Development issues in an agrarian economy—Bangladesh,* ed. W. Mahmud. Dacca, Bangladesh: Center for Administrative Studies.

Ahmed, R., and A. Barnard. 1986. "Seasonality of rice prices, effects of new technology and an approach to rice price stabilization in Bangladesh." Washington, D.C.: IFPRI (unpublished draft).

Ahmed, R., and S. Kumar. 1985. "Evaluation of food-for-work programs in Bangladesh." Washington, D.C.: IFPRI (study in progress).

Akong'a, J. J. 1982. *Famine, famine relief, and public policy in Kitui district.* IDS Working Paper no. 388. Nairobi: Institute for Development Studies, University of Nairobi.

Akrasanee, N., et al. 1983. *Rural off-farm employment in Thailand.* Bangkok: Industrial Management Co.

Alderman, H. 1986. "The effect of food price and income changes on the acquisition of food by low-income households." Washington, D.C.: IFPRI (mimeographed).
Allan, W. 1965. *The African husbandman*. Edinburgh: Oliver and Boyd.
Alverson, H. 1978. *Mind in the heart of darkness: Value and self-identity among the Tswana of southern Africa*. New Haven, Conn.: Yale University Press.
———. 1984. "The wisdom of tradition in the development of dry-land farming: Botswana." *Human Organization* 43 (no. 1):1-8.
Anderson, J., and P. B. R. Hazell. Forthcoming. *Variability in grain yields and implications for agricultural research and policy*. Baltimore, Md.: Johns Hopkins University Press.
Anderson, M. A., et al. 1981. *Nutrition interventions in developing countries, study I: Supplementary feeding*. Cambridge, Mass.: Oelgeschlager, Gunn, and Hain.
Andrae, G., and B. Beckman. 1985. *The wheat trap: Bread and underdevelopment in Nigeria*. London: Zed.
Annegers, J. F. 1973. "Seasonal food shortages in West Africa." *Ecology of Food and Nutrition* 2:251-57.
Anthony, K. R., B. Johnston, W. Jones, and V. Uchendu. 1979. *Agricultural change in tropical Africa*. Ithaca, N.Y.: Cornell University Press.
Antonsson-Ogle, B. 1984. "Wild plant resources." *Ceres* 17 (no. 5):38-40.
Apeldoorn, G. J. van. 1981. *Perspectives on drought and famine in Nigeria*. London: Allen and Unwin.
Arhem, Kaj, K. M. Homewood, and A. Rogers. 1981. *A pastoral food system: The Ngorongoro Maasai in Tanzania*. BRALUP Research Paper no. 70. Dar es Salaam: Bureau of Resource Assessment and Land Use Planning, University of Dar es Salaam.
Arjyal, P. C., and K. R. Poudyal. 1982. *Nepal system of price support and procurement of selected agricultural commodities*. Kathmandu, Nepal: Agricultural Projects Services Center.
Azariadis, C. 1975. "Implicit contracts and underemployment equilibrium." *Journal of Political Economy* 83:1183.
Azariadis, C., and J. Stiglitz. 1983. "Implicit contracts and fixed price equilibrium." *Quarterly Journal of Economics* 48 (no. 392):1-22.
Bailey, R. C., and N. R. Peacock. 1984. "Efe pygmies of Northeast Zaire: Subsistence strategies in the Ituri forest." In *Uncertainty in food supply*, ed. G. Harrison. Cambridge: Cambridge University Press.
Bardhan, K. 1970. "Price and output response of marketed surplus of foodgrains: A cross sectional study of some North Indian villages." *American Journal of Agricultural Economics* 52:51-61.
Bardhan, P. 1979. "Wages and unemployment in a poor agrarian economy: A theoretical and empirical analysis." *Journal of Political Economy* 87 (no. 3): 479-500.
———. 1980. "Interlocking factor markets and agrarian development: A review of issues." *Oxford Economic Papers* 32:82-98.
Bardhan, P., and K. Bardhan. 1971. "Price response of marketed surplus of foodgrains." *Oxford Economic Papers* 23:255-67.

Bardhan, P., and A. Rudra. 1978. "Interlinkage of land, labor, and credit relations: An analysis of village survey data in East India." *Economic and Political Weekly* 13:367-84.

———. 1981. "Terms and conditions of labour contracts in agriculture: Results of a survey in West Bengal, 1979." *Oxford Bulletin of Economics and Statistics* 43:89-111.

Barker, R., and R. W. Herdt. 1985. *The rice economy of Asia.* Washington, D.C.: Resources for the Future; and Los Baños, Philippines: IRRI.

Barnum, H. N., and L. Squire. 1979. "An econometric application of the theory of the farm household." *Journal of Development Economics* 6:79-102.

Barrett, D. E. 1984. "Nutrition and child behaviour: Conceptualization and assessment of social-emotional functioning and a report on an empirical study." In *Malnutrition and behaviour: Critical assessment of key issues.* Publication Series 4. Lausanne: Nestle Foundation.

Barrett, D. E., M. Radke-Yarrow, and R. E. Klein. 1982. "Chronic malnutrition and child behaviour: Effects of early caloric supplementation on social and emotional functioning at school age." *Developmental Psychology* 18:541-56.

Basson, P. 1981. "Women and traditional food technologies: Changes in rural Jordan." *Ecology of Food and Nutrition* 11:17-23.

Bates, D., and S. Lees. 1977. "The role of exchange in productive specialization." *American Anthropology* 79.

Bayliss-Smith, T. 1981. "Seasonality and labour in the rural energy balance." In *Seasonal dimensions to rural poverty,* pp. 30-38, ed. R. Chambers, R. Longhurst, and A. Pacey. London: Frances Pinter.

Beals, R. E., and C. F. Menezes. 1970. "Migrant labour and agricultural output in Ghana." *Oxford Economic Papers* 22 (no. 1):109-27.

Beaton, G. 1983. "Energy in human nutrition: Perspectives and problems." *Nutrition Review* 41:325-40.

Beaton, G., and H. Ghassemi. 1982. "Supplementary feeding programs for young children in developing countries." *American Journal of Clinical Nutrition* 35 (May):1280.

Beeny, J. M. 1975. *Agricultural mechanization study.* UNDP Report to the United Republic of Tanzania. Rome: FAO.

Behnke, R. H. 1985. "Measuring the benefits of subsistence versus commercial livestock production in Africa." *Agricultural Systems* 16:109-35.

Behrman, J. R. 1988a. "Intrahousehold allocation of nutrients in rural India: Are boys favored? Do parents exhibit inequality aversion?" *Oxford Economic Papers* 40:1.

———. 1988b. "Nutrition, health, birth order, and seasonality: intra-household allocation in rural India." *Journal of Development Economics* 28 (no. 7):43-63.

Behrman, J. R., and A. B. Deolalikar. 1987. "Will developing country nutrition improve with income? A case study for rural South India." *Journal of Political Economy* 95 (no. 3):492-507.

———. 1988a. "Health and Nutrition." In *Handbook of development economics,* vol. 1, pp. 631-711, ed. H. B. Chenery and T. N. Srinivasan. Amsterdam: North Holland.

———. 1988b. "How do food prices affect individual nutrient consumption and health status? A latent variable analysis." Philadelphia: University of Pennsylvania (mimeographed).

———. 1988c. "Is variety the spice of life? Implications for nutrient responses to income." Philadelphia: University of Pennsylvania (mimeographed).

———. 1989. "Agricultural wages in India: The role of health, nutrition, and seasonality." In this volume.

Behrman, J. R., R. A. Pollak, and P. Taubman. 1982. "Parental preferences and provision for progeny." *Journal of Political Economy* 90 (no. 1):52-73.

———. 1986. "Do parents favor boys?" *International Economic Review* 27 (no. 1):31-52.

Behrman, J. R., and P. Taubman. 1986. "Birth order, schooling, and earnings." *Journal of Labor Economics* 4 (no. 3, pt. 2.):S121-45.

Behrman, J. R., and B. L. Wolfe. 1984. "More evidence on nutrition demand: Income seems overrated and women's schooling underemphasized." *Journal of Development Economics* 14 (nos. 1-2):105-28.

Bell, C., and T. Srinivasan. 1985a. "Agricultural credit markets in Punjab: Segmentation, rationing, and spillover." Washington, D.C.: World Bank (unpublished paper).

———. 1985b. "An anatomy of transactions in rural credit markets in Andhra Pradesh, Bihar, and Punjab." Washington, D.C.: World Bank (unpublished paper).

Benefice, E., S. Chevassus-Agnes, and H. Barral. 1984. "Nutritional situation and seasonal variations for pastoralist populations of the Sahel (Senegalese Ferlo)." *Ecology of Food and Nutrition* 14:229-47.

Berg, A. 1984. "Nutrition review." Washington, D.C.: World Bank, Population, Health, and Nutrition Department.

Berg, E. 1965. "The economics of the migrant labour system." In *Urbanization and migration in West Africa*, pp. 160-81, ed. H. Kuper. Berkeley and Los Angeles: University of California Press.

Berio, A. J., P. Francois, and A. Odovafa. 1985. "Seasonal activity patterns and household food supply in the rural areas of Ivory Coast." Paper presented at IFPRI/FAO/AID workshop on seasonal causes of household food insecurity, policy implications, and research needs, Maryland Inn, Annapolis, Md., December 10-13.

Bernard, F. 1972. *East of Mount Kenya: Meru agriculture in transition*. IFO-Institut Afrika-Studien no. 75. Munich: Weltforum.

Bernstein, H. 1977. "Notes on capital and peasantry." *Review of African Political Economy* 10:44-62.

Berry, S. 1984. *Fathers work for their sons*. Berkeley and Los Angeles: University of California Press.

Beyer, J. L. 1980. "Africa." In *World systems of traditional resource management*, pp. 5-37, ed. G. A. Klee. New York: Wiley.

Bhalla, S. 1979. "The measurement of permanent income and its application to savings behavior." *Journal of Political Economy* 88:722-43.

———. 1980. "Measurement errors and the permanent income hypothesis: Evidence from rural India." *American Economic Review* 69:295-307.

Bharati, P., and A. Basu. 1982. "Uncertainties of food supply and nutritional deficiencies in relation to economic condition in a village population of southern West Bengal, India." Paper presented at the symposium on coping with uncertainty of food supply, Bad Homburg, West Germany.

Bidinger, P. D., B. Nag, and P. Babu. 1986. "Nutritional and health consequences of seasonal fluctuations in household food availability." *Food and Nutrition Bulletin* 8 (no. 1):36.

Billewicz, W. Z., and I. A. McGregor. 1982. "A birth-to-maturity longitudinal study of heights and weights in two West African (Gambian) villages, 1951-1975." *Annals of Human Biology* 9 (no. 4):309.

Binswanger, H. P. 1984. *Agricultural mechanization: A comparative historical perspective.* World Bank Staff Working Paper no. 673. Washington, D.C.: World Bank.

Binswanger, H. P., and N. S. Jodha. 1978. *Manual of instructions for economic investigators in ICRISAT's village level studies 2.* Hyderabad, India: ICRISAT.

Binswanger, H. P., and M. Rosenzweig, eds. 1984. *Contractual arrangements, employment, and wages in rural labor markets in Asia.* New Haven, Conn.: Yale University Press.

Birdsall, N., and R. Sabot, eds. 1988. *Labor market discrimination in developing economies.* Washington, D.C.: World Bank.

Bittenbender, H. C. n.d. "The home garden: An important horticultural farming system in less developed countries." East Lansing: Michigan State University (mimeographed).

Black, R. E., K. H. Brown, and S. Becker. 1984a. "Longitudinal studies of infectious diseases and physical growth of children in rural Bangladesh II: Incidence of diarrhoea and association with known pathogens." *American Journal of Epidemiology* 115:315.

———. 1984b. "Malnutrition as a determining factor in diarrhoeal duration but not in incidence among young children in a longitudinal study in rural Bangladesh." *American Journal of Clinical Nutrition* 37:87-94.

Black, R. E., K. H. Brown, S. Becker, A. R. M. A. Alim, and I. Huq. 1982. "Longitudinal studies of infectious diseases and physical growth of children in rural Bangladesh." *American Journal of Epidemiology* 115 (no. 3).

Blanc, J. 1969. "Nutrition et développement—Réflexion sur les aspects économiques et l'alimentation en Afrique de l'ouest." Ph.D. diss., Institut de Recherches et de Planification de Grenoble, Université des Sciences Sociales de Grenoble.

Bliss, C., and N. Stern. 1978. "Productivity, wages, and nutrition: Parts I and II: The theory," *Journal of Development Economics* 5 (no. 4):331-98.

Bloch, M. 1973. "The long and the short term: The economic and political significance of the morality of kinship." In *The character of kinship,* pp. 75-87, ed. J. Goody. Cambridge: Cambridge University Press.

Bohannan, P. 1954. *Tiv farm and settlement.* London: HMSO.

Bohdal, M., N. Gibbs, and W. Simmons. n.d. *Nutrition survey: A campaign against malnutrition in Kenya, 1964-1968.* Report to the Ministry of Health of Kenya on the WHO/FAO/UNICEF-assisted project. Nairobi, Kenya.

The Botswana Society. 1979. *Proceedings of the symposium on drought in Botswana*, ed. M. T. Hinchey. Hanover, N.H.: University Press of New England for the Botswana Society, with Clark University Press.

Bouis, H. E. 1983. "Seasonal rice price variation in the Philippines: Measuring the effects of government intervention." *Food Research Institute Studies* 19 (no. 1):81-92.

Boutillier, J. L., P. A. Cantrelle, J. Causse, and C. Levant. 1962. *La moyenne vallée du Senegal*. Paris: Ministère de la Cooperation.

Bouwkamp, J. C. 1984. "The potential of sweet potatoes and their problems." *World Crops* (March/April):59-62.

Boxall, R., M. Greeley, D. Tyagr, M. Lipton, and J. Nedkarta. 1978. *The prevention of farm level foodgrain storage losses in India: A social cost benefit analysis*. Brighton, England: Institute of Development Studies, University of Sussex.

Braun, J. von, and E. Kennedy. 1986. *Commercialization of subsistence agriculture: Income and nutritional effects in developing countries*. Working Paper no. 1. Washington, D.C.: IFPRI.

Braverman, A., and T. N. Srinivasan. 1984. "Agrarian reforms and developing rural economies characterized by interlinked credit and tenancy markets." In *Contractual arrangements, employment, and wages in rural labor markets in Asia*, pp. 63-81, ed. H. P. Binswanger and M. Rosenzweig. New Haven, Conn.: Yale University Press.

Breman, J. 1984. "Seasonal migration and cooperative capitalism: The Bardoli, South Gujarat." In *Contractual arrangements, employment, and wages in rural labor markets in Asia*, ed. H. P. Binswanger and M. Rosenzweig. New Haven, Conn.: Yale University Press.

Brokensha, D., and D. Riley. 1980. "Mbeere knowledge of their vegetation and its relevance for development: A case study from Kenya." In *Indigenous knowledge systems and development*, ed. D. Brokensha, D. Warren, and O. Werner. Washington, D.C.: University Press of America.

Brokensha, D., D. Warren, and O. Werner, eds. 1980. *Indigenous knowledge systems and development*. Washington, D.C.: University Press of America.

Brown, A. 1978. "The impact of economic development on health in the Papalopan." Paper presented at the annual meeting of the American Anthropological Association, Los Angeles, Calif.

Brown, K. H., R. E. Black, and S. Becker. 1982. "Seasonal changes in nutritional status and the prevalence of malnutrition in a longitudinal study of young children in rural Bangladesh." *American Journal of Clinical Nutrition* 36:303-13.

Brown, K. H., R. E. Black, A. D. Robertson, and S. Becker. 1985. "Effects of season and illness on the dietary intake of weanlings during longitudinal studies in rural Bangladesh." *American Journal of Clinical Nutrition* 41:343-55.

Bureau of Agricultural Economics (BAEcon). 1985. Estimates of palay area and production, unpublished data. Quezon City: Ministry of Agriculture, Government of the Philippines.

Burton, M., and D. White. 1984. "Sexual division of labor in agriculture." *American Anthropologist* 86:568-83.

Campbell, D. J. 1978. *Coping with drought in Kenya Masailand: Pastoralists and farmers of the Loitokitok area.* IDS Working Paper no. 337. Nairobi: Institute for Development Studies, University of Nairobi.

Campbell-Platt, G. 1980. "African locust bean (*parkia* spp) and its West African fermented food product, dawa-dawa." *Ecology of Food and Nutrition* 9:123-32.

Carloni, A. 1984. *The impact of maternal employment and income on the nutritional status of children in rural areas of developing countries.* Rome: United Nations Subcommittee on Nutrition.

Cassidy, C. 1980. "Nutrition and health in agriculturalists and hunter-gatherers." In *Nutritional anthropology*, ed. N. Jerome, R. Kandel, and G. Pelto. Pleasantville, N.Y.: Redgrave.

Cecelski, E. 1984. *The rural energy crisis, women's work, and family welfare: Perspectives and approaches to action.* World Employment Programme Research Working Paper (mimeographed).

Center for Research on Economic Development (CRED). 1977. *Marketing, price policy, and storage of foodgrain in the sahel: A survey,* vols. 1 and 2. Ann Arbor: University of Michigan.

Chambers, R. 1981. Introduction. In *Seasonal dimensions to rural poverty,* pp. 1-8, ed. R. Chambers, R. Longhurst, and A. Pacey. London: Frances Pinter.

———. 1982. "Health, agriculture, and rural poverty: Why seasons matter." *Journal of Development Studies* 18 (no. 2):217-38.

———. 1983a. "Bad times for rural children: Countering seasonal deprivation." *Journal of the Society for International Development* 1:45-49.

———. 1983b. "Seasonality, poverty and nutrition: A professional frontier." Paper presented in the proceedings of the National Workshop on Poverty and Nutrition, Tamil Nadu University, Coimbatore, India, February 7-10.

Chambers, R., R. Longhurst, and A. Pacey, eds. 1981. *Seasonal dimensions to rural poverty.* London: Frances Pinter.

Charsombat, P. 1981. *Labor migration from agriculture in Thailand.* SEAPRAP Research Report no. 55. Bangkok: Faculty of Economics and Business Administration, Kasetsart University.

Chaudhury, R. H. 1980. "Seasonal dimensions of rural poverty in Bangladesh: Employment, wages, and consumption patterns." *Social Action* 30 (January-March):1-27.

Chen, L. C., A. K. M. A. Chowdhury, and S. L. Huffman. 1979. "Seasonal dimensions of energy protein malnutrition in rural Bangladesh: The role of agriculture, dietary practices, and infection." *Ecology of Food and Nutrition* 8 (no. 3):175-87.

———. 1980. "Anthropometric assessment of energy-protein malnutrition and subsequent risk of mortality among preschool-aged children." *American Journal of Clinical Nutrition* 33:1836-45.

Chen, L. C., E. Huq, and S. D'Souza. 1981. "Sex bias in the family allocation of food and health care in rural Bangladesh." *Population Development Review* 7 (no. 1).

Chernichovsky, D., and M. O. Astra. 1984. *Patterns of food consumption and nutrition in Indonesia: An analysis of the national socioeconomic survey.* World Bank Staff Working Paper no. 670. Washington, D.C.: World Bank.

Chowdhury, A. K. M. A., S. L. Huffman, and L. C. Chen. 1981. "Agriculture and nutrition in Matlab Thana, Bangladesh." In *Seasonal dimensions to rural poverty,* ed. R. Chambers, R. Longhurst, and A. Pacey. London: Frances Pinter.
Chowdhury, A. K. M. A., S. L. Huffman, and G. T. Carlin. 1978. "Malnutrition, menarche, and marriage in rural Bangladesh." *Social Biology* 24.
Clay, E. 1986. "Rural public works and food-for-work: A survey." *World Development* 14 (nos. 10–11):1237–52.
Clayton, E. 1973. "Mechanization and employment in East African agriculture." In *Mechanization and employment in agriculture: Case studies from four continents,* pp. 19–44. Geneva: International Labour Office.
Cleave, J. H. 1974. *African farmers: Labor use in the development of smallholder agriculture.* New York: Praeger.
Cliffe, L. 1978. "Labor migration and peasant differentiation: The Zambian experience." *Journal of Peasant Studies* 5:268.
Cole, T. J. 1974. "Bronchitis, smoking, and obesity in an English and a Danish town—Male death after a 10-year following." *Bulletin de physio-pathologie respiratoire* 10:657–59.
Colson, E. 1979. "In good years and bad: Food strategies of self-reliant societies." *Journal of Anthropological Research* 35:18–29.
Conant, F. P. 1982. "Thorns paired, sharply recurved: Cultural controls and rangeland quality in East Africa." In *Desertification and development,* pp. 111–22, ed. B. Spooner and H. S. Mann. New York: Academic.
Condon, R. G., and R. Scaglion. 1982. "The ecology of birth seasonality." *Human Ecology* 10:495–511.
Connell, J., B. Dasgupta, R. Laishley, and M. Lipton. 1977. *Migration from rural areas.* Delhi: Oxford University Press.
Cooper, F. 1982. "Peasants, capitalists, and historians: A review article." *Journal of Southern African Studies* 7:284–314.
Coppock, D. L., et al. 1985. "Traditional tactics of resource exploitation and allocation among nomads in an arid African environment." In *Proceedings of the international rangelands resources development symposium, Salt Lake City, Utah, USA, February 13–14,* pp. 87–96, ed. L. White and J. Tiedeman. Pullman: Cooperative Extension Service, Washington State University.
Coughenour, M. B., J. E. Ellis, D. M. Swift, D. L. Coppock, K. Galvin, J. T. McCabe, and T. C. Hart. 1985. "Energy extraction and use in a nomadic pastoral ecosystem." *Science* 230 (no. 4726):619–25.
Coward, W. A., A. M. Prentice, P. R. Murgatroyd, H. L. Davies, T. J. Cole, M. Sawyer, G. R. Goldberg, D. Halliday, and J. P. Macnamara. 1984. "Measurement of CO_2 and water production rates in man using 2H, 180 labelled H_2O: Comparisons between calorimeter and isotope values." Paper presented at European nutrition workshop on human energy metabolism (physical activity and energy expenditure measurements in epidemiological research based upon direct and indirect calorimetry), Wageningen, Netherlands, October.
Dahl, G., and A. Hjort. 1976. *Having herds.* Stockholm Studies in Social Anthropology no. 2. Stockholm: Department of Social Anthropology, University of Stockholm.

Dalton, G. 1968. "Economic theory and primitive society." In *Economic anthropology: Readings in theory and analysis*, pp. 143-67, ed. E. LeClair and H. Schneider. New York: Holt, Rinehart, and Winston.

Dandekar, K., and M. Sathe. 1980. "Employment guarantee scheme and food-for-work program." *Economic and Political Weekly* 15:707-13.

Day, R. H., and I. Singh. 1977. *Economic development as an adaptive process: The green revolution in the Indu Punjab*. New York: Cambridge University Press.

Deaton, A., and J. Muellbauer. 1980. *Economics and consumer behavior*. Cambridge: Cambridge University Press.

Delgado, C. L. 1979. *The southern Fulani farming system in Upper Volta: A model for the integration of crop and livestock production in the West African savannah*. African Rural Economy Paper no. 20. East Lansing: Michigan State University.

———. 1986. "A variance components approach to food grain market integration in northern Nigeria." *American Journal of Agricultural Economics* 68 (no. 4).

Delgado, C. L., and P. J. Matlon. n.d. "Seasonal spreads in Burkina Faso grain prices: Predictability and related policy issues." Washington, D.C.: IFPRI (unpublished draft).

Delgado, C. L., and C. P. J. Miller. 1985. "Changing food patterns in West Africa: Implications for policy research." *Food Policy* 10 (no. 1):55-62.

Delgado, C. L., and C. G. Ranade. 1987. "Technological change and agricultural labor use." In *Accelerating food production in sub-Saharan Africa*, ed. J. W. Mellor, C. L. Delgado, and M. J. Blackie. Baltimore, Md.: Johns Hopkins University Press.

Deolalikar, A. B. 1988. "Nutrition and labor productivity in agriculture: Econometric estimates for rural South India." *Review of Economics and Statistics* 70 (no. 4).

Devitt, P. 1979. "Drought and poverty." In *Proceedings of the symposium on drought in Botswana*, pp. 121-27, ed. M. T. Hinchey. Hanover, N.H.: University Press of New England for the Botswana Society, with Clark University Press.

DeWalt, K. M. 1983. "Income and dietary adequacy in an agricultural community." *Social Science Medicine* 17:1877-86.

———. 1984. *Nutritional and agricultural change in a Mexican community*. Ann Arbor, Mich.: UMI Research.

DeWalt, K. M., P. Kelly, and G. Pelto. 1980. "Nutritional correlates of economic micro-differentiation in a highland Mexican community." In *Nutritional anthropology*, ed. N. Jerome, R. Kandel, and G. Pelto. Pleasantville, N.Y.: Redgrave.

DeWalt, K. M., and G. Pelto. 1977. "Food use and household ecology in a Mexican community." In *Nutrition and anthropology in action*, ed. T. Fitzgerald. Assen, Netherlands: Van Gorcum.

Dewey, K. G. 1980. "The impact of agricultural development on child nutrition in Tabasco, Mexico." *Medical Anthropology* 4:55-78.

———. 1981. "Nutritional consequences of the transformation from subsistence to commercial agriculture in Tabasco, Mexico." *Human Ecology* 9:151-87.

Dirks, R. 1978. "Resource fluctuations and competitive transformations in West Indian slave societies." In *Extinction and survival in human populations*, ed. C. D. Laughlin and I. A. Brady. N.Y.: Columbia University Press.
———. 1980. "Social responses during severe food shortages and famine." *Current Anthropology* 21:21-44.
Diskin, M. 1978. "Discussion in symposium on Mexican food systems." Paper presented at the 77th annual meeting of the American Anthropological Association, Los Angeles, Calif.
Donge, J. K. van. 1982. "Politicians, bureaucrats, and farmers: A Zambian case study." *Journal of Development Studies* 19 (no. 1):88-107.
Donham, D. 1981. "Beyond the domestic mode of production." *Man* 16 (no. 4):515-41.
Douglas, M. 1962. "Lele economy compared with the Bushong." In *Markets in Africa*, pp. 211-33, ed. P. Bohannan and G. Dalton. Evanston, Ill.: Northwestern University Press.
Draper, H. H. 1978. "The aboriginal Eskimo diet in modern perspective." *American Anthropology* 79:309-16.
DuBois, C. 1941. "Food and hunger in Alor." In *Language, culture, and personality: Essays in memory of Edward Sapir*, pp. 272-81, ed. L. Spier, A. I. Hallowell, and S. S. Newman. Menasha, Wisc.: Sapir Memorial Publication Fund.
Due, J. M. 1983. *Beans in the farming systems in Langali and Kibaoni villages, Mgeta area, Morogoro region, Tanzania*. Morogoro: Department of Agricultural Economics, Faculty of Agriculture, University of Dar es Salaam.
Dugdale, A. E. 1985. "Family anthropometry: A new strategy for determining community nutrition." *Lancet* 2(8456):672-73.
Dugdale, A. E., and J. Eaton-Evans. 1985. *The seasonality of infant growth in Brisbane, Queensland, Australia*. Human Nutrition Research Group Working Paper Q.4067. Brisbane, Australia: University of Queensland, Department of Child Health.
Durnin, J. V. G. A., and R. Passmore. 1967. *Energy, work, and leisure*. London: Heinemann.
Dynarski, M., and S. Sheffrin. 1985. "Housing purchase and transitory income: A study with panel data." *Review of Economics and Statistics* 67:195-204.
Edirisinghe, N. 1987. *The food stamp scheme in Sri Lanka: Costs, benefits, and options for modification*. Research Report 58. Washington, D.C.: IFPRI.
Eicher, C. K., and D. C. Baker. 1982. *Research on agricultural development in sub-Saharan Africa: A critical survey*. MSU International Development Paper. East Lansing: Michigan State University.
Ejiga, N. 1977. "Economic analysis of storage, distribution, and consumption of cowpeas in northern Nigeria." Ph.D. diss., Cornell University, Ithaca, N.Y.
Elling, M. 1981. *Background to agricultural development in central province, Zambia*. Rome: FAO/Agriplan.
Ellis, F. 1982. "Report on the West African grain stabilization project." Washington, D.C.: U.S. Agency for International Development (mimeographed).
Engle, P. L. 1984. "Intrahousehold allocation of resources: Perspectives from psychology." San Luis Obispo: California Polytechnic State University (mimeographed).

Epstein, T. S. 1973. *South India: Yesterday, today and tomorrow*. New York: Holmes and Meier.
Evans-Pritchard, E. E. 1940. *The Nuer*. Oxford: Oxford University Press.
Eveleth, P. B., and J. M. Tanner. 1976. *Worldwide variation in human growth*. Cambridge: Cambridge University Press.
Evenson, R., and H. Binswanger. 1984. "Estimating labor demand functions for Indian agriculture." In *Contractual arrangements, employment, and wages in rural labor markets in Asia*, pp. 263-79, ed. H. P. Binswanger and M. Rosenzweig. New Haven, Conn.: Yale University Press.
Evenson, R., and Y. Kislev. 1976. *Agricultural research and productivity*. New Haven, Conn.: Yale University Press.
Falcon, W., W. Jones, S. Pearson, J. Dixon, G. Nelson, F. Roche, and L. Unnevehr. 1984. *The cassava economy of Java*. Stanford, Calif.: Stanford University Press.
Farmer, B. H., ed. 1977. *Green revolutions?* p. 429. Boulder, Colo.: Westview.
Firth, R. 1959. *Social change in Tikopia*. London: Allen and Unwin.
Firth, R. 1966. *Housekeeping among Malay peasants*. London School of Economics Monograph on Social Anthropology no. 7. 2d ed. N.Y.: Humanities.
Fitt, A. B. 1924. *The human energy-rhythm through the year*. From the report of the 16th meeting of the Australian Association for the Advancement of Science, vol. 16, pp. 704-42. Wellington, New Zealand: W. A. G. Skinner, Government Printer.
Flannery, K. V. 1973. "The origins of agriculture." *Annual Review of Anthropology* 2:271-310.
Fleuret, A. 1979a. "Methods for evaluation of the role of fruits and wild greens in Shambaa diet: A case study." *Medical Anthropology* 3:249-69.
―――. 1979b. "The role of wild foliage plants in the diet: A case study from Lushoto, Tanzania." *Ecology of Food and Nutrition* 8:87-93.
Flores, M., and R. Flores. 1984. "Effects of dependence on seasonally available food." *Current Topics in Nutrition and Disease* 10:207-19.
Flowers, N. 1983. "Seasonal factors in subsistence, nutrition, and child growth in a central Indian community." In *Adaptive responses of native Amazonians*, pp. 357-90. New York: Academic.
Folbre, N. 1984. "Comment on 'market opportunities, genetic endowments, and intrafamily resource distribution.'" *American Economic Review* 74:518-20.
―――. 1986. "Cleaning house: New perspectives on households and economic development." *Journal of Development Economics* 22 (no. 1):5-40.
Food and Agriculture Organization of the United Nations (FAO). 1973. *Energy and protein requirements*. Nutritional Meetings Report Series no. 522. Rome: FAO.
―――. 1974. *The fourth world food survey*. Rome: FAO.
―――. 1979. *A Zambian handbook of pasture and fodder crops*. Rome: FAO.
―――. 1980. *Food balance sheets*. Rome: FAO.
―――. 1983. *Integrating crops and livestock in West Africa*. FAO Animal Production and Health Paper no. 41. Rome: FAO.
―――. 1984. *Animal energy in agriculture in Africa and Asia*. FAO Animal Production and Health Paper no. 42. Rome: FAO.

———. 1986. *Irrigation in Africa south of the Sahara*. FAO Investment Centre Technical Paper no. 5. Rome: FAO.

Food and Agriculture Organization-World Health Organization-United Nations University (FAO-WHO-UNU). 1985. *Energy and protein requirements*. Geneva: WHO.

Forbes, M. H. 1977. "Farming and foraging in prehistoric Greece: The nutritional ecology of wild resource use." In *Nutrition and anthropology in action*, pp. 46-61, ed. T. Fitzgerald. Assen, Netherlands: Van Gorcum.

Ford Foundation. 1982. *A report on the regional workshop on seasonal variations in the provisioning, nutrition, and health of rural families (31 March-2 April, 1982)*. Nairobi: Ford Foundation and AMREF.

Ford, R. I. 1972. "An ecological perspective on the eastern Pueblos." In *New perspectives on the Pueblos*, ed. A. Ortiz. Albuquerque: University of New Mexico Press.

Fortes, M., and S. L. Fortes. 1936. "Food in the domestic economy of the Tallensi." *Africa* 9:237-76.

Fowler, A. 1982. "The seasonal aspects of education in east and southern Africa." In *A report on the regional workshop on seasonal variations in the provisioning, nutrition, and health of rural families*, pp. 58-78. Nairobi: Ford Foundation and AMREF.

Fox, R. H. 1953. "A study of the energy expenditure of Africans engaged in various activities, with special reference to some environmental and physiological factors which may influence the efficiency of their work." Ph.D. diss., University of London.

Frank, D. A. 1984. "Malnutrition and child behaviour: A view from the bedside." In *Malnutrition and behaviour: Critical assessment of key issues*. Publication Series 4. Lausanne: Nestle Foundation.

Franke, R. W., and B. H. Chasin. 1980. *Seeds of famine: Ecological destruction and the development dilemma in the western Sahel*. Totowa, N.J.: Allanheld, Osmun.

Franzel, S. 1984. "Modeling farmers' decisions in a farming systems research exercise: The adoption of an improved maize variety in Krinyaga district, Kenya." *Human Organization* 43 (no. 3):199-207.

Galvin, K. A. 1985. "Food procurement, diet, and nutrition of Turkana pastoralists in an ecological and social context." Ph.D. diss., State University of New York, Binghamton.

Garcia, M., and P. Pinstrup-Andersen. 1987. *Consumer food subsidy experiment in the Philippines*. Research Report no. 61. Washington, D.C.: IFPRI.

Gathee, J. W. 1982. "Farming systems economics: Fitting research to farmers' conditions." In *Intercropping*, pp. 136-40, ed. C. L. Keswani and B. J. Ndunguru. Ottawa: International Development Research Centre.

Gladwin, C. H. 1982. "The role of a cognitive anthropologist in a farming systems program that has everything." In *The role of anthropologists in interdisciplinary teams developing improved food production technology*, pp. 73-92. Los Baños, Philippines: IRRI.

Glaeser, B. 1984. *Ecodevelopment in Tanzania*. Amsterdam: Mouton.

Goering, J. 1979. *Tropical root crops and rural development.* World Bank Staff Working Paper no. 324. Washington, D.C.: World Bank.
Goldman, R. H. 1974. "Seasonal rice prices in Indonesia, 1953-69: An anticipatory price analysis." *Food Research Institute Studies in Agricultural Economics, Trade and Development* 13 (no. 2):99-143.
Goldman, R. H., and L. Squire. 1978. *Technical change, labor use, and income distribution in the Muda irrigation project.* Development Discussion Paper no. 35. Cambridge, Mass.: Harvard Institute for International Development, Harvard University.
Goody, E. 1982. *Parenthood and social reproduction: Fostering and occupational roles in West Africa.* Cambridge: Cambridge University Press.
Gooneratne, W. 1982. *Labor absorption in rice-based agriculture: Case studies from Southeast Asia.* Bangkok: Asian Employment Programme, Asian Regional Team for Employment Promotion (ARTEP).
Gopalan, C. 1980. *Thirteenth Jawarharlal Nehru memorial lecture: Jawarharlal Nehru memorial lectures.* Patancheru, India: Teen Murti House, Jawarharlal Nehru Memorial Fund.
Gopalan, C., B. V. R. Sastry, and S. C. Balasubramanian. 1971. *Nutritive value of Indian foods.* Hyderabad, India: National Institute of Nutrition (rpt. 1976).
Government of Tamil Nadu, Commissioner of Statistics. Various issues. *Season and crop report.*
Grantham-McGregor, A. 1984. "The social background of childhood malnutrition." In *Malnutrition and behaviour: Critical assessment of key issues.* Publication Series 4. Lausanne: Nestle Foundation.
Greenberg, J. 1981. *Santiago's sword.* Berkeley and Los Angeles: University of California Press.
Griffiths, M. 1985. Personal communication.
Grivetti, L. E. 1978. "Nutritional success on a semi-arid land: Examination of Tswana agro-pastoralists of the eastern Kalahari, Botswana." *American Journal of Clinical Nutrition* 31:1204-20.
Gudeman, S. 1978. *The demand of a rural economy: From subsistence to capitalism in a Latin American village.* Boston: Routledge and Kegan Paul.
Guyer, J. 1980. "Household and community in African studies." *African Studies Review* 24:87-137.
———. 1984. *Family and farm in southern Cameroon.* Boston: African Studies Center.
Haessel, W. 1976. "The price and income elasticities of home consumption and marketed surplus of foodgrains." *American Journal of Agricultural Economics* 58 (no. 2):341-45.
Hall, M. 1968. "Mechanization in East African agriculture." In *Agricultural planning in East Africa,* pp. 81-116, ed. G. K. Helleiner. Nairobi: East African Publishing House.
Hall, R., and F. Mishkin. 1982. "The sensitivity of consumption to transitory income: Estimates from panel data on households." *Econometrica* 50:461-81.

Hankins, T., A. Larsen, R. Hulls, and J. Finucane. 1971. *Sukumaland interdisciplinary report*. BRALUP Research Report no. 40. Dar es Salaam: Bureau of Resource Assessment and Land Use Planning, University of Dar es Salaam.

Hansen, B. 1969. "Employment and wages in rural Egypt." *American Economic Review* 59:298-313.

———. 1971. "Employment and wages in rural Egypt: Reply." *American Economic Review* 61 (no. 3):500-508.

Hanson, J. 1971. "Employment and rural wages in Egypt: A reinterpretation." *American Economic Review* 61 (no. 3):492-99.

Hart, A. G. 1940. "Risk, uncertainty, and the unprofitability of compounding probabilities." In *Anticipations, uncertainty, and dynamic planning*. Chicago, Ill.: University of Chicago Press.

Hart, G. 1980. "Patterns of household labor allocation in a Javanese village." In *Rural household studies in Asia*, ed. H. P. Binswanger et al. Singapore: Singapore University Press.

———. 1986. "Interlocking transactions: Obstacles, precursor or instrument of agrarian capitalism." *Journal of Development Economics* 23 (no. 1):177-202.

Hartmann, B., and J. K. Boyce. 1983. *A quiet violence: View from a Bangladesh village*. London: Zed.

Haswell, M. R. 1953. *Economics of agriculture in a savannah village*. Colonial Research Study no. 8. London: HMSO.

———. 1981a. *Energy for subsistence*. London: Macmillan.

———. 1981b. "Food consumption in relation to labor output." In *Seasonal dimensions to rural poverty*, pp. 38-41, ed. R. Chambers, R. Longhurst, and A. Pacey. London: Frances Pinter; and Totowa, N.J.: Allanheld, Osmun.

Hayami, Y. 1978. *Anatomy of a peasant economy*. Los Baños, Philippines: IRRI.

Hays, H. M. 1985. *The marketing and storage of food grains in northern Nigeria*. Miscellaneous Paper no. 50. Zaria, Nigeria: Ahmadu Bello University, Institute of Agricultural Research.

Hays, H. M., and J. H. McCoy. 1984. "Foodgrain marketing in northern Nigeria: Spatial and temporal performance." *Journal of Development Studies* 14 (no. 2):182-92.

Hazell, P. B. R., and C. Ramasamy, eds. Forthcoming. *Technological change and rural welfare*. Baltimore, Md.: Johns Hopkins University Press.

Heckman, J. J. 1976. "The common structure of statistical models of truncation, sample selection, and limited dependent variables and a simple estimator for such models." *Annals of Economic and Social Measurement* 5 (no. 4):475-92.

Herdt, R. W. Forthcoming. "A retrospective view of technological and other change in Philippine rice farming, 1965 to 1982." *Economic Development and Cultural Change*.

Herdt, R. W., and C. Capule. 1983. "Adoption, spread, and production impact of modern rice varieties in Asia." Los Baños, Philippines: IRRI.

Herdt, R. W., and L. A. Gonzales. 1980. *The impact study of rapidly changing prices on rice policy objectives and instruments in the Philippines, 1980*. IRRI Agricultural Economics Department Paper no. 80-16. Los Baños, Philippines: IRRI.

Heyer, J. 1981. *Rural development in tropical Africa*. New York: St. Martin's.
Heyer, J., P. Roberts, and G. Williams. 1971. "A linear programming analysis of constraints on peasant farms in Kenya." *Food Research Institute Studies* 10 (no. 1):55-67.
Hill, A. G. 1985. *Population, health, and nutrition in the Sahel*. London: KPI.
Hitchcock, R. K. 1979. "The traditional response to drought in Botswana." In *Proceedings of the symposium on drought in Botswana*, pp. 91-97, ed. M. T. Hinchey. Hanover, N.H.: University Press of New England for the Botswana Society, with Clark University Press.
Holmberg, A. R. 1950. *Nomads of the long bow*. Washington, D.C.: Smithsonian.
Holmström, B. 1983. "Equilibrium long-term labor contracts." *Quarterly Journal of Economics* 48 (no. 392 supp.):23-54.
Horton, S. 1988. "Birth order and child nutrient status: Evidence on the intrahousehold allocation of resources in the Philippines." *Economic Development and Cultural Change* 36 (no. 2):341-54.
Hoskins, M. D. 1984. *The green revolution and cropping intensity*. Brighton, England: Institute for Development Studies.
Houghton, J. 1986. "Cereals policy in Burkina Faso." Report to CILSS/Club de Sahel by Elliot Berg Associates. Paris: OECD/Club de Sahel.
Huss-Ashmore, R. 1982. "Seasonality in rural highland Lesotho: Method and policy." In *A report on the regional workshop on seasonal variations in the provisioning, nutrition, and health of rural families*. Nairobi: Ford Foundation and AMREF.
———. 1984. "Seasonal cycles of nutrition and resource procurement in Lesotho." *American Journal of Physical Anthropology* 63:173.
Huss-Ashmore, R., and F. E. Johnston. 1985. "Bioanthropological research in developing countries." *Annual Review of Anthropology* 14:475-528.
Hyden, G. 1980. *Beyond Ujamaa in Tanzania: Underdevelopment and an uncaptured peasantry*. London: Heinemann.
Ibnouf, M. A. O. 1985. "An economic analysis of the mechanized food production schemes in the central plains of the Sudan." Ph.D. diss., Michigan State University, East Lansing.
ICRISAT/Burkina Faso. 1981. *Economics program annual report*. Kamboinse, Burkina Faso: ICRISAT.
Immink, M., and F. Viteri. 1981. "Energy intake and productivity of Guatemalan sugarcane cutters: An empirical test of the efficiency wage hypothesis." *Journal of Development Economics* 9:251-57.
Institute of Nutrition and Food Science (INFS). 1977. *Nutrition survey of rural Bangladesh: 1975-76*. Dacca, Bangladesh: University of Dacca.
———. 1983. *Nutrition survey of rural Bangladesh: 1981-82*. Dacca, Bangladesh: University of Dacca.
———. 1985. Unpublished anthropometric data. Dacca, Bangladesh: University of Dacca.
Jahnke, H. E. 1982. *Livestock production systems and livestock development in tropical Africa*. Kiel, West Germany: Kieler Wissenschaftsverlag Vauk.
James, W. P. T., and P. S. Shetty. 1982. "Metabolic adaptations and energy requirements in developing countries." *Human Nutrition: Clinical Nutrition* 36C:331-36.

de Janvry, A., and K. Subbarao. 1984. "Agricultural price policy and income distribution in India." *Economic and Political Weekly* 19:A166-78.
Jochim, M. 1981. *Strategies for survival.* New York: Academic.
Jodha, N. S. 1983. "Market forces and erosion of common property resources." Paper presented at the international workshop on agricultural markets in the semiarid tropics, ICRISAT, Hyderabad, India, October 24-28.
Jodha, N. S., and A. C. Mascarenhas. 1983. *Adjustment to climatic variability in self-provisioning societies: Some evidence from India and Tanzania.* Economics Program Progress Report 48. Patancheru, India: ICRISAT.
Johnson, A. 1980. "The limits of formalism in agricultural decision research." In *Agricultural decisionmaking: Anthropological contributions to rural development,* pp. 19-43, ed. P. Bartlett. New York: Academic.
Johnson, A., and C. A. Behrens. 1982. "Nutritional criteria in Machiguenga food production decisions: A linear-programming analysis." *Human Ecology* 10:167-89.
Johnston, B. 1980. "Socioeconomic aspects of improved animal-drawn implements and mechanization in semi-arid East Africa." In *Proceedings of the international workshop on socioeconomic constraints to development of semi-arid tropical agriculture,* pp. 221-33. Patancheru, India: ICRISAT.
Johnston, B., and P. Kilby. 1975. *Agriculture and structural transformation: Economic strategies in late-developing countries.* New York: Oxford University Press.
Jones, C. W. 1983. "The impact of the SEMRY I irrigated rice production project on the organization of production and consumption at the intrahousehold level." Washington, D.C.: U.S. Agency for International Development (mimeographed).
Jones, S. 1963. *A study of Swazi nutrition: Report of the Swaziland nutrition survey, 1961-62.* Paper prepared for the Swaziland Administration. Durban: Institute of Social Research, University of Natal.
Jones, W. 1972. *Marketing staple food in tropical Africa.* Ithaca, N.Y.: Cornell University Press.
———. 1984. Economic tasks for food marketing board in tropical Africa. *Food Research Institute Studies* 19:113-38.
Jones, W., and R. Egli. 1984. *Farming systems in Africa: The great lakes highlands of Zaire, Rwanda, and Burundi.* Technical Paper no. 27. Washington, D.C.: World Bank.
Josserand, H. P. 1984. "Farmers' consumption of an imported cereal and the cash/foodcrop decision: An example from Senegal." *Food Policy* 9 (no. 1):27-35.
Judd, M., J. Boyce, and R. Evenson. 1983. *Investing in agricultural supply.* Economic Growth Center Discussion Paper no. 442. New Haven, Conn.: Yale University.
Kahn, J. S. 1980. *Minangu social formations: Indonesian peasants and the world economy.* New York: Cambridge University Press.
Katz, R. W., and M. H. Glantz. 1977. "Rainfall statistics, droughts, and desertification in the Sahel." In *Desertification,* pp. 81-102, ed. M. H. Glantz. Boulder, Colo.: Westview.

Kennedy, E., and H. Alderman. 1987. *Comparative analyses of nutritional effectiveness of food subsidies and other food-related interventions.* Washington, D.C.: IFPRI.

Kennedy, E., and B. Cogill. 1988. *Income and nutritional effects of the commercialization of agriculture in southwestern Kenya.* Research Report. Washington, D.C.: IFPRI.

Keswani, C. L., and B. J. Ndunguru, eds. 1982. *Intercropping.* IDRC-186e. Ottawa: International Development Research Centre.

Keys, A. 1980. "Overweight, obesity, coronary heart disease, and mortality." *Nutrition Reviews* 38:297–306.

Kielmann, A. A., and C. McCord. 1978. "Weight-for-age as an index of risk of death in children." *Lancet* 1:1247–50.

Kline, C. K., D. A. G. Green, R. C. Donahue, and B. A. Stout. 1969. *Agricultural mechanization in equatorial Africa.* East Lansing: Michigan State University.

Krishnamurty, K. 1970. "Marketing and storage of foodgrains." *Bulletin of Grains Technology* 8:121–26.

Kuckertz, H. 1985. "Organizing labour forces in Mpondoland: A new perspective on work parties." *Africa* 55 (no. 2):115–32.

Kumar, S. 1979. *Impact of subsidized rice on food consumption and nutrition in Kerala.* Research Report no. 5. Washington, D.C.: IFPRI.

———. 1985. "Household production and consumption strategies for meeting seasonal labor needs." Paper presented at IFPRI/FAO/AID workshop on seasonal causes of household food insecurity, policy implications, and research needs, Maryland Inn, Annapolis, Md., December 10–13.

———. 1987. "The nutrition situation and its food policy links." In *Accelerating food production in sub-Saharan Africa,* ed. J. W. Mellor, C. L. Delgado, and M. J. Blackie. Baltimore, Md.: Johns Hopkins University Press.

Kyereme, S. S. 1984. "Food consumption and poverty patterns in Ghana." Ph.D. diss., Cornell University, Ithaca, N.Y.

Lagemann, J. 1977. *Traditional African farming systems in eastern Nigeria.* IFO-Institut Afrika-Studien no. 98. Munich: Welforum.

Lamb, G., and L. Muller. 1982. *Control, accountability, and incentives in a successful development institution: The Kenya Tea Development Authority.* World Bank Staff Working Paper no. 550. Washington, D.C.: World Bank.

Lau, L. J., Wuu-Long Lin, and P. A. Yotopoulos. 1978. "The linear logarithmic expenditure system: An application to consumption-leisure choice." *Econometrica* 46 (no. 4):843–68.

Lawrence, M., W. H. Lamb, F. Lawrence, and R. G. Whitehead. 1984. "Maintenance energy cost of pregnancy in rural Gambian women and influence of dietary status." *Lancet* 2:363–65.

Lawrence, M., J. Singh, F. Lawrence, and R. G. Whitehead. 1985. "The energy cost of common daily activities in African women: Increased expenditure in pregnancy?" *American Journal of Clinical Nutrition* 42:753–63.

Ledesma, A. J. 1982. *Landless workers and rice farmers: Peasant subclasses under agrarian reform in two Philippine villages.* Los Baños, Philippines: IRRI.

Lee, R. 1969. "!Kung Bushman subsistence: An input-output analysis." In *Environment and cultural behavior,* pp. 47–79, ed. A. Vayda. Garden City, N.Y.: Natural History Press.

———, ed. 1979. *The !Kung San: Men, women, and work in a foraging society.* New York: Cambridge University Press.
Leibenstein, H. 1957. *Economic backwardness and economic growth.* New York: Wiley.
Lele, U. 1971. *Foodgrain marketing in India, private performance, and public policy.* Ithaca, N.Y.: Cornell University Press.
Le Moigne, M. 1980. "Animal-draft cultivation in Francophone Africa." In *Proceedings of the international workshop on socioeconomic constraints to development of semi-arid tropical agriculture,* pp. 213–20. Patancheru, India: ICRISAT.
Lewis, J. 1981. "Domestic labor intensity and the incorporation of Malian peasant farmers into localized descent groups." *American Ethnologist* 8:53–73.
Linares, O. 1976. "Garden hunting in the American tropics." *Human Ecology* 4:331–49.
Lindenbaum, S. 1987. "Loaves and fishes in Bangladesh." In *Food and evolution: Toward a theory of human habits,* pp. 427–44, ed. M. Harris and E. Ross. Philadelphia: Temple University Press.
Lipton, M. 1968. "The theory of the optimizing peasant." *Journal of Development Studies* 4 (no. 3):21–30.
———. 1983. *Labor and poverty.* World Bank Staff Working Paper 616. Washington, D.C.: World Bank.
———. 1985. *The place of agricultural research in the development of sub-Saharan Africa.* IDS Discussion Paper 202. Brighton, England: Institute for Development Studies, University of Sussex.
Longhurst, R. 1984. *The energy trap: Work, nutrition, and child malnutrition in northern Nigeria.* Cornell International Monograph Series no. 13. Ithaca, N.Y.: Cornell University.
———. 1985. "Secondary crops, seasonality, and women's work: Implications for household food security." Rome: FAO (mimeographed).
Longhurst, R., and P. Payne. 1979. *Seasonal aspects of nutrition: Review of evidence and policy implications.* IDS Discussion Paper 145. Brighton, England: Institute for Development Studies, University of Sussex.
Lunn, P. G., A. M. Prentice, R. G. Whitehead, and S. Austin. 1980. "Influence of maternal diet on plasma-prolactin levels during lactation." *Lancet* 1:623–25.
Lunven, P. 1985. "Comments on Philip Payne's paper on public health and functional consequences of seasonal hunger and malnutrition." Prepared for the IFPRI/FAO/AID workshop on seasonal causes of household food insecurity, policy implications, and research needs, Maryland Inn, Annapolis, Md. December 10–13, 1985 (mimeographed).
Madden, P., and M. Yoder. 1972. *Program evaluation: Food stamps and commodity distribution in rural areas of central Pennsylvania.* Agricultural Experiment Station Bulletin no. 780. State College: Pennsylvania State University.
Malina, R. M., and H. H. Himes. 1977. "Seasonality of births in a rural Zapotec municipio, 1945–70." *Human Biology* 49:415–28.
Marshall, L. 1976. *The !Kung of Nyae Nyae.* Cambridge, Mass.: Harvard University Press.

Masseyeff, R., A. Cambon, and B. Bergeret. 1958. *Le groupement d'evodoula (Cameroun): Etude de l'alimentation.* Paris: ORSTOM.

Matlon, P. J. 1987. "The West African semiarid tropics." In *Accelerating food production in sub-Saharan Africa,* ed. J. W. Mellor, C. L. Delgado, and M. J. Blackie. Baltimore, Md.: Johns Hopkins University Press.

Mauss, M. 1979. *Seasonal variations of the Eskimo: A study in social morphology.* Boston: Routledge and Kegan Paul.

Maxwell, S., C. Stutley, and A. Bojanic. 1982. *Report on a case study programme in San Pedro.* Working Paper no. 27. Santa Cruz, Bolivia: Centro Internacional de Agricultura Tropical.

Mazumdar, D. 1959. "The marginal productivity theory of wages and disguised unemployment." *Review of Economic Studies* 26:190-97.

McCabe, J. T. 1984. "Livestock management among the Turkana: A social and ecological analysis of herding in an East African pastoral population." Ph.D. diss., State University of New York, Binghamton.

Mears, L. A. 1981. *The new rice economy in Indonesia.* Yogyakarta, Indonesia: Gadjah Mada University Press.

Mears, L. A., M. Agabin, T. Anden, and R. Marquez. 1974. *Rice economy of the Philippines.* Quezon City: University of the Philippines Press.

Mellor, J. W., C. L. Delgado, and M. J. Blackie, eds. 1987. *Accelerating food production in sub-Saharan Africa.* Baltimore, Md.: Johns Hopkins University Press.

Mellor, J. W., and G. M. Desai, eds. 1985. *Agricultural change and rural poverty.* Baltimore, Md.: Johns Hopkins University Press.

Mencher, J. 1981. "More food for whom?" *Culture and Agriculture* 11:3-5.

Messer, E. 1972. "Patterns of 'wild' plant consumption in Oaxaca, Mexico." *Ecology of Food and Nutrition* 1:325-32.

———. 1977. "The ecology of vegetarian diet in a modernizing Mexican community." In *Nutrition and anthropology in action,* pp. 117-24, ed. T. Fitzgerald. Assen, Netherlands: Van Gorcum.

———. 1978. *Zapotec plant knowledge: Classification, uses, and communications about plants in Mitla, Oaxaca, Mexico.* Mem. Mus. Anthropology, University of Michigan, no. 10, pt. 2. Ann Arbor: University of Michigan Museum of Anthropology.

———. 1981. "Hot-cold classifications: Theoretical and practical implications of a Mexican study." *Social Science Medicine* 15B:133-45.

———. 1982. "Getting through (three meals) a day: Diet, domesticity, and cash income in a Mexican community." Paper presented at the annual meeting of the American Anthropological Association, Los Angeles, Calif.

———. 1983. "The household focus in nutritional anthropology: An overview." *Food and Nutrition Bulletin* 5 (no. 4):2-12.

———. 1984a. "Perspectives on diet." *Annual Review of Anthropology* 13:205-49.

———. 1984b. "Sociocultural aspects of nutrient intake and behavioral responses to nutrition." In *Nutrition and behavior: Human nutrition* 5, ed. J. Galler. New York: Plenum.

———. 1987. "The green revolution: A retrospective with a focus on hunger." Providence, R.I.: Alan Shawn Feinstein World Hunger Program, Brown University.

———. Forthcoming. "The relevance of time allocation studies in nutritional anthropology." In *Methods in nutritional anthropology,* ed. G. Pelto, P. Pelto, and E. Messer.

Messer, E., and H. Kuhnlein. 1986. "Traditional foods." In *Training manual in nutritional anthropology,* pp. 60-81, ed. S. Quandt and C. Ritenbaugh. Special Publication of the American Anthropological Association no. 20. Washington, D.C.: American Anthropological Association.

Metrick, H. 1975. "Mechanization of peasant agriculture in East Africa." In *Change of agriculture,* pp. 555-66, ed. A. H. Bunting. London: Duckworth.

Migot-Adholla, S. E. 1975. "The politics of a growers' cooperative organization." In *Rural cooperation in Tanzania,* pp. 221-53, ed. L. Cliffe, P. Lawrence, W. Luttrell, S. Migot-Adholla, and J. Saul. Dar es Salaam: Tanzania Publishing House.

Mincer, J. B. 1974. *Schooling, experience, and earnings.* New York: National Bureau of Economic Research.

Ministry of Agriculture and Cooperatives, Office of Agricultural Economics. Various issues. *Agricultural Statistics of Thailand,* Bangkok.

Miracle, M. 1961. "Seasonal hunger: A vague concept and an unexplored problem." *Bulletin de l'Inst. Franc. Afrique Noire* 23, ser. B (nos. 1-2):273-83.

———. 1966. *Maize in tropical Africa.* Madison: University of Wisconsin Press.

Mongkolsmai, D. 1985. "The distributional impacts of irrigation." Bangkok: Faculty of Economics, Thammasat University (mimeographed).

Moris, J. R. 1970. "The agrarian revolution in central Kenya: A study of farm innovation in Embu district." Ph.D. diss., Northwestern University, Evanston, Ill.

Moris, J. R., and D. Thom. 1985. *African irrigation overview, summary.* WMS Report no. 37. Logan: Department of Agricultural and Irrigation Engineering, Utah State University.

Moris, J. R., D. Thom, and R. Norman. 1984. *Prospects for small-scale irrigation development in the Sahel.* WMS Report no. 26. Logan: Department of Agricultural and Irrigation Engineering, Utah State University.

Morrison, T. K. 1984. "Cereal imports by developing countries: Trends and determinants." *Food Policy* 9 (no. 1):13-26.

Moussie, M. 1985. Untitled report on OFNACER, p. 64. Lincoln, Mo.: Department of Agriculture and Office of International Agriculture, Lincoln University (mimeographed).

Mundle, S. 1983. "Effect of agricultural production and prices on incidence of rural poverty: A tentative analysis of interstate variations." *Economic and Political Weekly* 18:A48-53.

Musgrove, P. 1980. "Permanent household income and consumption in urban South America." *American Economic Review* 69:355-68.

Nambiar, R. G. 1983. "Comparative prices in a developing economy: The case of India." *Journal of Development Economics* 12:19-25.

National Research Council (NRC). 1982. *Nutritional analysis of P.L. 480 title II commodities.* Washington, D.C.: National Academy Press.
Nestle, P. 1986. "A society in transition: Developmental and seasonal influence on the nutrition of Maasai women and children." *Food and Nutrition Bulletin* 8 (no. 1):2-18.
Netting, R. McC. 1968. *Hill farmers of Nigeria: Cultural ecology of the Kofyar of the Jos Plateau.* Seattle: University of Washington Press.
New ERA. 1981. "Study on internal migration in Nepal." Ayaneshwar, Nepal (mimeographed).
Newman, J. L. 1970. *The ecological basis for subsistence change among the Sandawe of Tanzania.* Washington, D.C.: National Academy of Sciences.
Newman, M., I. Quedraego, and D. Norman. 1980. "Farm-level studies in the semi-arid tropics of West Africa." In *Proceedings of the international workshop on socioeconomic constraints to development of semi-arid tropical agriculture,* pp. 241-63. Patancheru, India: ICRISAT.
Nghiep, L. 1979. "The structure and changes of technology in pre-war Japanese agriculture." *American Journal of Agricultural Economics* 61:687-93.
Nimis, M. M. 1982. "The contemporary role of women in lowland Maya livestock production." In *Maya subsistence: Studies in memory of Dennis E. Puleston,* pp. 313-25, ed. K. V. Flannery. New York: Academic.
Niñez, V. 1985. "Introduction: Household gardens and small-scale food production." *Food and Nutrition Bulletin* 7 (no. 3):1-5.
Nnamyelugo, D., J. King, H. Ene-Obong, and P. Ngoddy. 1985. "Seasonal variations and the contribution of cowpea (*vigna unguiculata*) and other legumes to nutrition intake in Anambra state, Nigeria." *Ecology of Food and Nutrition* 17:271-87.
Norman, D. W. 1969. "Labour inputs of farmers: A case study of the Zaria province of the north-central state of Nigeria." *Nigerian Journal of Economic and Social Studies* 11 (no. 1):9-10.
———. 1973. *Economic analysis of agricultural production and labor utilization among the Hausa in the north of Nigeria.* African Rural Employment Paper no. 4. East Lansing: Department of Agricultural Economics, Michigan State University.
Norman, D. W., P. Beeden, W. J. Kroeker, D. H. Pryor, B. Huizinga, and H. M. Hays. 1976. *The feasibility of improved sole crop maize production for the small-scale farmer in the northern Guinea savanna zone of Nigeria.* Miscellaneous Paper no. 59. Zaria, Nigeria: Ahmadu Bello University, Institute of Agricultural Research.
Norman, D. W., E. Simmons, and H. M. Hays. 1982. *Farming systems in the Nigerian savanna.* Boulder, Colo.: Westview.
Nurse, G. T. 1975. "Seasonal hunger among the Ngoni and Ntumba of central Malawi." *Africa* 45:1-11.
Nylin, G. 1929. "Periodical variations in growth, standard metabolism and oxygen capacity of the blood in children." *Acta Medica Scandinavica* 31 (supp.).
Odell, M. F. 1982. "The domestic control of production and reproduction in a Guatemalan community." *Human Ecology* 10:47-69.

Ogbu, J. U. 1973. "Seasonal hunger in tropical Africa as a cultural phenomenon." *Africa* 43:317-22.

Ogle, B. M., and L. E. Grivetti. 1985. "Legacy of the chameleon: Edible wild plants in the kingdom of Swaziland, southern Africa. A cultural ecological, nutritional study." *Ecology of Food and Nutrition* 16:193ff.

Okigbo, B. 1984. "Improved permanent production systems as an alternative to shifting intermittent cultivation." In *Improved production systems as an alternative to shifting cultivation,* pp. 1-100. FAO Soils Bulletin no. 31. Rome: FAO.

O'Leary, M. F. 1980. "Response to drought in Kitui district, Kenya." *Disasters* 4 (no. 3):314-27.

———. 1984. "Ecological villains or economic victims: The case of the Rendille of northern Kenya." *Desertification Control Bulletin* 11:17-21.

Olsen, R. J. 1980. "The economic value of children in peasant agriculture." In *Population and development,* ed. R. Ridker. Baltimore, Md.: Johns Hopkins University Press.

Omawale. 1980. "Nutrition problem identification and development policy implications." *Ecology of Food and Nutrition* 9:113-22.

Oppen, M. von, V. T. Raju, and S. L. Bapna. 1979. "Foodgrain marketing and agricultural development in India." Paper presented at the international workshop on socioeconomic constraints to development of semi-arid tropic agriculture, ICRISAT, Patancheru, India, February 19-23.

Orr, J. B., and M. L. Clark. 1930. "A report on seasonal variation in the growth of school children." *Lancet* 2:365-67.

Pagezy, H. 1982. "Seasonal hunger as experienced by the Oto and the Twa of a Ntomba village in the equatorial forest (Lake Tumba, Zaire)." *Ecology of Food and Nutrition* 12:139-53.

———. 1984. "Seasonal hunger as experienced by the Oto and Twa women in the equatorial forest (Lake Tumba, Zaire)." *Ecology of Food and Nutrition* 15:13-27.

Palmer, C. E. 1933. "Seasonal variation of average growth in weight of elementary school children." *Public Health Reports* 48 (no. 9):211.

Palmer-Jones, R. 1981. "How not to learn from pilot irrigation projects: The Nigerian experience." *Water Supply and Management* 5 (no. 1).

Panda, H. 1985. "Impact of irrigation on farmers' acreage response to price: Case of Andhra Pradesh." *Economic and Political Weekly* 20 (no. 13).

Pardy, C. 1987. "Cereal sales behavior among farm households in four villages of Burkina Faso." In *The dynamics of grain marketing in Burkina Faso,* ed. J. Sherman, K. Shapiro, and E. Gilbert. Ann Arbor: Center for Research on Economic Development, University of Michigan.

Pasternak, B. 1978. "Seasons of birth and marriage in two Chinese localities." *Human Ecology* 6:299-323.

Patton, M. 1971. *Dodoma region, 1929-1959: A history of famine.* BRALUP Research Report no. 44. Dar es Salaam: Bureau of Resource Assessment and Land Use Planning, University of Dar es Salaam.

Payne, P. R. 1985. "The nature of malnutrition." In *Nutrition and development,* pp. 1-19, ed. M. Biswas and P. Pinstrup-Andersen. New York: Oxford University Press.

———. 1989. "Public health and functional consequences of seasonal hunger and malnutrition." In this volume.
Payne, P. R., and P. Cutler. 1984. "Measuring malnutrition: Technical problems and ideological perspectives." *Economic and Political Weekly*, August 25, pp. 1485-92.
Payne, P. R., and A. E. Dugdale. 1977. "A model for the prediction of energy balance and body weight." *Annals of Human Biology* 4:525-35.
Penning de Vries, F. W. T. 1983. *The productivity of Sahelian rangelands—A summary report*, p. 33. ODI Pastoral Network Paper 15b. London: Agricultural Administration Unit, Overseas Development Institute.
Peterson, J. T. 1978. "Hunter-gatherer/farmer exchange." *American Anthropologist* 80:335-51.
Pinckney, T. C., and C. H. Gotsch. 1987. "Simulation and optimization of price stabilization policies: Maize in Kenya." *Food Research Institute Studies* 20 (no. 3).
Pingle, V. 1975. "Some studies of two tribal groups of Central India: Part 2. Nutritive importance of foods consumed in two different seasons." *Plant Foods for Man* 1:195-208.
Pinstrup-Andersen, P. 1981. *Nutritional consequences of agricultural projects: Conceptual relationships and assessment approaches*. World Bank Staff Working Paper no. 456. Washington, D.C.: World Bank.
———. 1982. *Agricultural research and technology in economic development*. New York: Longman.
———. 1985. "Food prices and the poor in developing countries." *European Review of Agricultural Economics* 12 (nos. 1-2).
———, ed. 1988. *Consumer-oriented food subsidies: Costs, benefits, and policy options for developing countries*. Baltimore, Md.: Johns Hopkins University Press.
Pinstrup-Andersen, P., and M. Garcia. 1987. "Household vs. individual food consumption as indicators of the nutritional impact of food policy." Washington, D.C.: IFPRI (mimeographed).
Pinstrup-Andersen, P., and P. B. R. Hazell. 1985. "The impact of the green revolution and prospects for the future." *Food Reviews International* 1:1-25.
Pinstrup-Andersen, P., and M. Jaramillo. Forthcoming. "The impact of technological change in rice production on food consumption and nutrition in North Arcot, India." In *Technological change and rural welfare*, ed. P. B. R. Hazell and C. Ramasamy. Baltimore, Md.: Johns Hopkins University Press.
Pitt, M., and M. R. Rosenzweig. 1984. "Agricultural prices, food consumption and the health and productivity of farmers." *University of Minnesota Economic Development Center Bulletin* 84 (no. 1).
———. 1985. "Health and nutrient consumption across and within farm households." *Review of Economics and Statistics* 67 (no. 2):212-23.
———. 1986. "Agricultural prices, food consumption, and the health and productivity of farmers." In *Agricultural household models: Extensions, applications, and policy*, ed. I. J. Singh, L. Squire, and J. Strauss. Washington, D.C.: World Bank.
Popkin, B. 1980. "Time allocation of the mother and child nutrition." *Ecology of Food and Nutrition* 9:1-14.

Prentice, A. M. 1980. "Variations in maternal dietary intake, birthweight, and breastmilk output in the Gambia." In *Maternal nutrition during pregnancy and lactation,* pp. 167-83, ed. H. Aebi and R. G. Whitehead. Bern: Hans Huber.

Prentice, A. M., S. B. Roberts, A. Prentice, A. A. Paul, M. Watkinson, A. Watkinson, and R. G. Whitehead. 1983a. "Dietary supplementation of lactating Gambian women: I. Effect on breastmilk output and quality." *Human Nutrition: Clinical Nutrition* 37C:53-64.

Prentice, A. M., R. G. Whitehead, M. Watkinson, W. H. Lamb, and T. J. Cole. 1983b. "Pre-natal dietary supplementation of African women and birthweight." *Lancet* 1:489-92.

Price, E. C. 1982. "Adoption and impact of new cropping systems in Iloilo and Pangasinan, Philippines." In *Cropping systems research in Asia.* Los Baños, Philippines: IRRI.

Price, M. R. S. 1981. "The rangelands: Pastoralism and ranching." In *The development of Kenya's semi-arid lands,* pp. 150-75, ed. D. Campbell and S. E. Migot-Adholla. IDS Occasional Paper no. 36. Nairobi: Institute for Development Studies, University of Nairobi.

Pryor, F. 1977. *The origins of the economy.* New York: Academic.

Psacharopoulos, G. 1982. "Returns to education: An updated international comparison." *Comparative Education* 17 (no. 3):321-41.

Quetelet, L. A. J. 1871. *Anthropometrie on mesure des differentes facultés des hommes,* p. 479. Brussels: Muquardt.

Ramasamy, C. 1985. Personal communication. Washington, D.C.: IFPRI.

Rand, W. M., N. S. Scrimshaw, and V. R. Young. 1979. "An analysis of temporal patterns in urinary nitrogen excretion of young adults receiving constant diets at two nitrogen intakes for 8 to 11 weeks." *American Journal of Clinical Nutrition* 32:1408-14.

Ray, S. K., R. W. Cummings, Jr., and R. W. Herdt. 1979. *Policy planning for agricultural development.* New Delhi, India: Tata McGraw Hill.

Reardon, T., and P. Matlon. 1985. "The seasonality of calorie intake in rural Burkina Faso." Paper presented at IFPRI/FAO/AID workshop on seasonal causes of household food insecurity, policy implications, and research needs, Maryland Inn, Annapolis, Md., December 10-13.

République-Unie du Cameroun (RUC). 1978. *National nutrition survey.* Yaoundé: RUC.

Richards, A. I. 1939. *Land, labor, and diet in northern Rhodesia: An economic study of the Bemba tribe.* London: G. Routledge.

Richards, P. 1985. *Indigenous agricultural revolution. Ecology and food production in West Africa.* London: Hutchinson.

de Ridder, N., L. Stroosnijder, A. M. Cisse, and H. van Keulen. 1982. *Productivity of Sahelian rangelands: A study of the soils, the vegetations, and the exploitation of that natural resource.* PPS Course Book, 2 vols. Wageningen, Netherlands: Department of Soil Science and Plant Nutrition, Wageningen Agricultural University.

Rizvi, N. 1986. "Seasonal variation in nutritional status among women of different occupational groups in Bangladesh." In *Proceedings of the XIIIth Interna-*

tional Congress of Nutrition, pp. 150-55, ed. T. G. Taylor and N. K. Jenkins. London: John Libbey.

Roberts, S. B., A. A. Paul, T. J. Cole, and R. G. Whitehead. 1982. "Changes in activity, birth weight, and lactational performance in rural Gambian women." *Trans. Roy. Soc. Trop. Med. Hyg.* 76 (no. 5):668.

Robson, J. R. K., and J. Elias. 1978. *The nutritional value of wild plants: An annotated bibliography.* Troy, N.Y.: Whitson.

Rogers, G. B. 1975. "Nutritionally based wage determination in the low-income labour market." *Oxford Economic Papers* 27:61-81.

Rosenzweig, M. R. 1984a. "Determinants of wage rates and labor supply behavior in the rural sector of a developing country." In *Contractual arrangements, employment, and wages in rural labor markets in Asia,* pp. 211-41, ed. H. P. Binswanger and M. Rosenzweig. New Haven, Conn.: Yale University Press.

———. 1984b. *Program interventions, intrahousehold allocation and the welfare of individuals: Economic models of the household.* Minneapolis: University of Minnesota.

Rosenzweig, M. R., and T. P. Schultz. 1982. "Market opportunities, genetic endowments, and intrafamily resource distribution: Child survival in rural India." *American Economic Review* 72 (no. 4):803-15.

———. 1984. "Market opportunities, genetic endowments, and intrafamily resource distribution: Reply." *American Economic Review* 74:521-22.

Ruthenberg, H. 1968. "Some characteristics of smallholder farming in Tanzania." In *Smallholder farming and smallholder development in Tanzania,* pp. 325-55, ed. H. Ruthenberg. IFO-Institut, Afrika-Studien no. 24. Munich: Weltforum.

———. 1976. *Farming systems in the tropics.* Oxford: Oxford University Press.

Ryan, J. G. 1982. *Wage functions for daily labor market participants in South Asia.* Progress Report 38. Patancheru, India: ICRISAT.

Ryan, J. G., P. D. Bidinger, N. P. Rao, and P. Pushpamma. 1984. *The determinants of individual diets and nutritional status in six villages of southern India.* ICRISAT Research Bulletin no. 7. Patancheru, India: ICRISAT.

Ryan, J. G., and R. D. Ghodake. 1984. "Labor market behavior in rural villages in South India: Effect of season, sex, and socioeconomic status." In *Contractual arrangements, employment, and wages in rural labor markets in Asia,* pp. 169-83, ed. H. P. Binswanger and M. Rosenzweig. New Haven, Conn.: Yale University Press.

Ryan, J. G., R. D. Ghodake, and R. Sarin. 1980. "Labor use and labor markets in semi-arid tropical rural villages of peninsular India." In *Socioeconomic constraints to development of semi-arid tropical agriculture.* Patancheru, India: ICRISAT.

Ryan, J. G., and T. D. Wallace. 1986. "Determinants of labor market wages, participation, and supply in rural South India." Canberra, Australia: Australian Centre for International Agricultural Research (mimeographed).

Sabean, D. 1978. "Small peasant agriculture in Germany at the beginning of the nineteenth century: Changing work patterns." *Peasant Studies* 7 (no. 4):218-24.

Safilios-Rothschild, C. 1980. *The role of the family—A neglected aspect of poverty.* World Bank Staff Working Paper no. 403. Washington, D.C.: World Bank.

Sahlins, M. 1972. *Stone age economics.* Chicago, Ill.: Aldine.

Sahn, D. 1986. "Measures to ensure access to food by the poor." Rome: FAO (mimeographed).

Sahn, D., and H. Alderman. 1987. "The role of the foreign exchange and commodity auctions in trade, agriculture, and consumption in Somalia." Washington, D.C.: IFPRI (mimeographed).

———. 1989. "The effects of human capital on wages and the determinants of labor supply in a developing country." *Journal of Development Economics* 29 (no. 2).

Sahn, D., and J. von Braun. 1987. "The relationship between food production and consumption variability: Policy implications for developing countries." *Journal of Agricultural Economics* 38 (no. 1).

Sahn, D., B. Rogers, and D. Nelson. 1981. "Assessing the uses of food aid: P.L. 480 title II program in India." *Ecology of Food and Nutrition* 10 (no. 3).

Sandford, S. 1983. *Management of pastoral development in the third world.* New York: Wiley.

Saul, M. 1987. "The marketing of grain in Upper Volta." In *The dynamics of grain marketing in Burkina Faso,* ed. J. Sherman, K. Shapiro, and E. Gilbert. Ann Arbor: Center for Research on Economic Development, University of Michigan.

Schlage, C. 1969. "Analysis of some important foodstuffs of Usambara." In *Investigations into health and nutrition in East Africa,* pp. 55–70, ed. H. Kraut and H. D. Cremer. IFO-Institut, Afrika-Studien no. 42. Munich: Weltforum.

Schneider, H. K. 1968. "Economics in East African aboriginal societies." In *Economic anthropology: Readings in theory and analysis,* pp. 425–45, ed. E. LeClair and H. Schneider. New York: Holt, Rinehart, and Winston.

Schofield, S. 1974. "Seasonal factors affecting nutrition in different age groups and especially preschool children." *Journal of Development Studies* 2 (no. 1):22–40.

Scott, C. 1980. "Practical problems in conducting surveys on living standards." In *Conducting surveys in developing countries: Practical evidence and experience in Brazil, Malaysia, and the Philippines,* ed. C. Scott, P. de Andre, and P. Chander. Living Standards Measurement Study Working Paper no. 5. Washington, D.C.: World Bank.

Scott, E. P., ed. 1984. *Life before the drought.* Boston: Allen and Unwin.

Scrimshaw, M., and S. Cosminsky. n.d. "The impact of women's health on food procurement." In *Diet and domesticity in social life,* ed. A. Sharman, J. Theophano, K. Curtis, and E. Messer (under press review).

Scudder, T. 1962. *The ecology of the Gwenbe Tonga.* Manchester: Manchester University Press.

———. 1980. "African river basin development and local initiatives in savanna environments." In *Human ecology in savanna environments,* pp. 383–405, ed. D. Harris. New York: Academic.

Seckler, D. 1980. "Malnutrition: An intellectual odyssey." *Western Journal of Agricultural Economics* 5:219.
———. 1982. "Small but healthy: A basic hypothesis in the theory, measurement and policy of malnutrition." In *Newer concepts in nutrition and their implications for policy,* ed. P. V. Sukhatme. Pune, India: Maharashtra Association for the Cultivation of Science Research Institute.
Sen, A. K. 1981. *Poverty and famines: An essay on entitlement and deprivation.* Oxford: Clarendon.
Serdula, M. K., J. Sward, J. S. Marks, N. S. Staehling, O. Galal, and F. Trowbridge. 1986. "Seasonal differences in breastfeeding in rural Egypt." *American Journal of Clinical Nutrition* 44 (no. 3):405-9.
Service, E. 1966. *The hunters.* Englewood Cliffs, N.J.: Prentice-Hall.
Shack, D. N. 1969. "Nutritional processes and personality development among the Gurage of Ethiopia." *Ethnology* 8:292-300.
Shack, W. 1971. "Hunger, anxiety, and ritual: Deprivation and spirit possession among the Gurage of Ethiopia." *Manual of Natural Science* 6 (no. 1):30-43.
Sharman, A. 1970. "Social and economic aspects of nutrition in Padhold Dukedi district, Uganda." Ph.D. diss., University of London.
Sharman, A., J. Theophano, K. Curtis, and E. Messer, eds. n.d. *Diet and domesticity in social life* (under press review).
Sharp, J. S., and A. D. Spiegel. 1985. "Vulnerability to impoverishment in South African rural areas: The erosion of kinship and neighbourhood as social resources." *Africa* 55 (no. 2):133-52.
Shepherd, A. 1981. "Agrarian change in northern Ghana: Public investment, capitalist farming, and famine." In *Rural development in tropical Africa,* pp. 168-92, ed. J. Heyer, P. Roberts, and G. Williams. New York: St. Martin's.
Shepherd, J. 1975. *The politics of starvation.* New York: Carnegie Endowment for International Peace.
Shetty, P. S. 1984. "Adaptive changes in basal metabolic rate and lean body mass in chronic undernutrition." *Human Nutrition: Clinical Nutrition* 38C:443.
Siamwalla, A. 1984. "Public stock management and its implications for prices and supply," pp. 9-12. Washington, D.C.: IFPRI (mimeographed).
Siamwalla, A., and A. Valdes. 1984. "Food security in developing countries: International issue." In *Agricultural development in the third world,* ed. C. Eicher and J. Staatz. Baltimore, Md.: Johns Hopkins University Press.
Sidhu, J. S., and D. S. Sidhu. 1985. "Price support versus fertilizer subsidy: An evaluation." *Economic and Political Weekly* 20:A17-22.
Simmons, E. B. 1975. "The small-scale rural food processing industry in northern Nigeria." *Food Research Institute Studies* 14:147-61.
———. 1976a. *Calorie and protein intakes in three villages of Zaria province, May 1970-June 1971.* Miscellaneous Paper no. 55. Zaria, Nigeria: Ahmadu Bello University, Institute of Agricultural Research.
———. 1976b. *Rural household expenditures in three villages of Zaria province, May 1970-June 1971.* Miscellaneous Paper no. 56. Zaria, Nigeria: Ahmadu Bello University, Institute of Agricultural Research.

———. 1981. "A case study in food production, sale, and distribution." In *Seasonal dimensions to rural poverty*, pp. 73–80, ed. R. Chambers, R. Longhurst, and A. Pacey. London: Frances Pinter.
Simpson, J. R., and P. Evangelou, eds. 1984. *Livestock development in sub-Saharan Africa*. Boulder, Colo.: Westview.
Singh, I. J., L. Squire, and J. Strauss, eds. 1986. *Agricultural household modes: Extensions, applications, and policy*. Washington, D.C.: World Bank.
Société d'Etudes pour le Développement Economique et Social (SEDES). 1964–65. *Le niveau de vie des populations de la zone Cacaoyére du centre Cameroun.* Paris: SEDES.
Soemarwoto, O., et al. 1986. "The Javanese home garden as an integrated agroecosystem." *Food and Nutrition Bulletin* 7 (no. 3):44–47.
Somerville, C. M. 1986. *Drought and aid in the Sahel*. Boulder, Colo.: Westview.
Southworth, V. R., W. O. Jones, and S. R. Pearson. 1979. "Food crop marketing in Atebubu district, Ghana." *Food Research Institute Studies* 17 (no. 2).
Spencer, D. 1976. *African women in agricultural development: A case study in Sierra Leone*. African Rural Economy Program Working Paper no. 11. East Lansing: Michigan State University.
Spencer, T., and P. Heywood. 1983. "Seasonality, subsistence agriculture, and nutrition in a lowlands community of Papua New Guinea." *Ecology of Food and Nutrition* 13:221–29.
Spitz, P. 1982. "Drought and self-provisioning." In *A report on the regional workshop on seasonal variations in the provisioning, nutrition, and health of rural families*, pp. 22–40. Nairobi: Ford Foundation and AMREF.
Squire, L. 1981. *Employment policy in developing countries*. New York: Oxford University Press.
Srinivasan, T. N. 1981. "Malnutrition: Some measurement and policy issues." *Journal of Development Economics* 8 (no. 1):3–19.
Stiglitz, J. 1976. "The efficiency wage hypothesis, surplus labour, and the distribution of income in LDCs." *Oxford Economic Papers* 28.
Strauss, J. 1982. "Determinants of food consumption in rural Sierra Leone: Application of the quadratic expenditure system to the consumption-leisure component of a household farm model." *Journal of Development Economics* 11:327–53.
———. 1985. "The impact of improved nutrition in labor productivity and human resource development: An economic perspective." Paper prepared for the IFPRI workshop on the political economy of nutrition improvements, West Virginia (mimeographed).
———. 1986. "Does better nutrition raise farm productivity?" *Journal of Political Economy* 94 (April):297–320.
Sukhatme, P. V. 1961. "The world's hunger and future needs in food supplies." *Journal of the Royal Statistical Society* 124, ser. A:463–525.
———, ed. 1982. *Newer concepts in nutrition and their implications for policy*. Pune, India: Maharashtra Association for the Cultivation of Science Research Institute.
Svedberg, P. 1988. *Undernutrition in sub-Saharan Africa: Is there a sex bias?* Stockholm: Institute for International Economic Studies, University of Stockholm.

Swanberg, K., and E. Hogan. 1981. *Implications of the drought syndrome or agricultural planning in East Africa: The case of Tanzania.* Development Discussion Paper no. 120. Cambridge, Mass.: Harvard Institute for International Development, Harvard University.

Swift, J. 1981. "Labor and subsistence in a pastoral economy." In *Seasonal dimensions to rural poverty,* pp. 80–87, ed. R. Chambers, R. Longhurst, and A. Pacey. London: Frances Pinter.

Szarleta, E. 1987. "Increasing the marketed surplus of cereal grains in Burkina Faso: An identification of farmer response." In *The dynamics of grain marketing in Burkina Faso,* ed. J. Sherman, K. Shapiro, and E. Gilbert. Ann Arbor: Center for Research on Economic Development, University of Michigan.

TAC. 1984. "An analysis of research priorities in the CGIAR system." TAC/CGIAR 34th Meeting, Addis Ababa, Ethiopia, June 1984.

Tanzania Ministry of Agriculture and Livestock Development. 1984. *Tanzania national food strategy: Main report,* vol. 1. Rome: FAO, for the Tanzania Government.

Teitelbaum, J. M. 1977. "Human versus animal nutrition: A development project among the Fulani cattle-keepers in the Sahel of Senegal." In *Nutrition and anthropology in action,* ed. T. Fitzgerald. Assen, Netherlands: Van Gorcum.

Teokul, W., P. Payne, and A. Dugdale. 1986. "Seasonal variations in nutritional status in rural areas in developing countries: A review of the literature." *Food and Nutrition Bulletin* 8 (no. 4):7–10.

Thomas, J. W. 1985. "Food-for-work: An analysis of current experience and recommendations for further performance." Cambridge, Mass.: Harvard Institute for International Development, Harvard University (mimeographed).

Tiffen, M. 1985. *Land tenure issues in irrigation planning, design and management in sub-Saharan Africa.* Working Paper no. 16. London: Overseas Development Institute.

Timberlake, L. 1985. *Africa in crisis.* London: IIED, Earthscan.

Timmer, C. P. 1974. "A model of rice marketing margins in Indonesia." *Food Research Institute Studies* 13 (no. 2):145–67.

———. 1981. "Is there a 'curvature' in the Slutsky matrix?" *Review of Economics and Statistics* 63:395–402.

Timmer, C. P. and H. Alderman. 1979. "Estimating consumption parameters for food policy analysis." *American Journal of Agricultural Economics* 61:982–87.

Tin-May-Than and Ba-Aye. 1985. "Energy intake and energy output of Burmese farmers at different seasons." *Human Nutrition: Clinical Nutrition* 39C:7–15.

Toksoz, S. 1981. *An accelerated irrigation and land reclamation program for Kenya: Dimension and issues.* Development Discussion Paper no. 114. Cambridge, Mass.: Harvard Institute for International Development, Harvard University.

Tolley, A. S., V. Thomas, and C. M. Wong. 1982. *Agricultural price policies and the developing countries.* Washington, D.C.: World Bank; and Baltimore, Md.: Johns Hopkins University Press.

Tomkins, A. 1985. "Protein-energy malnutrition and the risk of infection." *Proceedings of the Nutrition Society* 15:289.

Toulmin, C. 1986. "Access to food, dry season strategies and household size amongst the Bambara of central Mali." *IDS Bulletin* 17 (no. 3):58-66.

Trairatvorakul, P. 1984. *The effects on income distribution and nutrition of alternative rice price policies in Thailand.* Research Report 46. Washington, D.C.: IFPRI.

Tripp, R. 1981. "Farmers and traders: Some economic determinants of nutritional status in northern Ghana." *Journal of Tropical Pediatrics* 27:15-22.

Tully, D. 1982. "The decision to migrate: Conflicts between the individual and the household in Dar Masalit Sudan." Paper presented at the symposium on household level analysis in economic and political research (a reevaluation with African evidence). Also presented at the 81st annual meeting of the American Anthropological Association, Washington, D.C.

Tukey, J. W. 1982. "The problem of multiple comparison." In *SAS user's guide: Statistics,* pp. 169-75. Raleigh, N.C.: SAS Institute.

Turnbull, C. 1962. *The forest people.* New York: Simon and Schuster.

———. 1972. *The mountain people.* New York: Simon and Schuster.

Turton, D. 1977. "Response to drought: The Mursi of Southwest Ethiopia." In *Human ecology in the tropics,* ed. J. P. Garlick and R. W. J. Keay. Symposia of the Society for the Study of Human Biology. London: Taylor and Frances.

Tyagi, D. S. 1982. "Farm level storage: Orderly marketing, the public distribution system and labour absorption." *IDS Bulletin* 13 (no. 3):45-48.

Tyson, P. D. 1979. "Southern African rainfall: Past, present, and future." In *Proceedings of the symposium on drought in Botswana,* pp. 45-52, ed. M. T. Hinchey. Hanover, N.H.: University Press of New England for the Botswana Society, with Clark University Press.

UNICEF. 1982. *The UNICEF home garden handbook—For people promoting mixed gardening in the humid tropics.* New York: UNICEF.

United Nations University. 1985. "Household-level food production." *Food and Nutrition Bulletin* 7 (no. 3).

Unnevehr, L. J. 1983. *The effect and cost of Philippine government intervention in rice markets.* Working Paper no. 9 on Rice Policies in Southeast Asia Project. Washington, D.C.: IFPRI, International Fertilizer Development Center; and Los Baños, Philippines: IRRI.

———. 1985. "The costs of squeezing marketing margins: Philippine government intervention in rice markets." *Journal of Developing Economies* 23 (no. 2).

U.S. Department of Health, Education, and Welfare (US-DHEW). 1966. *Nutrition survey of East Pakistan: 1962-64.* Washington, D.C.: Government Printing Office.

Valverde, V., H. Delgado, R. Flores, and R. Sibrian. 1985. "Nutritional and health consequences of seasonal fluctuations in household food availability." Presented at United Nations University conference on the impact of the choice of agricultural and food policies on nutrition and health status, Bellagio, Italy, March 1985. Guatemala City, Guatemala: Instituto Nutricion de Centro America y Panama (mimeographed).

Waaler, H. T. 1984. "Height, weight, and mortality: The Norwegian experience." *Acta Medica Scandinavica* (no. 679 supp.).

Walden, T. 1980. "Entrepreneurial illiquidity preference and the African extended family." In *Development planning in Kenya,* pp. 119-39, ed. T. Pinfold and G. Norcliffe. Ontario: Department of Geography, Atkinson College, York University.
Wallace, T. 1981. "The Kano river project, Nigeria: The impact of an irrigation scheme on productivity and welfare." In *Rural development in tropical Africa,* pp. 281-305, ed. J. Heyer, P. Roberts, and G. Williams. New York: St. Martin's.
Walsh, R. P. D. 1981. "The nature of climatic seasonality." In *Seasonal dimensions to rural poverty,* pp. 11-29, ed. R. Chambers, R. Longhurst, and A. Pacey. London: Frances Pinter.
Warman, A. 1980. *We come to object.* Baltimore, Md.: Johns Hopkins University Press.
Wasserstrom, R. F. 1983. *Class and society in central Chiapas.* Berkeley and Los Angeles: University of California Press.
Waterbury, J. 1983. *The Senegalese peasant: How good is our conventional wisdom?* Princeton, N.J.: Princeton University Press.
Waterlow, J. C. 1972. "Classification and definition of protein-calorie malnutrition." *British Medical Journal* 3:566-69.
Waterlow, J. C., R. Buzina, W. Keller, J. M. Lane, and M. Z. Nichaman. 1977. "The presentation and use of height and weight data for comparing the nutritional status of groups of children under the age of 10 years." *WHO Bulletin* 55:489-98.
Watts, M. 1983. *Silent violence: Food, famine, and peasantry in northern Nigeria.* Berkeley and Los Angeles: University of California Press.
West, D., and D. W. Price. 1976. "The effects of income, assets, food programs, and household size on food consumption." *American Journal of Agricultural Economics* 58:725-30.
Wharton, C. R. 1962. "The economic meaning of subsistence." *Malayan Economic Review* 8 (no. 2).
White, D. R., M. Burton, and M. Dow. 1981. "Sexual division of labor in African agriculture: A network autocorrelation analysis." *American Anthropologist* 83:824-49.
White, G. F., ed. 1974. *Natural hazards: Local, national, global.* New York: Oxford University Press.
Wienpahl, J. 1984. "The role of women and small stock among the Turkana." Ph.D. diss., University of Arizona, Tucson.
Wiessner, P. 1981. "Measuring the impact of social ties on nutritional status among the !Kung San." *Social Science Information* 20:641-78.
Wijga, A. 1983. "The nutritional impact of food-for-work programs." Paper prepared for Netherlands Universities Foundation for International Cooperation, Wageningen, Netherlands.
de Wilde, J. C. 1967. "Implements and machinery." In *Experiences with agricultural development in tropical Africa: The synthesis,* vol. 1, pp. 95-131. Baltimore, Md.: Johns Hopkins University Press.
Wilmsen, E. N. 1978. "Seasonal effects of dietary intake on the Kalahari San." *Fed. Proc.* 37:65-72.

———. 1982. "Studies in diet, nutrition, and fertility among a group of Kalahari bushmen in Botswana." *Social Science Information* 21:95-125.

Wilson, R. T., A. Diallo, and K. Wagenaar. 1985. "Mixed herding and the demographic parameters of domestic animals in arid and semi-arid zones of tropical Africa." In *Population, health, and nutrition in the Sahel,* pp. 116-38, ed. A. G. Hill. London: KPI.

Winikoff, B., ed. 1975. *Nutrition and national policy.* Cambridge: Massachusetts Institute of Technology.

Winterhalder, B., and E. A. Smith, eds. 1981. *Hunter-gatherer foraging strategies.* Chicago, Ill.: University of Chicago Press.

Wisner, B. 1977. "The human ecology of drought in eastern Kenya." Ph.D. diss., Clark University, Worcester, Mass.

Wisner, B., and P. Mbithi. 1974. "Drought in eastern Kenya: Nutritional status and farmer activity." In *Natural hazards: Local, national, global,* pp. 87-97, ed. G. F. White. New York: Oxford University Press.

Wolpin, K. 1982. "A new test of the permanent income hypothesis: The impact of weather on income and consumption of farm households in India." *International Economic Review* 23:583-94.

Working, H. 1958. "A theory of anticipatory prices." *American Economic Review* 48 (no. 2):188-99.

Wright, T. L., R. C. Carter, M. K. V. Carr, and M. G. Kay. 1982. *Nigeria: Manpower needs and training requirements for irrigation.* London: British Council.

Contributors

Mohammed Abdullah, a nutritionist, is currently a professor at the Institute of Nutrition and Food Science of the University of Bangladesh. His research has focused on intrahousehold allocation of nutrients and determinants of dietary adequacy.

Harold Alderman, a senior research associate at the Cornell University Food and Nutrition Policy Program and a research fellow at IFPRI, has focused his work on the influence of various aspects of food policy on consumption and nutrition, including a major study of the Egyptian food ration and subsidy system. He is presently undertaking a study of household-level food security in Pakistan.

Jere R. Behrman is the William R. Kenan, Jr., Professor of Economics, co-director of the Center for Analysis of Developing Economies and of the Center for Household and Family Economics, and a member of the Population Center and the South Asia Regional Studies Program at the University of Pennsylvania. He has published research on a wide range of topics primarily related to economic development in 14 books and about 140 professional articles.

Timothy J. Cole has been a senior statistician at the MRC Dunn Nutrition Unit, Cambridge, United Kingdom.

W. Andrew Coward has been a member of the scientific staff of the MRC Dunn Nutrition Unit, Cambridge, United Kingdom. Currently he is in charge of the unit's stable isotope research program.

Christopher Delgado is a research fellow and co-coordinator for African research at IFPRI. He also teaches at the Johns Hopkins University School of Advanced International Studies. Previously he worked at the Center for Research on Economic Development, University of Michigan, and at the University of Ouagadougou, Burkina Faso. He has written on West African farming and pastoral systems, agricultural price policy, food grain marketing issues, the political economy of West Africa, and long-run policies to support African agricultural production. He is the co-editor (along with John W. Mellor and Malcolm J. Blackie) of *Accelerating Food Production in Sub-Saharan Africa.*

Anil B. Deolalikar, a visiting associate professor of economics at Harvard University during 1987–88, has been an assistant professor of economics at the University of Pennsylvania since 1983. He also is a member of the Population Studies Center and the South Asia Regional Studies Program at the University of Pennsylvania. Besides being engaged in research on several aspects of development, including agriculture and rural development, economic demography, human resource development, and technology transfer, he has been a consultant to several international organizations.

Lynn Ellsworth is a program officer in the social sciences division of the International Development Research Centre, based at the center's office for West Africa in Dakar, Senegal. She received her Ph.D. in agricultural economics from the University of Wisconsin-Madison.

Jane Guyer is an associate professor of anthropology at Boston University. She has spent considerable time doing field work in Nigeria and Cameroon, focusing on the history of the division of labor in the household. Among her publications is the edited volume *Feeding African Cities: Studies in Regional Social History.*

Robert W. Herdt is director of Agricultural Sciences at the Rockefeller Foundation. Over the past two decades he has actively worked with biological and social scientists to understand how production and income of small-scale farmers in developing countries can be improved. He received the award for the outstanding paper in the *American Journal of Agricultural Economics* in 1977 (co-authored with Yujiro Hayami). He has co-authored 3 books and over 60 papers in the field of agricultural development. His most recent book, *The Rice Economy of Asia,* co-authored with Randolph Barker, is the culmination of their three decades of experience working on that subject. It received the 1986 award for outstanding research communication from the American Agricultural Economics Association.

Mauricio Jaramillo was formerly connected with IFPRI as a research assistant.

Françoise Lawrence is a member of the scientific staff of the MRC Dunn Nutrition Unit, Cambridge, United Kingdom, based at the unit's field station in Gambia, West Africa.

Mark Lawrence is a member of the scientific staff of the MRC Dunn Nutrition Unit, Cambridge, United Kingdom, based at the unit's field station in

Gambia, West Africa. Currently he is a lecturer in nutrition and physiology at the University of Glasgow, United Kingdom.

Michael Lipton is director of the IFPRI Food Consumption and Nutrition Policy Program. Previously he was a professorial fellow in economics at the Institute of Development Studies at Sussex University, and has taught or researched at All Souls College, Oxford, and the Institute of Social Studies, The Hague. He has been a consultant to a number of international organizations and governments. His main interest is agriculture and rural development in South Asia and in eastern and southern Africa. Of his many publications, the most recent is a series of working papers for the World Bank on characteristics of poverty groups.

Richard Longhurst is a program officer and representative in Sudan for the Ford Foundation. He prepared this chapter while at the Institute for Development Studies at Sussex University, drawing extensively on his work while a consultant and staff member of the Food Policy and Nutrition Division of the Food and Agriculture Organization.

Peter J. Matlon is director of research for the West Africa Rice Development Association (WARDA) in Bouake, Ivory Coast. During 1979-88 he served as principal economist with the International Crops Research Institute for the Semi-Arid Tropics (ICRISAT). He has written extensively on farming systems in the West African semiarid tropics with emphasis on an assessment of current and prospective production technologies. The present chapter was written while he was a visiting scholar at the Food Research Institute, Stanford University.

Ellen Messer is an associate professor in the Alan Shawn Feinstein World Hunger Program at Brown University and a visiting scholar at the Center for International Studies, International Food and Nutrition Program, Massachusetts Institute of Technology. A cultural and nutritional anthropologist, she works on the ecology, culture, and socioeconomics of food production and consumption patterns, and the nutrition and health consequences of changing food habits.

Dow Mongkolsmai is an associate professor at the Faculty of Economics, Thammasat University, Bangkok, Thailand. Her research interests include irrigation economics and policy and public finance. She has written extensively on development issues in Thailand. Among her publications are "Status and Performance of Irrigation in Thailand," Rice Policies in Southeast Asia Project Working Paper no. 8; and "The Distributional Impacts of Irrigation in Thailand," which will appear in a forthcoming IFPRI research report on irrigation and equity in Southeast Asia.

Contributors

Jon R. Moris edits the pastoral development network from London's Overseas Development Institute while on leave from Utah State University. He has spent much of his career in Africa, at universities in Uganda, Kenya, and Tanzania, where he was professor of agricultural education and extension. His *Managing Induced Rural Development* remains a major reference work in its field. His recent research has included a USAID-funded review of irrigation throughout Africa.

Philip Payne is professor of applied nutrition and head of the Department of Human Nutrition at the London School of Hygiene and Tropical Medicine. His interests include protein quality assays, mathematical models of whole-body protein, and energy metabolism. He is concerned with the use of nutrition data at the national and international level and with the development of nutrition data systems for government planning and policymaking in the agriculture and food sector. In 1977 he set up the Nutrition Policy Unit at the London School of Hygiene and Tropical Medicine.

Per Pinstrup-Andersen is a professor of food economics at Cornell University and director of the Cornell Food and Nutrition Policy Program. He was formerly the director of the Food Consumption and Nutrition Policy Program at IFPRI. He has researched and published extensively in the areas of the economics of technological change in agriculture and the effect of public policy, particularly food subsidies, on income distribution, food consumption, and nutrition. He recently co-edited *Nutrition and Development*.

Thomas Reardon is a research fellow at IFPRI. His research focuses on marketing behavior of farmers and price and consumption policy issues.

Mark W. Rosegrant is co-coordinator of IFPRI's research on technology policy. His current work includes research analyzing the productivity of alternative irrigation systems and management methods to assess the appropriate strategies for investment in irrigation in Southeast Asia, examination of the effects of irrigation on income distribution, assessment of government efforts in the Indonesian food crop sector, and analysis of the effects of risk on input use and production.

David E. Sahn is currently the deputy director of the Cornell University Food and Nutrition Policy Program. Previously, he was a research fellow at IFPRI, where he prepared this book. He has served as a consultant with numerous international organizations, including the World Bank, Food and Agricultural Organization, World Health Organization, World Food Council, and United Nations University. His research on the effects of agricultural and macroeconomic policies on incomes, consumption, and nutrition has focused on issues related to poverty alleviation and distributional concerns. He has worked in numerous African, Asian, and Latin

American countries, and has written extensively on his research and experiences.

Kenneth Shapiro is a professor of agricultural economics and associate dean for international agricultural programs at the University of Wisconsin-Madison. He directed the four village studies under the Burkina Faso Grain Marketing Research Project. His prior research in Africa includes work throughout the Sahel on grain marketing and on livestock production and marketing, as well as work in Tanzania on farmer efficiency and modernization. Dr. Shapiro received his B.S. from Cornell University, his M.S. from the University of Maryland, and his M.A. and Ph.D. from Stanford University.

Jane Singh is a member of the scientific staff of the MRC Dunn Nutrition Unit, Cambridge, United Kingdom, based at the unit's field station in Gambia, West Africa.

Roger G. Whitehead has been the director of the MRC Dunn Nutrition Unit, Cambridge, United Kingdom.

Index

Abdullah, Mohammed, 5, 7, 57, 58
Adams, W. M., 231n
Adaptability model, for biological regulation, 22, 23-24
Africa
—food market deficiencies, 137
—price incentives to increase production, 138
—relationship between cash and food crop production, 92
—seasonality in food production: effectiveness of methods to reduce, 231-34; indigenous solutions to, 210, 213-14, 214-16, 216-20, 220-21, 232; introduced, 223-24, 224-25; introduced solutions to, 226-28, 228-29, 229-31, 306; recommended policies and programs to reduce, 316; social adjustments to, 221-22, 222-23
—seasonality in rainfall, 209
—seasonal stress on food availability, 12, 13; determinants of, 209-10; from interferences with home production, 160-62
Agricultural earnings, 3, 8; and consumption, 84; periodicity of, 82. *See also* Agricultural income; Wages
Agricultural income: diversification of, 162-65; seasonality, 8, 82, 139; seasonally targeted projects to improve, 96-97, 310, 311-16; sources, 82
Agricultural production: Africa, 138, 226-28, 228-29, 229-31; cropping intensity and, 242-43, 252; home, 160-62; strategies for reducing seasonal variability in, 91-93; technological change effect on, 12-13, 15, 91, 92, 93, 264

Ahmed, I., 228n
Ahmed, R., 96, 184, 238
Akong'a, J. J., 209n, 223
Alderman, H., 8, 9, 67, 81, 83, 84, 107, 108, 172n, 309n, 310, 311n
Alverson, H., 165, 166, 168, 226
Anderson, M. A., 312
Andrae, G., 230n, 231n
Annegers, J. F., 161
Anthony, K. R., 227
Anthropological studies
—on food strategies to meet dwindling food supplies, 155-58; adjustment of household size and composition, 167-68; feasting and gifts, 166-67; foraging to diversify diet, 159; home production, 160-62; household rationing, 167; income diversification, 162-66
—on household and time management, 172, 174-75
—on seasonal food insecurity: conclusions, 173-74; ethnographic information on, 153-58; findings, 151-53
Anthropometric indicators: for children, 20, 21, 27-29; to measure health status, 28-29, 68; to measure nutritional status, 20; for relation of body size to work output, 45
Antonsson-Ogle, B., 289n
Apeldoorn, G. J. van, 209n, 227
Arbitrage: intertemporal, 189, 314; temporal 13, 194
Arjyal, P. C., 238
Asia, seasonality in food security: cropping intensity and, 242-43; factors affecting, 237; government policies influencing, 237-38; marketing interventions and, 239; measurement, 235-

355

Asia, seasonality in food security (cont'd) 36; new crop varieties and, 239; production timing and, 240-42
Asian Vegetable Research and Development Center (AVRDC), 292
Australia, seasonality in children's growth rates, 28
AVRDC. *See* Asian Vegetable Research and Development Center
Azariadis, C., 85, 87

Ba-Aye, 34, 40
Babu, P., 45
Bailey, R. C., 153
Baker, D. C., 187
Balasubramanian, S. C., 68, 111
Banana production, Africa, 211, 214
Bangladesh
—food availability: cropping pattern and, 63; seasonality in, 57-58
—food-for-work program, 97
—intrahousehold food distribution, 57, 58-59
—relationship between children's body weight and death risk, 25-26
—seasonal migration for employment, 94
—seasonal variation in food intake, 59; by children, 61, 62, 63, 64, 65; and growth pattern, 20; by men, 60, 64; in poor rural communities, 301; by women, 60, 62, 63, 64; by young girls, 64-65
—wholesale price spread for foodgrains, 184, 185
Bardhan, K., 198
Bardhan, P., 85, 86, 87, 88n, 89
Barker, R., 238
Barnum, H. N., 83
Barral, H., 154, 166, 289
Basson, P., 165
Bates, D., 153, 154
Bayliss-Smith, T., 210
Beals, R. E., 94
Beaton, G., 103, 312n
Becker, S., 27, 57
Beckman, B., 230n, 231n
Beeny, J. M., 229
Behnke, R. H., 217
Behrens, C. A., 169
Behrman, J. R., 5, 7, 9, 66, 68n, 70n, 76, 84, 107, 108n, 109, 119
Bell, C., 85, 88
Benefice, E., 154, 166, 289
Berg, A., 312n

Berg, E., 94
Bergeret, B., 140
Berio, A. J., 290
Bernard, A., 184
Bernard, F., 211, 214
Bernstein, H., 137n
Berry, S., 138, 224n
Beyer, J. L., 213n
Bhalla, S., 98
Bidinger, P. D., 45
Billewicz, W. Z., 20, 28
Binswanger, H. P., 68n, 83, 85, 228
Biological regulation to maintain nutritional status, 5; adaptability model for, 23-24; genetic potential model for, 22-23; policy implications of models, 24-27
Biomass, seasonal fluctuations in Sahel, 217, 218
Birdsall, N., 108, 110, 116
Bittenbender, H. C., 310n
Black, R. E., 20, 25, 27, 28, 57
Blackie, M. J., 305n
Blanc, J., 104
Bliss, C., 85n, 107
Bloch, M., 139n
Body growth: environmental conditions and, 27; as index of nutritional status, 20; seasonality of, 20-21, 25
Bohannan, P., 143
Bojanic, A., 29
Botswana: specialized commercial food production in, 226-27; World Food Program recipient, 97n
Botswana Society, 209n
Bouis, H. E., 188, 192, 193, 238
Boutillier, J. L., 104
Bouwkamp, J. C., 288
Boxall, R., 43
Boyce, J. K., 158, 164, 293
Braun, J. von, 15n, 138
Braverman, A., 88n
Breman, J., 94, 95, 161
Brokensha, D., 159
Brown, A., 161
Brown, K. H., 27, 57
Burkina Faso, 5, 9, 10, 11
—crop versus livestock care in, 93
—food insecurity in Sahel savanna and Sudano-Sahel regions: chronic versus single-season, 131-33; overgrazed and overhunted lands and, 119-20; policy implications of, 134-36; poor transport and, 119; seasonal, 118, 123, 125, 126-

Index 357

27; seasonal activity patterns in, 121-22; subsistence-oriented cropping system and, 120-21; unstable labor market and, 119
—grain economies: government intervention in, 203-5; household marketing pattern in, 202-3; market versus nonmarket transfers, 201-2; seasonality in prices, sales, and purchases, 198-200; villages surveyed, 197-98
—household vulnerability in Sahel-savanna and Sudano-Sahel regions, 122-23, 124; dependence on common property resources and, 131; dependence on food aid and, 130-31; from market dependence for food, 127, 129-30; by seasonal food and caloric intake, 127-29
—national cereal marketing organization, 196
—seasonal price spread for sorghum, 186
—secondary food crops consumption, 289
Burma, farmers of: body mass index, 42, 43; body size and energy expenditure rate, 34-35; reliance on rice, 42; seasonal changes in body weight, 40, 41
Burton, M., 152

Caloric intake, 4, 5, 6
—Burkina Faso, 127-29
—India: income and total expenditures and, 276-82; price effects on, 72-73, 282, 283; seasonal distribution, 68-69, 76; for small versus large households, 74-76, 266-71; technological change and, 269, 283, 284; and wages, 114-15; and worker productivity, 111, 115
—secondary food crops as source of, 289-91
Cambon, A., 140
Cameroon, Beti peoples of southern: food consumption, 140-42; income earnings, 142, 143-44, 144-47, 148-49, 150; nutritional quality of diet, 140-41; seasonal food expenditures, by product, 147-48; seasonality, 139-40
Campbell, D. J., 209n
Campbell-Platt, G., 288
Carlin, G. T., 153
Carloni, A., 103
Cassava: key role in peasant agriculture, 213; production, 214; as secondary food crop, 288
Cassidy, C., 154

Center for Research on Economic Development (CRED), 193
CGIAR. See Consultative Group on International Agricultural Research
Chambers, R., 19, 45, 66, 90, 107, 118, 138, 151, 210, 225, 235, 264
Chasin, B. H., 162, 209n
Chaudhury, R. H., 89, 93, 95
Chen, L. C., 26, 57, 64, 103, 160, 165
Chevassus-Agnes, S., 154, 166, 289
Children: body growth differences, 24, 25-26, 27-28, 29-30; nutritional status and mothers' work in trade and agriculture, 165; seasonal variations in work by, 102-3; supplementary feeding program for, 311-12
Chowdhury, A.K.M.A., 26, 57, 103, 153, 160, 165
Clark, M. L., 27
Clay, E., 310
Clayton, E., 228n
Cleave, J. H., 90, 92, 102
Cliffe, L., 164
Cocoa, 140, 141-42
Cogill, B., 101n
Cole, T. J., 47
Colson, E., 164, 165, 166, 168, 170
Condon, R. G., 153
Connell, J., 94
Consultative Group on International Agricultural Research (CGIAR), 292, 293
Consumption: credit and, 98-99; current versus future, 97, 100; earnings and, 84; income and, 97-98; relationship between leisure and, 101-3; seasonal variability in, 101; work efficiency and, 85. See also Food consumption
Cooper, F., 163
Cosminsky, S., 158, 160
Coughenour, M. B., 219, 221n
Coward, W. A., 47, 52
Cowpeas, seasonal price variability in Nigeria, 187
CRED. See Center for Research on Economic Development
Credit: employment link to, 85; food security and access to, 4, 11; relationship between seasonal migration for employment and, 95; to smooth out variations in consumption, 98-99
Crops: diversification of, to counteract seasonality, 210-13; intensity in production, 242-43, 252; new varieties to

Crops (cont'd)
improve food security, 239; production of root, 313–14. *See also* Secondary food crops
Cummings, R. W., Jr., 238
Cutler, P., 22, 108
Cyprus, World Food Program recipient, 97n

Dahl, G., 216, 220
Dalton, G., 221
Dandekar, K., 96
Deaton, A., 102
Delgado, C. L., 9, 10, 11, 13, 14, 93, 119n, 138, 179, 182, 243, 305n
Deolalikar, A. B., 5, 7, 9, 66, 68n, 70n, 76, 84, 107, 108, 109, 110n, 118
de Ridder, N. L., 217
Desai, G. M., 15n
Developing countries. *See* Third World
Devitt, P., 220
DeWalt, K. M., 158, 160, 170
de Wilde, J. C., 228n, 229
Diallo, A., 219
Diarrhea, 45–46
Dirks, R., 171
Disease, infectious: effects of seasonal, 29–30; immunization and control program for, 46; limiting effect on labor supply, 89–90; relationship between body growth, nutrition, and, 26–27; vector-borne, 45
Diskin, M., 166
Donge, J. K. van, 210n
Donham, D., 144n
Dow, M., 152
D'Souza, S., 57, 64
DuBois, C., 152, 161
Dugdale, A. E., 28, 32, 44
Dunn Nutrition Unit, 56
Durnin, J.V.G.A., 50
Dynarski, M., 98

Earnings. *See* Agricultural earnings; Agricultural income; Wages
Eaton-Evans, J., 28
Edirisinghe, N., 8
Efe pygmies, 153–54
Efficiency wage models, 83, 97; conditions necessary for, 84; described, 83–84; nutrition-productivity link, 117
Egli, R., 216
Egypt: seasonal variation in wages, 89; seasonal variation in work by women and children, 102-3

Eicher, C. K., 187
Ejiga, N., 187
Elias, J., 159
Elling, M., 215
Ellsworth, Lynn, 8, 9, 10, 11, 13, 196
Employment: agricultural, 4, 85, 90, 91; cycles in, 81; food-for-work programs, 96, 97; public work programs to generate, 96–97, 310; seasonal migrations for agricultural, 93, 94, 95–96
Energy
—balance in, 4, 5; adaptation to seasonal changes in, 31–32; in pregnant Gambian women, 20, 47–48
—expenditure patterns, 5; body size and, 30–31, 35; as percent of basal metabolic rate, 34
—intake requirements: defined, 30; of Gambian farmers, 36–37; reduced energy expenditures to compensate for, 31
—simulation model for changes in balance, 31–32; difficulties in measurements, 45; model farm, 33–34; model man, 32–33; policy implications, 43–46
Ethiopia, 12
Ethnographic studies on seasonal food insecurity, 151, 152; in foraging societies, 153–54; in pastoral societies, 154–55; in peasant households, 155–58
Evangelou, P., 216
Evans-Pritchard, E. E., 152
Evenson, R., 83, 293

FAO. *See* Food and Agriculture Organization
Fertilizer, 92, 226
Firth, Rosemary, 152
Fitt, A. B., 28
Flannery, K. V., 152
Fleuret, A., 159, 286, 289n
Fores, M. and R., 45
Flowers, N., 154
Food and Agriculture Organization (FAO), 42, 61–62, 64, 213, 214, 229, 230, 286n
Food and Agriculture Organization-World Health Organization-United Nations University, 30, 119
Food availability, 3, 4, 7; energy expenditure and, 137; from food storage and processing, 192, 209, 211, 214; and hospitality, 152, 166–67; strategies for improving, 155, 159, 160–62, 162–66, 166–67, 167–68; from technological

changes in farm production, 12-13, 15, 91, 92, 93, 264; trade to stabilize, 192-93
Food consumption: household, 7-9, 104-5, 167; indicators, 30-31; seasonal cycles in, 28; target levels, 4; withdrawal of food from storage for, 39-40
Food-for-work programs, 96
Foodgrains: Burkina Faso marketing of, 197-205; prices, 184-87, 189-90, 192-93; specialized commercial production in Africa, 227-28
Food insecurity, 4-5
—African seasonality in: determinants of stress from, 209-10; effectiveness of methods to reduce, 231-34; indigenous solutions to, 210-21; introduced, 223-25; introduced solutions to, 225-31; social adjustments to, 221-23
—agricultural growth and market development to reduce, 301-2, 304-5
—Burkina Faso, 118-22, 125-27, 131-36
—defined, 118-19
—food habits as determinant of, 168; cultural and symbolic factors influencing, 169-70; dietary structure and, 170-71; ecological and economic factors influencing, 168-69; seasonal supplies and, 171
—interyear and intrayear variability in, 14-15
—reciprocal sharing to avoid, 9, 201-2
—seasonal: anthropological findings on, 151-53; ethnographic documentation of, 151, 153-58
—seasonal price increases and, 179-80
Food prices
—anticipatory hypothesis for, 188; for Thai rice, 249-50, 251, 262-63
—government role in reducing seasonal variability in: costs of, 190-91; direct approach to, 189-90, 192-93; by improving quality of information, 304; indirect approach to, 189, 193-94, 195; by infrastructure development, 304-5; policy requirements for, 191-92; through trade policy, 304
—and nutrition and health status in rural South India, 66, 76-77; data for, 67-70; policy implications, 77-78; use of reduced-form functions to determine, 67, 70-76
—seasonal variations in, 3, 66; and food insecurity, 179-80; as function of supply and demand expectations, 9-10, 187-89; magnitude of, 180, 184, 194; relation to storage and transaction costs, 183-87; stabilization, 303-4
Food security
—Asian seasonality in, 235; factors affecting, 237; government policies influencing, 237-38; marketing interventions and, 239; measurement in, 235-36; new crop varieties and, 239; production timing and, 240-43
—government policy recommendations to improve, 313-16
—household access to food and, 4, 7-11, 167-68
—seasonal variations in, 86, 239-40, 301
—targeted projects to improve: adjustments in administrative procedures, 313; advantages of, 309; by employment and income generation, 310; by food price subsidies, 311; by food rations and supplementary feeding programs, 311-12; by income transfers to households, 310-11
—untargeted projects to improve: agricultural research for technological change, 305-8; infrastructure development, 304-5; price stabilization, 303-4
Food storage: costs, 183, 199n; and food availability, 13, 192, 209, 211, 214; on-farm, 10-11; seasonal food price variability relation to, 183-89; withdrawal of food from, 39-40
Food supplementation programs, 5, 47-48, 311-12
Foraging societies, 153-54
Forbes, M. H., 159
Ford, R. I., 166
Fortes, M. and S. L., 155, 170
Fowler, A., 223
Fox, R. H., 34, 36
François, P., 290
Franke, R. W., 162, 209n
Franzel, S., 212, 213
Fruits, as secondary food crops, 289

Galvin, K. A., 221n
Gambia, 5
—described, 47
—farmers: body size and energy expenditure rate, 34-35; food intake strategies, 36-37, 38; seasonal weight fluctuations, 36, 37, 39; work output, 38
—food supplementation program, 47-48
—Genieri efforts to improve food production and availability, 156-57

Gambia (cont'd)
—index of nutritional status, 20
—seasonality effect on pregnant women: on energy balance, 47, 53; on energy expenditures, 47, 48-52; on pregnancy outcome and lactation, 55, 56; on resting metabolic rate, 53-55
—seasonal role of secondary food crops, 228
—women's role in agriculture, 49
Garcia, M., 8, 104
Gathee, J. W., 226
Genetic potential model, for biological regulation, 22-24
Germany, domestic crop production pattern for 18th and 19th century peasant, 143, 150
Ghana, seasonal migration for employment, 94
Ghassemi, H., 312n
Ghodake, R. D., 91, 110, 116
Gladwin, C. H., 212n
Glaeser, B., 227
Glantz, M. H., 209n
Goering, J., 213
Goldman, R. H., 188, 249
Goody, E., 168
Gopalan, C., 21, 24-25, 68, 111
Gotsh, C. H., 191
Grantham-McGregor, A., 21
Greenberg, J., 166
Green leafy vegetables, as secondary food crops, 289
Griffiths, M., 28
Grivetti, L. E., 159
Grove, A. T., 231n
Guyer, Jane, 8, 9, 15, 137, 139n, 172n

Hall, Malcolm, 228n
Hall, Robert, 98n
Hankins, T., 224
Hansen, B., 89n, 103
Hanson, J., 89n
Hart, Gillian, 86n, 87, 89, 94, 102
Hartmann, B., 158, 164
Haswell, M. R., 156, 157, 160, 161, 227, 288
Hayami, Y., 244
Hays, H. M., 180, 185, 216n, 220, 228
Hazell, P.B.R., 265, 266
Health status: body size and growth relationship to, 24-25; nutritional status and, 23; and productivity, 107-8; seasonal effects on, 19, 25, 107

Health status, rural South India
—food price effects on, 66, 67, 73-74, 77-78
—reduced-form demand relations, 67, 70; for individuals, 73-74; for small versus large households, 74-76
—response to household assets, 66, 74, 78
—seasonal variability effects on, 67-68; by impact on wage rates, 108-10; policy implications of, 77-78; by season and sex, 111, 114, 116-17; by village-specific peak-period months, 110-11, 112-13, 114-15
Heckman, J. J., 110
Herdt, Robert W., 9, 10, 12, 235, 238, 243
Heyer, J., 89, 93
Heywood, Peter, 288
Hill, A. G., 219
Himes, H. H., 158
Hitchcock, R. K., 209n
Hjort, A., 216, 220
Hogan, E., 214
Holmberg, A. R., 152
Holmström, B., 86
Hoskins, M. D., 242
Houghton, J., 187
Households
—food consumption for, 7-9; individual versus, 104-5; to meet labor demands, 103-4
—food demands, 168
—food security: access to food and, 7-11; by adjusting household size and composition, 167-68; defined, 3; by eating behavior, 167; factors contributing to, 4
—seasonal changes in food distribution within, 7; effect on women and children, 104-5; seasonal pattern in labor and, 6, 85
—vulnerability in Burkina Faso region, 122-23, 124; defined, 119; dependence on common property resources, 131; dependence on food aid and, 130-31; household consumption deficit and, 118-19; from market dependence for food, 127, 129-30; by seasonal food and caloric intake, 127-29
Huffman, S. L., 26, 103, 153, 160, 165
Hug, E., 57, 64
Huss-Ashmore, R., 153, 156, 160, 161, 164

Huxley, E., 227
Hyden, G., 138n

Ibnouf, M.A.O., 228n
ICRISAT VLS. *See* International Crops Research Institute for the Semi-Arid Tropics Village Level Studies
IFPRI. *See* International Food Policy Research Institute
ILCs. *See* Implicit labor contracts
Immink, M., 103
Implicit labor contracts (ILCs), 85-86, 87-88
Income. *See* Agricultural income
India: inelasticity of rural labor demand and supply, 83; new varieties of rice and wheat, 242; relationship between children's body weight and risk of death, 25-26; seasonal migration for employment, 94, 96; seasonal variability in employment, 90-91; seasonal variability in grain prices, 186-87; secondary food crops proportion of dietary intake, 219. *See also* India, North Arcot district; India, rural South
India, North Arcot district: household energy consumption, 266-70, 283-84; rice production study, 264-66, 271-76, 277, 283
India, rural South: farm household models of demand for nutrient intakes and health status, 67-70, 72-78; health and nutrition status effect on productivity, 108-17; nutrient intake and health status, 66-78
Indonesia: analysis of seasonal rice price, 249-50; foodgrain imports to stabilize prices, 192, 193; government food price and market intervention, 191, 245; production instability effect on food prices, 188; rice harvested, by month, 240, 241; seasonal migration for employment, 94; seasonal variation in wages, 94
Institute of Nutrition and Food Science (INFS), 57
International Center for Agricultural Research in Dry Areas, 293
International Crops Research Institute for the Semi-Arid Tropics Village Level Studies (ICRISAT VLS), 70n; daily wage data, by district and village in India, 110-15; food intake survey for Burkina Faso, 122; survey data on

India's nutrient intake-health status relationship, 67-68
International Food Policy Research Institute (IFPRI), 122, 265
International Institute for Tropical Agriculture, 293
Intertemporal price equilibrium, 183, 188
Intertemporal utility theory models: access to credit assumption, 99-100; permanent income hypothesis, 98; research on food acquisition in terms of, 106
Irrigation, 12; and cropping intensity, 242-43, 252; in Nigeria, 230, 231; to reduce seasonal variability in production, 91, 229-31, 239, 306; for rice in Thailand, 248-49, 251-57, 260-62
Ivory Coast, secondary food crops, 290

Jahnke, H. E., 216
Japan, 83
Jaramillo, M., 11, 12, 15, 101n, 264, 273, 307
Jochim, M., 168, 169
Jodha, N. S., 68n, 91, 291
Johnson, A., 169
Johnston, B., 137n, 229
Johnston, F. E., 153
Jones, C., 161
Jones, W., 92, 183, 187, 193, 216
Jones, W. O., 186, 187
Josserand, H. P., 138n
Judd, M., 293

Kahn, J. S., 165
Katz, R. W., 209n
Kelly, P., 160
Kennedy, E., 101n, 138, 309n, 310, 311n
Kenya: crop diversification, 211, 212-13; farm mechanization, 228; gender-linked division of household-farming activities, 222; irrigation, 229, 230; seasonal variation in wages, 89; specialized commercial food production, 226-27; vertisol areas, 216
Keswani, C. L., 211
Kielmann, A. A., 26
Kilby, P., 137n
Kinsey, B., 228n
Kline, C. K., 229
Krishnamurty, K., 43
Kuckertz, H., 139n
Kuhnlein, H., 159, 160
Kumar, S., 93, 96, 291

Kyereme, S. S., 101n

Labor demand, 66, 82-83, 94-95
Labor supply, 83, 89-90, 102
Lagemann, J., 211
Latin America: decreasing access to raw materials in, 163; effects of land and future crops mortgaging, 165; migrant labor, 163
Lawrence, F., 47
Lawrence, M., 5, 20, 47, 50, 52, 311
Ledesma, A. J., 244
Lee, R., 153, 220, 221n
Lees, S., 153, 154
Leibenstein, H., 83, 107
Lele, U., 186, 187
Le Moigne, M., 228
Lesotho, 97n, 156
Lewis, J., 114n
Liberia, 94
Lindenbaum, S., 164
Lipton, M., 12, 13, 91, 93, 285, 287
Livestock: to cope with seasonalities, 9, 217; exchange of food for, 222-23; integration of grain production and, 219-20; mixed herds of, 219; as source of energy, 217, 219
Longhurst, R., 12, 13, 45, 66, 90, 101n, 104, 107, 118, 138, 210, 225, 264, 285, 289
Lunn, P. G., 56
Lunven, P., 26

McCabe, J. T., 221n
McCord, C., 26
McCoy, J. H., 185
McGregor, I. A., 20, 28
Madden, P., 100
Malaria, 45, 46
Maize, 224-25
Malawi, efforts to meet seasonal food shortages, 157-58
Malaysia: family labor input, 90n; relationship of labor supply elasticity and wages, 83
Mali: farm mechanization in, 228-29; on-farm crop production, 92
Malina, R. M., 158
Malnutrition
—seasonal food insecurity and, 4, 20; causes, 6; food supplementation program to reduce, 5, 47-48; intrahousehold food distribution and, 6-7, 167-68
—seasonal growth retardation and, 20-21, 24, 25
—theories, 22-27
—uncertainty over meaning and cause of, 20-21
Marshall, L., 153
Mascarenhas, A. C., 91
Masseyeff, R., 140-41
Matlon, Peter, 5, 11, 118
Mauss, M., 152
Maxwell, S., 29
Mazumdar, D., 107
Mbithi, P., 209n
Mears, L. A., 99n, 187n, 192, 193, 238
Mellor, J. W., 15n, 305n
Men: seasonal variations in food intake, 60, 61, 63-64; task segmentation between women and, 116; wage rate and health and nutrition, 111-13, 116
Mencher, J., 170
Menezes, C. F., 94
Messer, Ellen, 8, 9, 15, 151, 154, 159, 160, 164, 165, 166, 168, 169, 170, 172, 289
Metrick, H., 228n
Mexico, 158, 289
Migot-Adholla, S. E., 229
Migration: for developing long-term contacts, 166; and household food demands, 168; for seasonal agricultural employment, 93-96
Miller, C.P.J., 138
Mincer, J. B., 108, 110
Miracle, M., 155, 224
Mishkin, F., 98n
Mongkolsmai, D., 9, 15, 246
Moris, J. R., 9, 11, 12, 13, 209, 212
Morrison, T. K., 138
Muellbauer, J., 102
Musgrove, P., 98

Nag, B., 45
National Academy of Sciences, 213n
National Nutrition Study, 140, 145
National Research Council, 312n
Ndunguru, B. J., 211
Nelson, D., 310
Nepal, 93
New ERA, 93, 95
Newman, J. L., 220, 224
Newman, M., 228
Nghiep, L., 83
Nigeria: agricultural production strategies, 92, 93; consumption to meet labor

demand, 103; irrigation, 230, 231; Onicha Ibo seasonal food shortages, 157; seasonal food price variability, 180, 182; seasonal variation in cowpeas prices, 187; seasonal variation in wages, 89; secondary food crops, 289-90; study on seasonal pattern of work, 90; trading and off-farm activities during dry season, 220, 221; vertisol areas in, 215, 216; wholesale price spread for foodgrains, 184-85
Nimis, M. M., 165
Niñez, V., 310n
Nnamyelugo, D., 290
Norman, D. W., 90, 216n, 220, 228
Nurse, G. T., 157, 161
Nutrient intake. *See* India, rural South
Nutritional status: anthropological attempts to quantify changes in, 153; anthropometry to measure, 20; and health, 23; and productivity, 107-8; seasonal patterns affecting, 152-53; social organization and, 153-55; theories of, 22-26; wages and, 85
Nylin, G., 27

Odell, M. F., 158, 163
Odovafa, A., 290
Office National des Céréales (OFNACER), Burkina Faso, 196
Ogbu, J. U., 157, 161, 162, 170
Okigbo, B., 213
O'Leary, M. F., 209n, 228
Olsen, R. J., 110, 114
Omawale, 170
Orr, J. B., 27

Pacey, A., 45, 66, 90, 118, 138, 210, 225, 264
Pagezy, H., 153
Palmer, C. E., 27, 28
Pardy, C., 198, 204
Passmore, R., 50
Pasternak, B., 153
Pastoral households, 154-55
Patton, M., 209n
Payne, Philip R., 5, 19, 22, 32, 44, 101n, 104, 107, 108
Peacock, N. R., 154
Pearson, S. R., 186, 187
Pelto, G., 160, 170
Penning de Vries, F.W.T., 217
Peterson, J. T., 154
Philippines: cropping intensity, 242-43; government intervention in rice market, 244-45; government price stabilization efforts, 190-91, 192; measurement of food security seasonality in, 236; seasonal pattern of harvested rice area, 240-41; trade effect on seasonal food price spread, 188, 193
Pinckney, T. C., 191
Pingle, V., 291
Pinstrup-Andersen, P., 4, 8, 11, 13, 15, 72n, 76, 101n, 104, 237, 238, 264, 273, 307, 311n, 315n
Pitt, M., 70n, 235
Popkin, B., 165
Poudyal, K. R., 238
Prentice, A. M., 48, 51, 55, 56
Price, D. W., 8
Price, E., 243
Price, M.R.S., 217
Prices. *See* Food prices
Production. *See* Agricultural production
Productivity, link between nutrition, health, and, 5, 84, 107-8; caloric intake and, 111, 114-15; wage rates and, 108-10, 114-16
Protein intake, 68-69, 72-73
Pryor, F., 221
Psacharopoulos, G., 108, 110, 116
Public health strategies, 46
Public works programs, to generate employment, 96-97, 310
Purdue University, 197, 198

Quedraego, I., 228

Ramasamy, C., 265, 266
Ranade, C. G., 243
Ray, S. K., 238
Reardon, T., 5, 11, 118
République-Unie du Cameroun (RUC), 140
Research: to improve food security, 174-75, 305-8; on seasonal patterns of employment and wage formation, 105-6; for secondary food crops, 285, 292-93, 294, 295, 296-97
Rhodesia, food production and management practices of Bemba tribe, 156
Rice
—Asian: area harvested, 240-41; government intervention in market for, 244-45; seasonal market pattern for, 242; technological change and, 239-40
—Bangladesh production, 57, 58

364 *Index*

Rice (cont'd)
—cost of African, 230
—North Arcot, India: intrayear fluctuations in prices, 282, 283; intrayear fluctuations in production, 271-73, 283
—Philippines: government intervention in market, 244-45; seasonal pattern of area harvested, 240-41
—Thailand: cropping intensity, 251, 252-53; double cropping, 249, 261; exports, 248, 261, 263; prices, 249-50, 251, 262-63
—Thailand's irrigation of: effect on rice availability and price, 248-49; and household income, 254-55, 256-57; and labor use, 252, 255, 260-61, 262; and production, 251-52, 253-54
Richards, A. I., 155, 156, 160, 170, 171, 210
Richards, P., 143, 215
Riley, D., 159
Rizvi, N., 163
Roberts, S. B., 20
Robson, J.R.K., 159
Rogers, B., 310
Rogers, G. B., 84n, 85, 97
Rosegrant, M. W., 9, 15, 246
Rosenzweig, M. R., 70n, 83, 85, 89, 235
RUC. *See* République-Unie du Cameroun
Rudra, A., 87
Ruthenberg, H., 212, 213
Rwanda, vertisol areas, 215, 216
Ryan, J. G., 68, 69n, 70n, 89, 91, 109, 110, 111, 116, 243

Sabean, D., 143, 150
Sabot, R., 108, 110, 116
SAFEGRAD. *See* Semi-Arid Food Grain Research and Development program
Safilios-Rothschild, C., 57
Sahel region. *See* Burkina Faso
Sahlins, M., 154
Sahn, D. E., 3, 8, 9, 10, 11, 13, 14, 15n, 81, 83, 84, 107, 108, 172n, 179, 301, 309n, 310, 311n
Sandford, S., 216
Sarin, R., 110
Sastry, B.V.R., 68, 111
Sathe, M., 96
Saul, M., 201
Savings: agricultural-based, 11; livestock as store of, 219; out of transitory income, 98; research on household behavior on, 105-6; to smooth out variations in consumption, 98-99
Scaglion, R., 153
Schlage, C., 211
Schneider, H. K., 219, 222
Schofield, S., 57, 66, 101n, 103
Scott, C., 99n
Scott, E., 160
Scrimshaw, M., 158, 160
Scudder, T., 160, 166, 220, 231
Seasonal cycles, 81
Seasonality: anthropometric indicators and, 27-28; from biological and sociocultural perspectives, 151; methods for reducing bad effects of, 16; predictability, 14, 15-16; role of agricultural research on, 13, 294-97; secondary food crops to reduce, 285-94; stochastic interyear fluctuations versus, 14, 15
Seckler, D., 25
Secondary food crops (SFCs): agricultural research on, 292-93, 294, 295, 296-97, 306; environmental conditions for, 295; impact on poverty, 285, 291, 295; micronutrients in, 286-87; negative antinutritional aspects of, 287, 296; propositions relating to, 285-86; role in reducing food insecurity, 12-13, 285, 289; seasonal role, 288-91; for subsidiary home consumption food source, 286
Semi-Arid Food Grain Research and Development (SAFGRAD), 197, 198
Sen, A. K., 27
Senegal: farm mechanization in, 228, 229; pattern of rural food purchases, 138n; seasonal migration for employment in, 93; use of women and children during peak labor demands, 103
Serdula, M. K., 153
Service, E., 221
SFCs. *See* Secondary food crops
Shack, D. N., 152
Shack, W., 152
Shapiro, K., 8, 9, 10, 11, 13, 196
Sharman, A., 155, 166, 167
Sharp, J. S., 139n
Sheffrin, S., 98
Shepherd, A., 227, 228n
Shepherd, J., 209n
Siamwalla, A., 191, 192
Sierra Leone, relationship of labor supply elasticity and wages, 83

Simmons, E. B., 100n, 103, 165, 216n, 220, 228, 289, 307
Simpson, J. R., 216
Singh, J., 47
Smith, E. A., 168, 169
Sociocultural studies, to evaluate food insecurity, 151, 152-53
Soemarwoto, O., 310n
Southworth, V. R., 186, 187
Spencer, D., 103
Spencer, T., 288
Spiegel, A. D., 139n
Squire, L., 83, 90n
Sri Lanka, relationship of labor supply elasticity and wages, 83
Srinivasan, T., 85, 88, 108
Stern, N., 85n, 107
Stiglitz, J., 87, 107
Storage. See Food storage
Strauss, J., 83, 84, 108, 109
Stutley, C., 29
Sudan, irrigation schemes, 230, 231
Sukhatme, P. V., 30, 108
Svedberg, P., 104
Swanberg, K., 214
Sweet potatoes, 213, 288, 291
Swift, J., 217
Szarleta, E., 198, 204

Tamil Nadu Agricultural University, 265
Tanzania: banana varieties produced, 211; farm mechanization in, 228, 229; specialized commercial food production in, 226, 227; study of crop diversification in, 212; wild sources of food, 220
Technological change: and caloric intake, 269, 283, 284; comparative performance of, 231-34; cropping intensity from, 242-43; effect on employment seasonality, 9, 243-44; effect on food security seasonality, 239-40; for food storage, 209; to improve food security, 305, 306, 307, 308; to increase crop production, 91-92, 264-65; labor-saving, 86; and rice production, 239-40; wage sensitivity to, 83, 86. See also Irrigation
Teitelbaum, J. M., 155
Teokul, W., 44
Thailand, rice
—cropping intensity, 251, 252-53
—double cropping, 249, 261
—exports, 248, 261, 263
—irrigation, 9; development of, 247-48; effect on household income, 254-55, 256-57; effect on labor use, 252, 255, 260-61, 262; effect on rice availability and price, 248-51; effect on rice production, 251-52, 253-54; use of cultivation inputs and, 255, 258-59
Third World: contributors to seasonal variations in food security, 302; determinants of nutrition, 107; food energy sources, 292; permanent income hypothesis for, 98; savings out of transitory income, 98; seasonal food price variations, 180, 181; wage equation studies for, 108-10
Thomas, J. W., 310
Thomas, V., 192n
Tiffen, M., 231n
Timberlake, L., 120
Timmer, C. P., 67, 190
Tin-May-Than, 34, 40
Toksoz, S., 216, 229
Tolley, A. S., 192n
Toulmin, C., 119
Trade: effects on seasonal food price changes, 188, 304; food-grain price stabilization from imports, 192-93; stabilization of food availability from, 14, 179, 192-93
Tripp, R., 165
Tully, D., 164
Turnbull, C., 171, 221
Turton, D., 171
Tyagi, D. S., 238, 242
Tyson, P. D., 209n

Uganda, 102, 155
UNICEF, 310n
Unnevehr, L. J., 188n, 191
U.S.-DHEW, 57

Valverde, V., 101n
Vertisol areas, 214-15, 215-16
Vietmeyer, Noel, 213n
Viteri, F., 103

Wagenaar, K., 219
Wages, agricultural
—alternative models of formation of, 82-83; efficiency wage model, 83-85; implicit labor contract model, 85-88
—floor level, 84, 85
—impact of seasonal migration on, 95-96
—nutritionally determined, 85
—seasonal variations in, 3; factors respon-

Wages, agricultural (cont'd)
  sible for, 88-89; need for additional research on, 105-6; from seasonal labor market changes, 15, 82; from seasonally targeted public works, 83; from shifts in labor supply and demand, 82-83
Wallace, T., 221, 231n
Walsh, R.P.D., 209n
Warman, A., 158, 163
Warren, D., 159
Wasserstrom, Robert F., 158, 163
Waterbury, John, 93
Watts, M., 228
Weight-for-height, rural South India, 68; for individuals, 71; price effects on, 73-74; seasonal distribution, 69; for small versus large households, 74-76; as short-run health status determinant, 77
Werner, O., 159
West, D., 8
WFP. *See* World Food Program
Wheat, 230, 242
Wheeler, E. F., 57, 58n
White, D., 152
Whitehead, R. G., 47
WHO. *See* World Health Organization
Wienpahl, J., 221n
Wiessner, P., 153, 221n

Wijga, A., 96, 97n
Wilmsen, E. N., 153
Wilson, R. T., 219
Winikoff, B., 164
Winterhalder, B., 168, 169
Wisner, B., 209n
Wolfe, B. L., 76
Wolpin, K., 98
Women: in Cameroon, 144-45, 145n, 146, 147; effect on child care of working, 165; in Gambia, 47, 49-52, 53-55; seasonal variations in food intake in Bangladesh, 60, 61, 62, 63; seasonal variations in work by, 103; small-scale food processing by, 165; supplementary feeding projects for, 47-48, 311-12; task segmentation between men and, 116; wage rates and health and nutrition variables in India, 111-13, 116
Wong, C. M., 192n
Working, H., 188
World Food Program (WFP), 97n
World Health Organization (WHO), 118
Wright, T. L., 216

Yoder, M., 100

Zambia: secondary food crops as source of caloric intake, 290-91; vertisol areas, 215; wild sources of food, 220